Lake Naivasha, Kenya

Developments in Hydrobiology 168

Series editor
H. J. Dumont

Lake Naivasha, Kenya

Papers submitted by participants at the conference "Science and the Sustainable Management of Shallow Tropical Waters" held at Kenya Wildlife Services Training Institute, Naivasha, Kenya, 11–16 April 1999, together with those from additional studies on the lake

Edited by

D.M. Harper[1], R.R. Boar[2], M. Everard[3] & P. Hickley[4]

[1] *University of Leicester, U.K.*
[2] *University of East Anglia, U.K.*
[3] *Environment Agency, Reading, U.K.*
[4] *The Environmental Agency, Kidderminster, U.K.*

Reprinted from Hydrobiologia, volume 488 (2002)

LAKE NAIVASHA RIPARIAN ASSOCIATION

EARTHWATCH INSTITUTE
FINDING SOLUTIONS FOR A SUSTAINABLE FUTURE

Springer-Science+Business Media, B.V.

Library of Congress Cataloging-in-Publication Data

A C.I.P. Catalogue record for this book is available from the Library of Congress.

ISBN 978-90-481-6272-7 ISBN 978-94-017-2031-1 (eBook)
DOI 10.1007/978-94-017-2031-1

Printed on acid-free paper

TABLE OF CONTENTS

Hydrobiologia **488:** vii, 2002.
D.M. Harper, R. Boar, M. Everard & P. Hickley (eds), Lake Naivasha, Kenya.

Preface

The 18 papers assembled in this volume arose from research conducted at Lake Naivasha over the past two decades. Most were presented orally at the international conference 'Science & The Sustainable Management of Shallow Tropical Waters' held at the Kenya Wildlife Services Training Institute, Naivasha, in April 1999. One-page summaries of all the oral presentations were published as a UNESCO IHP Technical Document (Harper & Zalewski, 2001) and 26 full papers in *Hydrobiologia*, volume 458. Some papers here were added after the conference to make a more complete picture of the ecology of this important freshwater lake.

The work achieved at Naivasha and its publication here is a testament to the forward thinking of three groups of people – few of them trained as research scientists. In the first instance, a group of visionaries along the eastern seaboard of the United States, led by Brian Rosborough, developed the idea of ordinary citizens investing in environmental knowledge by supporting, with their own time, efforts and at their own expense, field research leading to conservation. Thirty years on, the Earthwatch Institute supports 148 expeditions a year from a membership in 48 countries. The EWI has supported almost all of the research described in this volume, since 1987.

The second group of visionaries are those who were, or are now, resident in the Lake Naivasha area, who have been working tirelessly to conserve the lake's ecosystem intact. Forty years ago the late Roger Mennel made a laboratory on his lakeshore farm available to John Gaudet, Jap Kalff, John Melack and a succession of Kenyan students for limnological research. The Wardens of the former Elsamere Conservation Centre – Jock & Enid Dawson, John & Jane Carver, Angus & Jill Simpson and Velia & Tony Carn, took up this mantle from the mid-1980s to 1999. By the time of the conference, Lake Naivasha had been given Ramsar status by the Kenya government and the management of the site vested in the Lake Naivasha Riparian Association, due mainly to the work of its Chairman (Lord Enniskillen) and Honorary secretary (Sarah Higgins).

The third group of visionaries are those people, all over the globe, who see the degraded state that our planet is in and play an active part trying to arrest and sometimes reverse its degradation. For most of us in the developed world this means giving time and money but rarely changing our lifestyles. For those in the developing world it is often changing their way of life and sometimes even sacrificing it. There are millions of these visionaries, but until there are countless millions, the degradation will inexorably continue all over the world.

This work is dedicated to those hundred or so Naivasha residents who have made this work possible by help, encouragement and permissions and the thousand or so Earthwatch volunteers who have contributed by their hard work to the data collection that finds an outlet in these pages. Finally, it is dedicated to the tens of thousands of people in the Lake Naivasha region, who toil on the flower farms, struggle in the dukas, or stand for long hours to sell newspapers in the streets, depending on the lake yet knowing nothing about it. At the time of writing this editorial, Kenya has a new government which has already declared primary schooling will be free and appointed Wangari Maathai, a prominent environmentalist as Assistant Minister of Environment and Natural Resources. The prospects for all the 'wananchi' of Kenya to understand, care and protect their own resources for future generations have suddenly improved markedly.

DAVID M. HARPER
Department of Biology
University of Leicester
England

KENNETH M. MAVUTI
Department of Zoology
University of Nairobi
Kenya

Reference

Harper, D.M. & Zalewski, 2001. Science and the Sustainable Management of Shallow Tropical Waters. Technical Documents in Hydrology 46. UNESCO, Paris.

Hydrobiologia **488:** ix–xii, 2002.
D.M. Harper, R. Boar, M. Everard & P. Hickley (eds), Lake Naivasha, Kenya.

Introduction

The Lake Naivasha Management Plan – consensus-building to conserve an international gem

In this introduction, as I said in my address to the conference where many of the following papers were first presented, I want to take as my theme the synergy between the science and the management that we, the 'stakeholders' in present-day jargon, are using to conserve Lake Naivasha. It is not just the science, but also the turning of science into layman's language and offering practical solutions to the everyday problems and potential conflicts of interest facing the community in Naivasha, that is essential to the conservation process.

The definition of the word 'gem' is an object of great beauty or worth. Naivasha is both. Its conservation in its beautiful state for future generations whilst utilising its worth to benefit present humankind is what the Management Plan is all about. Like anything of great worth, its security could be achieved by ring-fencing, protecting it by 'putting it in the bank'. The alternative is to make it of such value to the stakeholders that they all play a part in its conservation and feel they own it; and thus have no need to exploit or steal it. Ours is that latter objective. Unfortunately to achieve it is a long-term task, and meanwhile damage can and will occur. That is why we advocate a practical common-sense and damage limitation approach based upon the 'precautionary principle'.

The Lake Naivasha Riparian Association (LNRA) was jointly-awarded the 1999 Ramsar Wetland Conservation Award (Ramsar, 1999a), given to 'those governments, organisations or individuals who have contributed significantly to the long-term and sustainable use of a wetland site, especially those initiatives which might serve as inspirational or practical examples for others'. Those are the significant buzzwords – 'conservation', 'sustainable wise use', 'practical', 'inspirational' and 'examples'. A brief history of the Association will help to understand why its efforts were rewarded this way (Ramsar, 1999b).

The LNRA was 70 years old in 1999. It was originally formed by the landowners around the lake to administer an agreement with the government over the use of riparian land – the 'ecotone' – exposed by water level decline. Naivasha is a shallow lake (see Hickley et al., 2002, for a depth map). Being a shallow lake with very gently sloping shores, and being susceptible to large variations in depth between wet and dry cycles (see Becht & Harper, 2002, for fluctuations this century), the extent of the ecotone varies greatly even on an annual basis. We can experience a water edge shift measured in kilometres and a variation in depth measured in metres, in the space of 6 months. The riparian land is naturally very fertile with volcanic/lacustrine soils and subsurface water, so the agreement with the government of the day was that the landowners around the lake could use the exposed land below their riparian boundary (arbitrarily chosen as the level the lake attained in 1906) (Fig. 1) for grazing, cultivation and access to the lake only. No permanent buildings were allowed, no claims against the government in the event of flooding would be entertained and the landowners themselves through their Association would arbitrate any disputes over boundaries between one and another.

Unfortunately the agreement has proved not to be sustainable in the long term because of certain shortcomings, though in other respects it was a unique and enlightened agreement, which gave a true sense of ownership of the resource to the immediate community. It is undoubtedly that which has sustained the lake to date and prevented the worst abuses that could otherwise have occurred. Its shortcomings were in the lack of definitions and the omissions. For example, what is a 'permanent building', what is 'cultivation'? 'Should there have been a buffer strip, beyond which no human activity is allowed?' 'Should livestock be allowed free access to the littoral zone and to water direct from the lake?' None of this was spelled out, but as the size of human population began to increase rapidly, livestock numbers increased. Horticultural technology, with its attendant use of pesticides and fertilisers has grown dramatically over the past two decades, so these errors and omissions became serious in the extreme. Landowners who interpreted cultivation to mean the establishment of long-term intensive crops with a 10–15-year life did so

Figure 1. Lake Naivasha in November 2001. The RAMSAR site boundary is the lake road.

apparently legitimately. Inevitably there followed problems of cultivation too close to the lake and the erection of structures to protect their investment from flooding by the rising lake, thus interfering with natural processes. Those who believed that plastic greenhouses were not permanent buildings put them up, apparently legitimately, and soon we had whole segments of the ecotone devoid of vegetation and covered with plastic. The labour-intensive flower industry developed and with it the need for housing, water and latrines – especially near the workplace on the shoreline!

Trying to solve these problems in the absence of effective environmental legislation on the one hand and in a developing country desperately short of jobs, foreign exchange and socio-economic development on the other, was potentially a recipe for ecological disaster. Fortunately, a combination of enhanced scientific knowledge, enlightened flower growers and agriculturalists in Naivasha, together with a more discerning market in the developed world which increasingly demands 'environmentally-friendly' production methods for flowers, fruits and vegetables, has since enabled us to begin to address these shortcomings in the agreement, by negotiation with the government.

In the late 80s we foresaw the problem that we were going to face and also recognised one major weakness of the situation. Although I have emphasised the sense of ownership imparted by the riparian agreement – the reality is that true ownership of the ecotone remains with the government. In a country such as this, where productive land is scarce and the population is rapidly increasing, 'government land' means available land, land to be parcelled out for political favours. In the absence then of much environmental awareness among politicians, little consensus among stakeholders about conservation, this was a dangerous situation which required delicate consensus-building to prevent either the land being declared a protected area or being invaded by land-hungry allotees, neither of which could be sustainable. The Association was a well-established organisation, with an interest in the resource, a diverse membership representing a cross-section of society, and a common interest in maintaining the quality and quantity of lake water. We set out to build on these strengths. I believe that we were also fortunate to have the ideal conditions of climate, soils and water for flower-growing. For all its potential to destroy the lake, the industry is contributing greatly towards sustainable utilization of such a valuable resource. In the absence of the industry, I believe more abuse and more fragmented development would have occurred and would have been more difficult to contain.

So, in the early 1990s, the LNRA resolved to put in place a management plan for the lake which would promote voluntarily adopted, sectoral, codes of conduct, within an overall strategy with the objective of ensuring that all the human activities then represented around the lake could be sustained (quite independently of Ramsar's 'wise use' principles of which we were not aware at that time).

Even so, there was a wealth of research on which to draw, so our first priority was to assemble a synopsis of that research in a single, readable, document, achieved in late 1993 (Goldson, 1993). The second phase was the writing of the management plan, which took approximately 2 years, involving a steering group representing a wide cross-section of interests from the LNRA and beyond, involving the community at large. During this second phase we worked closely with the Kenya Wildlife Service to have Lake Naivasha nominated as Kenya's second Ramsar site, achieved in April 1995.

We always knew that the third, implementation, phase of the process would be the most difficult. We were set back by nearly a year after the plan was written because some important stakeholders believed that their livelihoods were threatened not only by the plan but also by the designation as a Ramsar site. These difficulties were overcome by dialogue and continued consensus-building. The plan was finally approved by stakeholders and received the official approval of government through the District Development Committee, being subsequently presented to His Excellency The President in 1997 when the implementation phase officially began (LNRA, 1999).

Water must urgently attain its true worth as a limited and highly valuable natural resource. For the vast majority of humans, water is the very basis of their lives. The daily trudge for miles to fetch it from unreliable and often polluted sources in rural areas or broken taps and running sewers in urban sprawls is their reality. Society has to change that, as we have to ensure that Naivasha does not become just another such polluted source.

Conservation for its own sake without grassroots education is only a rich man's game. Education for all in environmental conservation, understanding the long-term effects of pollution, awareness of best practices and a sense of responsibility for the world around is a long-term programme. In Kenya, huge numbers of people are educated at village and community level where, although they may lack sophisticated knowledge, they perhaps have a better opportunity than their educated peers to appreciate the natural world. In future they will need to be raised relying more on the resources around them. We must educate the educators to emphasize environmental issues and give them due weight in school curricula. Dr Richard Leakey, KWS Director, in opening this conference, said that humans had a right to live in a healthy environment – equal to their right to education and good health. I would go further, as I do not believe that humans have rights without corresponding responsibilities. To attain the right to live in a healthy environment we – everybody – has a responsibility to care for it.

Our lessons must be realistic. We cannot teach the negative aspects of, or enforce control of, tree cutting and charcoal burning in developing countries where alternative energy sources are either unavailable or expensive. We heard during our conference about papyrus brickettes, fuelwood crops using fast-growing local trees such as *Sesbania sesban*, but they must be made more affordable and the communities must have a strong sense of ownership of the resource in order for it not to be over-exploited. The human population in Naivasha has expanded from 25 000 to 250 000 over the last 15 years, largely through immigrant labour for the horticultural industry. They need energy to cook with and food to eat – a practical problem. The forest cover remaining in the Naivasha catchment, the papyrus around the lake and the biodiversity of the lake itself are under threat unless we address these issues. The stakeholders must be allowed to benefit directly from their water resources – only then will they look after them. This is an integral part of the management plan, whose scope extends into the catchment.

The Management Plan has already shown progress, in the metering of irrigation water, voluntary retreat from riparian land, adoption of drip-feed irrigation amongst some stakeholders. But there is still a long way to go. We grapple with individual abuses where greed or lack of understanding damages the resource, while lack of management or financial resources still leaves public utilities polluting the lake. We badly need the new environmental legislation before Parliament to underpin our control of the abuses. Ultimately, the state of Lake Naivasha revealed in these pages will only be sustained by the adoption of the codes of conduct by all stakeholders, which itself will come from a sense of ownership of the wetland resources of this 'gem'.

References

Becht, R. & D. M. Harper, 2002. Towards an understanding of human impact upon the hydrology of Lake Naivasha. Hydrobiologia 488 (Dev. Hydrobiol. 168): 1–11.

Goldson, J., 1993. A Three Phase Environmental Impact Study of Recent Developments around Lake Naivasha. Lake Naivasha Riparian Owners' Association, Naivasha, December 1993: 109 pp.

Hickley, P. R. Bailey, D. M. Harper, R. Kundu, M. Muchiri, R. North & A. Taylor, 2002. The status and future of the Lake Naivasha fishery, Kenya. Hydrobiologia 488 (Dev. Hydrobiol. 168): 181–190.

LNRA, 1999. The Lake Naivasha Management Plan. Lake Naivasha Riparian Association, Naivasha, 77 pp.

Ramsar, 1999a. http://ramsar.org/key_awards99_index_e.htm

Ramsar, 1999b. http://ramsar.org/key_awards99_interview_lnra.htm

ENNISKILLEN
Chairman
Lake Naivasha Riparian Association
P.O. Box 1011
Naivasha
20117
Kenya

Hydrobiologia **488**: 1–11, 2002.
D.M. Harper, R. Boar, M. Everard & P. Hickley (eds), Lake Naivasha, Kenya.

Towards an understanding of human impact upon the hydrology of Lake Naivasha, Kenya

Robert Becht[1,*] & David M. Harper[2]
[1]*ITC, P.O. Box 6, 7500 AA Enschede, The Netherlands*
[2]*Department of Biology, University of Leicester, Leicester LE1 7RH, U.K.*
**Author for correspondence. E-mail: becht@itc.nl*

Key words: lake hydrology, modeling, water budget, evaporation, irrigation

Abstract

The water balance of Lake Naivasha, has been calculated from a model based upon the long-term meteorological data of rainfall, evaporation and river inflows. The lake is Kenya's second Ramsar site because of its international importance as a wetland, but supplies drinking water to Nakuru and irrigation water to the nationally important industries of horticulture and power generation. Groundwater flows into and out of the lake are estimated from the model's success in predicting water level fluctuations over the same period. The most accurate predictions of lake level were derived from the data sets of river discharges known to be from the most-reliable time period and gauging stations. The model estimated a current annual abstraction rate of 60×10^6 m^3 ann^{-1}, a figure perhaps six-times higher than that calculated as a 'safe' yield in the 1980s. There is an urgent need to accurately measure all abstractions and provide consistent, reliable, hydrological and meteorological data from the catchment, so that a 'safe' yield may be agreed upon by all stakeholders and sustainable use of the lake waters achieved.

Introduction

Lake Naivasha is situated on the floor of the Eastern Rift Valley, at its highest elevation 1890 m, in Kenya (0° 45′ S, 36° 20′ E). The faulting which produced the Rift Valley has led to extensive, and often intense, volcanic activity associated with it, evidence for which is widespread (Fig. 1.) The valley floor is composed of a complex stratigraphy of volcanic and fluvio-lacustrine deposits laid down in Pleistocene times from a larger lake (Butzer et al., 1972; Richardson & Richardson, 1972). The underlying rocks are a complex and fractured mosaic as a consequence of this tectonic activity, and include a deep and wide aquifer. Water from the lake seeps into this (Gaudet & Melack, 1981) and moves through it, probably both southwards towards Longonot and northwards towards Gilgil (Clarke et al., 1990). The lake has no surface outlet.

This underground outflow is maintained by inflow from the higher altitudes of the Rift's flanks (because on the Rift floor, evaporation exceeds rainfall). The catchment on the north and north-eastern side consists of two long axial river systems, draining into Naivasha as a result of past intra-rift faulting and of recent volcanic activity (Clarke et al., loc. cit.), the Gilgil and Malewa (formerly Melewa). Their high altitude origin (>2500 m; rainfall 1100 mm ann^{-1} compared to 600 mm ann^{-1} at Naivasha) makes them permanent, feeding Lake Naivasha mainly through surface flow (input calculated as 85% surface and 15% subsurface, Gaudet & Melack (1981)).

On the eastern, western and north-western and southern side of the catchment, no surface water reaches the lake (Fig. 2). Rivers, often clearly incised into the landscape, flow seasonally. Only the eastern stream, the Karati, reaches the lake in high rains. Others end before the lake, such as the Marmonet from the Mau Escarpment on the western Rift flanks, which recharges the Ndabibi Plains (Clarke et al., loc. cit.)

The outflow water from Lake Naivasha has been detected in the north (estimated 30% contribution to the warm springs on the northern edge of Lake Ele-

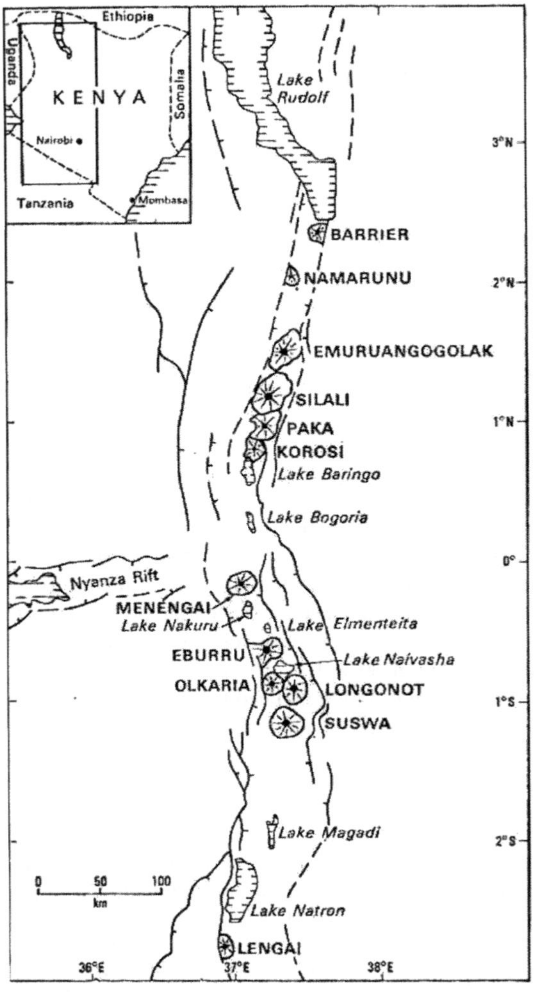

Figure 1. The Eastern (Gregory) Rift, showing the evidence of tectonic activity and the location of Lake Naivasha. Modified from Clarke et al. (1990).

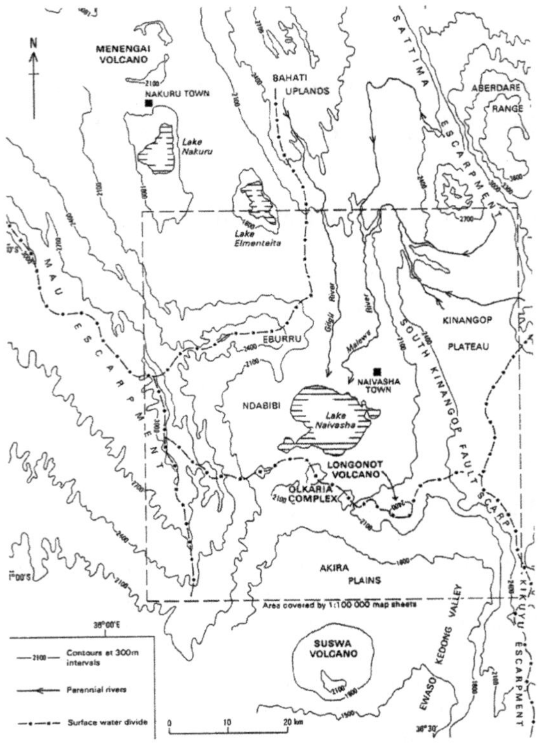

Figure 2. The catchment of Lake Naivasha showing the surface-water drainage pattern. Modified from Clarke et al. (1990).

menteita) and in the south (an estimated 60% contribution to the Olkaria well-field and a lower contribution to fumaroles east of this) (Clarke et al., loc. cit.).

The water balance of Lake Naivasha has been of wider interest for over 100 years; initially because of scientific curiosity about the causes of its extreme fluctuations but latterly for its economic value for irrigation and supply of potable water to Nakuru.

The scientific interest was strengthened by the archaeological evidence from the 1930s onwards from raised shorelines, which showed that Naivasha was part of a much larger lake (Leakey, 1931; Nilsson, 1932), (Fig. 3), either including present-day Elementeita and Nakuru, or just the former (Åse et al., 1986). This realization then led to suggestions of synchrony between the major ice ages in the Northern Hemisphere with pluvials (wet periods) in the tropics. Palaeolimnological analysis did not support this hypothesis (Richardson, 1972; Richardson & Richardson, 1972). The lake level fluctuations have attracted analysis because their dependence upon the high-altitude rainfall makes them also dependent upon the Inter-Tropical Convergence Zone (ITCZ), which brings annual rainfall to the East African highlands in two rainy seasons as a consequence of its march northwards then southwards in a pattern controlled by the meeting of the westerly winds from the Indian Ocean with the easterly winds from the Atlantic. This large-scale climatic influence was confirmed by Vincent et al. (1979), who showed the similarity of Naivasha and Turkana level fluctuations in contrast to Lake Victoria's. These fluctuations were correlated with the altitude fluctuations of the snout of the Lewis glacier on Mount Kenya.

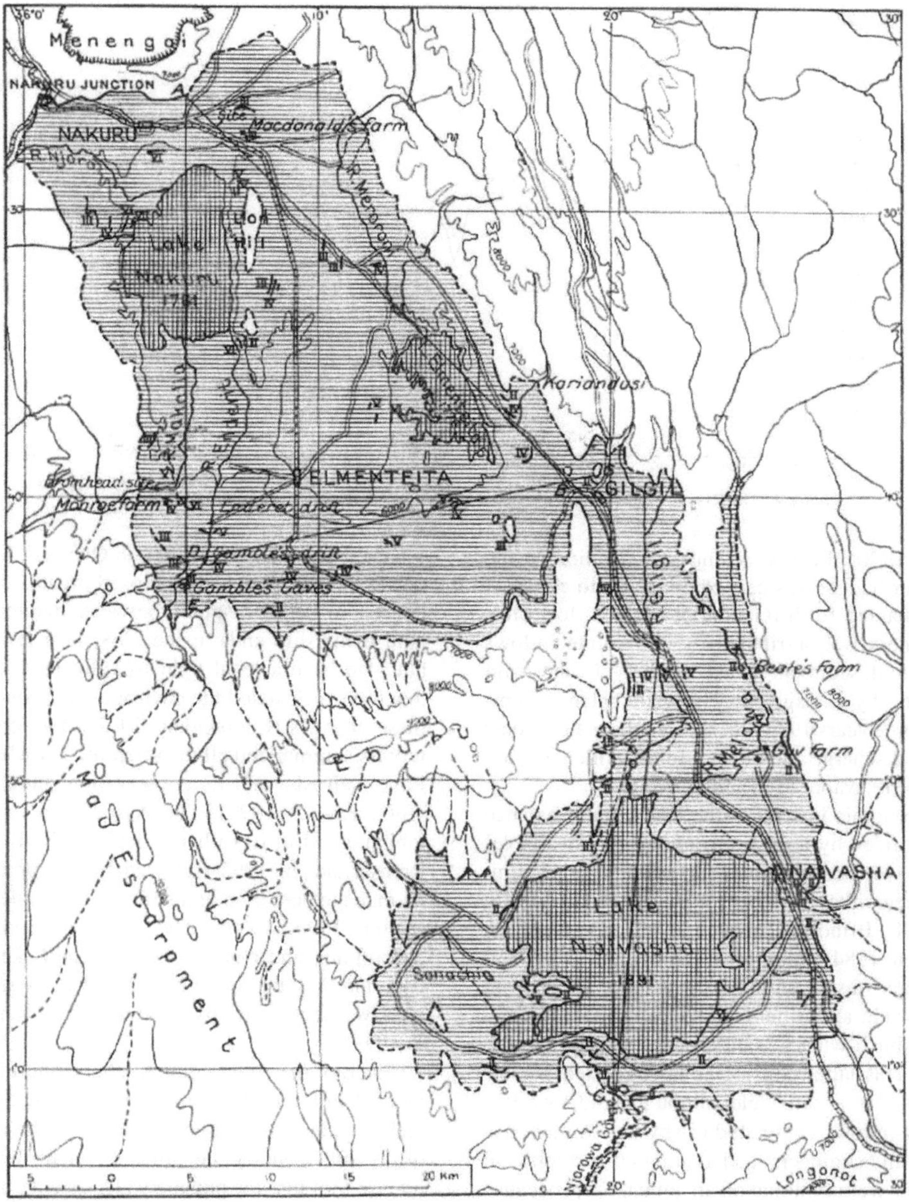

Figure 3. Pleistocene lakes in the Rift Valley. Modified from Nilsson (1932).

Searches for causes in lake level fluctuations have focused upon sun spot cycles, which showed a good correlation early in the 20th century but a weak one latterly (Åse et al., loc. cit.). Vincent et al., (loc. cit.) found strong evidence for an approximate 7-year periodicity, similar to the interval of El Nino events in the Southern Pacific. These apparent cycles of short-term lake level rises and falls have occurred against a longer-term pattern of lake level decline, which has been in evidence throughout the 20th Century (Fig. 4) with the exception of the decade 1955–65 when the decline was temporarily reversed.

Figure 4. The level of Lake Naivasha over the past 130 years (reconstructed from several sources).

In the 1970s the first scientific evaluation of the lake's water budget was carried out to try to establish why it remained fresh (Gaudet & Melack, loc. cit.). The freshness was attributed to very dilute inflows combined with sediment uptake and loss of water with some solutes by seepage out. The water balance, compiled for three years 1973–75, showed the major inputs to be river discharge and direct rainfall while the main outflows were evaporation and seepage.

The economic interest in the lake's water balance has been driven by a desire to utilize the 'available' freshwater with calculations appearing at approximately 10-year intervals over 60 years (Sikes, 1936; Tetley 1948; Brind & Robertson 1958; Oestergaard, 1974; Anon, 1984) although it was not until 1984 that any attention was given to how a yield of water from the lake would affect the ecology of the lake by trying to establish what a 'safe' level would be (Anon, loc. cit.). This latter consideration suggested 16.5×10^6 m^3 ann^{-1} as a 'safe yield' although it did not relate this to any inflow. These studies (by government hydrologists and engineers) had assumed that catchment abstractions were minimal on an annual basis because the existing licenced abstraction volume was small.

By the 1990s the nature of agriculture around the lake had changed substantially. Former stock-rearing, ranching and sisal-cultivation had given way to approximately 100 km^2 of irrigated horticulture (Johnson et al., 1995) whose output is air-freighted to Europe. This land use change, dramatic in itself, has brought even greater social changes in the rise in population of estate-labourers and their dependents.

The Lake Naivasha Riparian Owners' Association (LNROA) (until recently a closed organization which existed to settle disputes between members whose land ownership ends at the 1906 lake level contour but who have a legal right to cultivate lake bed below this), articulated the environmental concern about these changes (Enniskillen, 2002). Two consultants' reports summarized all the scientific knowledge about Lake Naivasha and its conservation status (Goldson 1993, Khroda 1994). The Association's subsequent lobbying led to Lake being declared Kenya's second Ramsar site in 1995 followed by the production of its management plan and strategy for implementation (Anon, 1996, 1997). The Association changed its constitution and name to become the Lake Naivasha Riparian Association (LNRA) and its organizational work was recognized by the award of the Ramsar Prize in 1999.

Each of the Association's reports and its Management Plan focused upon the water balance and water yield as the most important issue. At the conclusion of the 20th Century the demands for freshwater were intense, not just for potable water as envisaged half a century earlier but also for intensive irrigation and for water in the Olkaria Geothermal Power Station, 10 km to the south, which generates 15% of Kenya's power. The total calculated yield of freshwater from the lake plus catchment from these three uses in the late 1980s–early 1990s was estimated by Goldson (loc. cit.) at 37, 39 and 15 ×; 10^6 m^3 ann^{-1} respectively, a total of 91 $\times$ 10^6 m^3 ann^{-1}.

There is now an urgent need on the one hand to accurately measure what is actually happening to the water balance of the whole catchment and to support this with an hydrological model which allows hypothetical scenarios to be evaluated. One hydrological model was produced in the 1990s (Stuttard et al., 1999), but it was based on satellite imagery and sophisticated GIS packages, which make it less suitable for field use in Africa. The relatively short calibration period during a period that large-scale abstractions were increasing on a yearly base make it less suitable to evaluate the long-term waterbalance. This paper addressed the need for an hydrological model with a simple and usable one based upon a Microsoft Exel™ programme.

Methods

Model construction

The model is based upon the monthly change in a simplified water balance. Components used are inflow from rivers, rainfall on the lake surface, evaporation from the lake surface, a constant seepage (groundwater outflow) and a dynamic groundwater component to take into account the interactions with the aquifer surrounding the lake.

The lake Level–Area–Volume relationship is built into the model and allows the calculation of the rain and evaporation as a volume. The model uses a monthly time step, and is is expressed as:

$$\text{lake volume change} = \text{inflow} + \text{rainfall} - \text{evaporation} \pm Q_{aq} - Q_{out}, \quad (1)$$

where Q_{aq} is the inflow to or outflow from a hypothetical dynamic groundwater aquifer linked to the lake. It is derived as:

$$Q_{aq} = \mathrm{C}\,(\mathrm{H}_{lake} - \mathrm{H}_{aquifer})\ \mathrm{m}^3\ \mathrm{month}^{-1}, \quad (2)$$

where C is the hydraulic conductance of the aquifer ($\mathrm{m}^2\ \mathrm{month}^{-1}$) and H is the water level (m).

The water level in the aquifer is updated using the in/outflow calculated for the previous month:

$$H_{aquifer} = Q_{aq}\ /A \times \mathrm{Sy}\ (\mathrm{m}^3\ \mathrm{month}^{-1})$$

and

$$H_{aquifer^{new}} = H_{aquifer^{old}} + H_{aquifer}(m),$$

where A is the surface area and Sy is the specific yield (porosity) of the aquifer.

Q_{out} ($\mathrm{m}^3\ \mathrm{month}^{-1}$) is the water balance deficit, set to a constant for each model run. It lumps and to a certain extent balances out all missing parts and errors in the water balance. The major component is the outflow from the lake, but also the long-term unknown inflows from direct runoff and groundwater inflow, the evaporation of riparian vegetation and a systematic over or underestimate of the inflow, rainfall and evaporation are a part of this term.

Inflow is the river inflow, expressed as $\alpha \times$ gauged inflow ($\mathrm{m}^3\ \mathrm{month}^{-1}$), where α is a factor to modify the inflow for a systematic error in the data. *Rainfall and Evaporation* are as named, input in mm month^{-1} from the meteorological records.

The model was optimized by minimizing the sum of squared differences between observed and simulated monthly water levels. The optimizing model parameters were the constant outflow, the hydraulic conductance of the lake aquifer system and the specific yield of the groundwater reservoir.

Input data and limitations

Meteorological data were obtained from the Ministry of Environment and Natural Resources (MENR), Government of Kenya, which is the custodian of all meteorological data collected in the country, some dating back to the beginning of the 20th Century. Lake level data have been recorded at Naivasha throughout the century as feet or metres above sea level (discussed more fully in Åse, 1986). The latter data are derived from a confusing and conflicting range of absolute levels. For the model it was decided to use the old data collected by the erstwhile Ministry of Works and using the Cassini Projection after Laws & Flintoff (1950). Data since 1983 have been recorded by Sulmac, a large horticultural company, whose level has been authenticated by a professional surveyor on behalf of LNRA.

Lake bathymetry has been recorded on several occasions since the first government survey in 1928. The most recent are Åse, in 1983 (Åse 1986), Hickley in 1993 (Hickley et al., this volume) and WRAP/MNR&E, in 1997. The flow data for the rivers Malewa, Gilgil and Karati have been used, starting from 1932 with missing data infilled from the nearest data by statistical correlation to give a continuous monthly record to 1998.

The data are reliable until the mid-1970s, after which the frequency of missing data increases. Malewa data were not recorded after 1985 and so the flow has been calculated using the Turasha (Malewa main tributary) data, which had good quality data from 1950 to 1990 and the Gilgil from 1958–94.

The rainfall data of Naivasha (District Office). and Kinankop Forest Station have been recorded from 1901 and 1915, respectively, to present. Evaporation data are derived from the measured pan evaporation data have been used for the period 1960 to 1998. For the period 1901 to 1960, mean monthly values of the 1960–98 period have been used.

Model calibration

The model parameters which can be optimized are α, hydraulic conductance, specific yield, initial water level and most important the constant outflow, Q_{out}. In the calibration of the model α, the inflow correction factor, was set to one. The initial water level was set to the water level in January 1932.

None of the optimized parameters were time variant. This means that the parameters were not optimized for intervals to get a better fit but always for the full calibration period. The optimalization was carried out using the data from 1932 to 1978. because it was assumed that the abstraction rates during this period were low and that the overall quality of the flow data was good.

Water budget

An average annual natural water budget of the lake was estimated for a reliable period prior to industrial abstraction. In February 1932 and again in June 1981 the lake was at 1889.2 m and the area 164 km^2, and so the average of this period was taken. The purpose of understanding the water balance of Lake Naivasha is, as has been for 70 years, to estimate a 'safe yield' of water from the lake. An estimate of this 'safe yield' was made starting from this long-term annual average. An 'equilibrium lake area', Aeq$_w$ was calculated, where

$$(\text{rainfall} - \text{evaporation}) \times A_{\text{eq}_w} + \text{inflow} - \text{outflow} = 0. \qquad (3)$$

The relation between lake area and lake level defines the long term equilibrium lake level. This concept is important discussing the sustainable or safe yield. It should be realized that a constant abstraction from the system translates in a reduced lake area and therefore lake level. For every rate of abstraction (smaller than the total inputs) a long-term equilibrium level will be established and the system is in water balance terms in an equilibrium and therefore a sustainable state. The problem of what is the safe yield, is rather a political and ecological question than an hydrological one.

Results

The first model runs made it clear that simulation of the lake levels was possible, but that data after 1978 were not of high enough quality (Fig. 5). The model was re-run using only Malewa discharge data from 1932–1950 and Turasha from 1950–1990 as these were considered the most reliable sets. Figure 6 shows a better match between simulated and actual levels was achieved. The standard deviation of the difference between the two was 0.26, which means that 95% of all monthly levels differ 0.52 m or less.

The deviation of simulated from observed water levels in the past two decades is distinct and indicates the magnitude of industrial abstraction. Since the model (Equation (1)) does not contain an abstraction component the divergence of the curves from 1983 onwards is caused by the abstractions from the basin. Using the monthly differences between observed and simulated lake levels allows an estimate of lake volume used for industrial abstraction starting from January 1983. This has risen to 60×10^6 m^3 ann^{-1}, a value close to estimates derived from the area under irrigation and the irrigation requirement of the crop pattern. The calculated abstraction has resulted in a lake which might have been 3–4 m higher than was observed in November 1997 before the rapid rise caused by the 'El Nino' rains (Fig. 7).

Mean monthly local rainfall (Naivasha D.O.), lake inflow and lake level rises show an approximately monthly time lag (Fig. 8), indicating the natural catchment response for the time taken to restore soil moisture deficit after the dry season. Naivasha local rainfall does not directly influence the lake, because of the positive dominant influence of the high-altitude rainfall and the negative influence of evaporation. Nevertheless, there are four reasons why a lake level model based on local rainfall would be valuable:

1. Rainfall data exist from 1900 for Naivasha town and can be used to model the lake levels from 1900 to present.
2. If rainfall data can be used to accurately predict lake levels for the period before exploitation

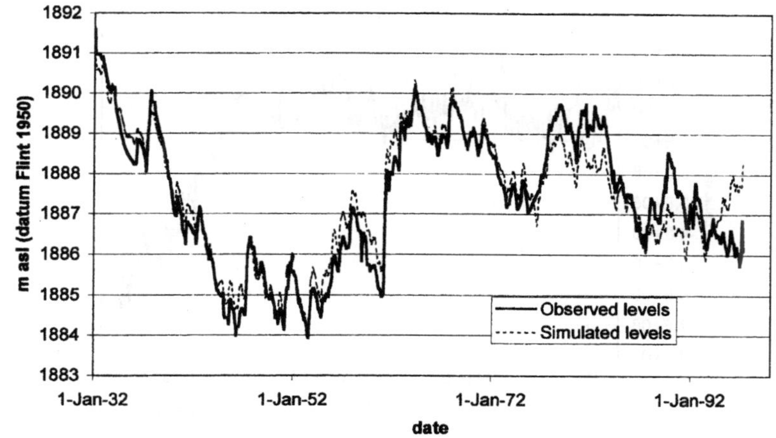

Figure 5. The simulation of Lake Naivasha water levels from 1932 to present, based upon measured inflows from the Malewa, Gilgil and Karati rivers compared to recorded levels.

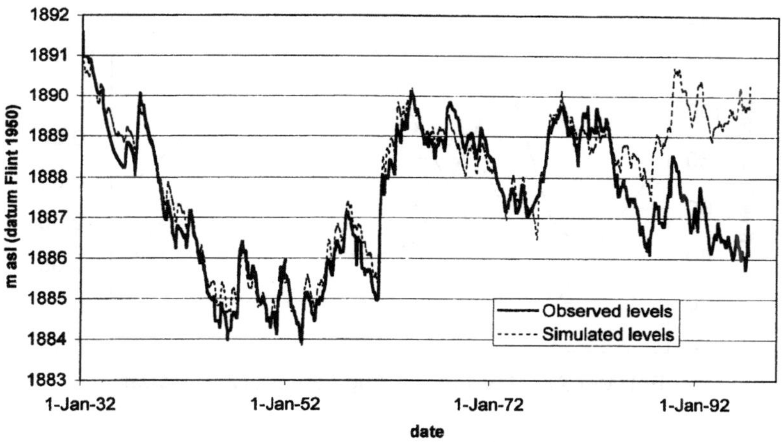

Figure 6. The simulation of Lake Naivasha water levels from 1932 to present based upon measured inflows from the Turasha river compared to recorded levels.

of water resources started, this model can show the effect of abstractions in the whole Naivasha catchment.

3. If a reliable rainfall runoff model exists it renders the model independent of the availability of flow data.
4. If the model is physically based it can be used to predict the hydrological effect of changes in the catchment and climate.

Figure 9 shows the output of this model, constructed through correlation of monthly inflow volume with monthly rainfall. The accuracy of this model is lower than the fuller hydrological model (a lake level confidence of 2 m compared with 0.52 m).

It is peculiar that the rainfall data from Naivasha town should give slightly better results than the rainfall data from Kinankop Forest Station, which is in the middle of the water-generating upper catchment. This indicates that the large fluctuations are driven by long-term overall drier and wetter periods.

Sustainable water resource exploitation

The average annual water balance of the lake for the period February 1932 to June 1981 (Table 1) was estimated with an error of 1.36×10^6 m^3 (equivalent to

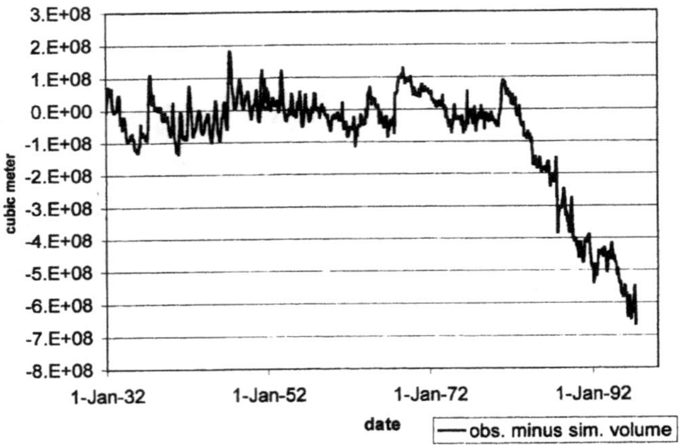

Figure 7. The difference in lake volume between observed and simulated levels, based upon a composite flow series (1932–1952 Malewa; 1952–1997 upscaled Turasha).

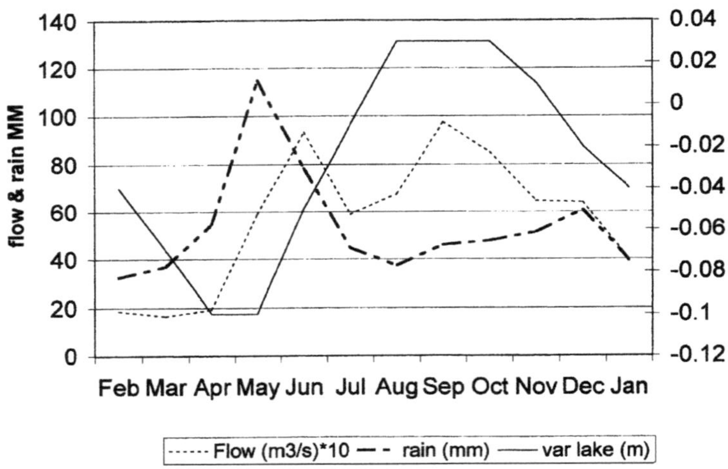

Figure 8. The mean monthly rainfall at Naivasha, inflow of the Malewa river and level of lake Naivasha, 1932–1997.

Table 1. The long-term average annual water balance from the model, prior to irrigation abstraction

Surface water inflow	217.4 $\times 10^6$ m^3
Rainfall	93.9 $\times 10^6$ m^3
Evaporation	256.3 $\times 10^6$ m^3
Water loss	56 $\times 10^6$ m^3
Sum (error)	1.36 $\times 10^6$ m^3

8 mm of water level). The equilibrium water level was estimated at 1886.5 m, corresponding to a long-term equilibrium lake area of 140 km^2. Abstraction from the lake will result in a reduction of this equilibrium lake level and area. Simulating the effect of estimated abstractions of 60 $\times$ 10^6 m^3 ann^{-1} on the new equilibrium lake level and area, using Equation (4), gave a level of 1883.4 m, which corresponds to an area of 82 km^2. Figure 10 shows the hypothetical effect of an exploitation of 60 $\times$ 10^6 m^3 ann^{-1} from 1932 to present. The average lake level has been reduced from approximately 1988 to approximately 1984.5 m. The simulation also shows that it takes some 10 years before a new equilibrium has reached.

The rise of the simulated water level after 1993, and the almost equal simulated and observed water levels at the end of 1997, are due to three effects:

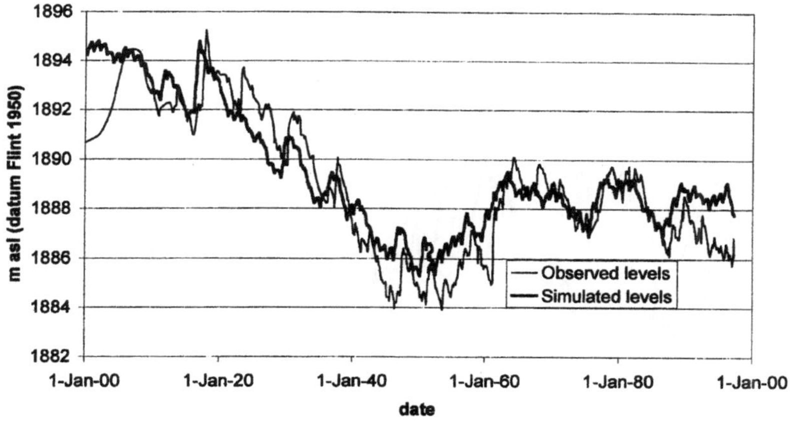

Figure 9. The simulation of lake Naivasha water levels through the 20th Century based upon rainfall at Naivasha D.O. compared to recorded levels.

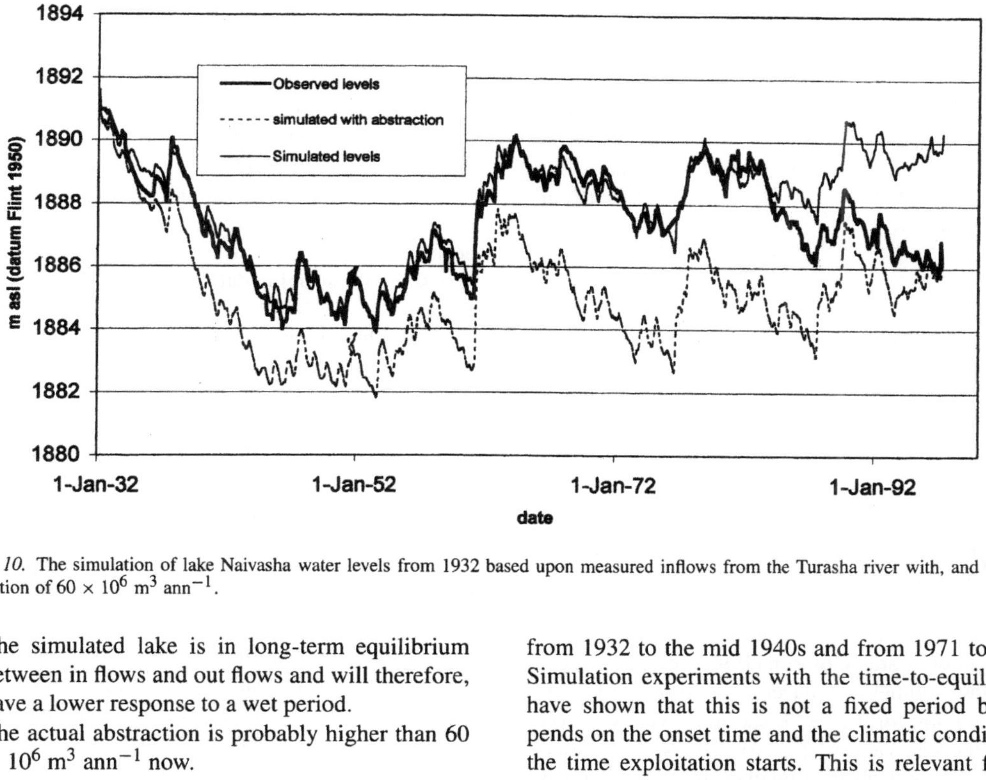

Figure 10. The simulation of lake Naivasha water levels from 1932 based upon measured inflows from the Turasha river with, and without, abstraction of 60×10^6 m^3 ann^{-1}.

1. The simulated lake is in long-term equilibrium between in flows and out flows and will therefore, have a lower response to a wet period.
2. The actual abstraction is probably higher than 60 $\times 10^6$ m^3 ann^{-1} now.
3. If all assumptions are correct the two curves will always converge to the new equilibrium level.

Comparison of exploitation of 60×10^6 m^3 ann^{-1} commenced in 1932 with a similar exploitation in 1961 as is shown in Figure 11. Here again the system needs a considerable time span to finds its equilibrium, from 1932 to the mid 1940s and from 1971 to 1990. Simulation experiments with the time-to-equilibrium have shown that this is not a fixed period but depends on the onset time and the climatic condition at the time exploitation starts. This is relevant for the interpretation of the present situation since it is not possible to predict how close or distant the lake is from a new equilibrium. However, since the simulations use a constant rate of abstraction, whereas in the real situation the abstraction rate has been constantly increasing over the last 15 years due to the expanding

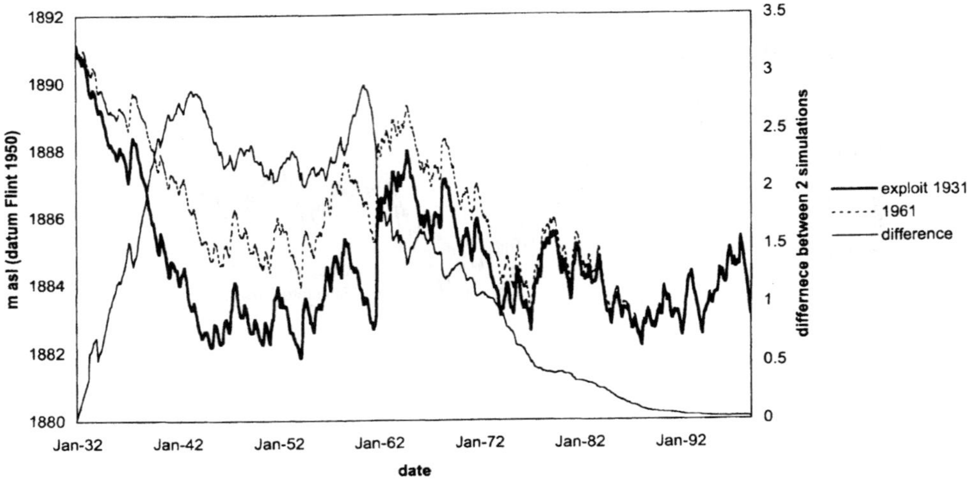

Figure 11. The simulation of Lake Naivasha levels under exploitation starting in 1961 compared to 1932 demonstrating the time taken to reach an 'equilibrium level'.

Table 2. The long-term average water balance figures used for the calculation of equilibrium lake levels and areas

Naivasha DO rainfall 1900–1998	648 mm
Evaporation Naivasha MoWD 1959–1990	1788 mm
Inflow Malewa 1932–1980	217.4×10^6 m^3
Groundwater outflow	56×10^6 m^3

area under irrigation we may safely assume that the lake is not in an equilibrium state.

Discussion

The abstractions from the Naivasha catchment are not known and can only be estimated from abstraction permits or from the known area of cropland and known requirements of the crops. Groundwater, abstracted from boreholes, may or may not be contiguous with lake or with inflow water. Goldson (1993), after discussing all the uncertainties and inaccuracies, estimated an abstraction figure for the 1990s six times what was considered the safe yield in the early 1980s. The present model suggests an abstraction in the mid-1990s at two-thirds of Goldson's estimate, but still three to four times the original concept of 'safe'. Some of the permits for river abstraction may well have influenced the long-term flow data, since the total permitted abstraction accounts for approximately 17% of the estimated long-term surface inflows to the lake (Table 2). There is thus an urgent need to accurately measure all the abstractions from river, lake and groundwater, in order to build an accurate water balance. Equally, there is an important requirement to scientifically and logistically establish an accurate and reliable hydrological and meteorological data base, as is evidenced by the accuracy of the model when run with only the time series and the gauging data considered to be the most accurate.

The other important need is to understand what a 'safe yield' means in terms of a sustainable lake, which can be interpreted in two ways. The first is the lake in terms of its water quality and quantity for irrigation. As the lake level (and area) decrease, water may increase in the short-term (as it becomes closer to dilute river water) but may decrease in the long-term (from greater algal activity in a more nutrient-rich environment, for example). The costs of using lake water for sensitive horticultural crops increase. The second is in terms of the biodiversity, ecology and conservation of the lake. No one has yet tried to predict the way in which the myriad aspects of the lake's ecology will change with either a declining lake level or an 'equilibrium', more predictable, lake level.

Encompassing both the uncertainty over the economic use of the lake and the unpredictable ecological response, is the concept of the 'Tragedy of the Commons' (Hardin, 1968) and the real prospect of its appearance at Lake Naivasha. At present, the waters in the Naivasha catchment, like the waters in many parts of the developing as well as developed world, are treated as common property, to be exploited by each and every individual, company and para-statal according only to their needs and ability to utilize them (i.e., wealth). With no measurement of 'how much', no concept of what is a sustainable/safe yield, and no policy on how to match what is used with what is sustainable, Lake Naivasha has a bleak long-term future in a continent with a bleak future (Ashton, 2002) unless drastic action is taken soon. This paper is a small step in trying to define a sustainable yield for Lake Naivasha and facilitate that action.

Acknowledgements

The work presented here is one of the results of a collaboration program between the Ministry of Environment and Natural Resources (Water Resources Assessment Program), Kenya Wildlife Services (Kenya–Netherlands Wetlands Training and Conservation Program) in Kenya and ITC in the Netherlands. In the framework of this program ITC students carried out their M.Sc. research in the lake Naivasha basin as part of a broader cooperation between the authors and the stakeholders of Lake Naivasha.

References

Anon, 1984. Lake Naivasha Studies. File WD/4/4/231, 23/10/84. Ministry of Water Development, Nairobi: 4 pp.

Anon, 1996. The Lake Naivasha Management Plan. Lake Naivasha Riparian Owners' Association, Naivasha, July 1996: 48 pp.

Anon 1997. A Proposal for the Implementation of the Lake Naivasha Management Plan. Lake Naivasha Riparian Owners' Association, Naivasha, April 1997: 19 pp.

Åse, L-E., Sernbo, K. & Syren, P., 1986. Studies of Lake Naivasha, Kenya and its Drainage Area. Forkningsrapporter 63, Stockholms Universitet Naturgeografiska Institutionen, Stockholm: 75 pp.

Ashton, P. J., 2002. Avoiding conflicts over Africa's water resources. Ambio 31: 231–235.

Brind, W. & J. Robertson, 1958. The Hydrology of Lake Naivasha. Hydrology Section, Ministry of Works, Nairobi: 9 pp.

Butzer, K. W., G. L. Isaac, J. L. Richardson, & C. Washbourne-Kamau, 1972. Radiocarbon dating of East African lake levels. Science 175: 1069–1076.

Clarke, M. C. G., D. G. Woodhall, D. Allen & G. Darling, 1990. Geological, Volcanic and Hydrological Controls on the Occurrence of Geothermal Activity in the area surrounding Lake Naivasha, Kenya. Minstry of Energy, Nairobi: 138 pp.

Enniskillen, 2002. The Lake Naivasha Management Plan – consensus-building to conserve an international gem. Hydrobiologia 488 (Dev. Hydrobiol. 168): ix–xii.

Gaudet, J. & J. M. Melack, 1981. Major ion chemistry in a tropical African lake basin. Freshwat. Biol. 11: 309–333.

Goldson, J. 1993. A Three Phase Environmental Impact Study of Recent Developments around Lake Naivasha. Lake Naivasha Riparian Owners' Association, Naivasha, December 1993: 109 pp.

Leakey, L. S. B., 1932. East African lakes. Geog. J. 77: 497–512.

Nilsson, E., 1932. Quarternary Glaciations and Pluvial lakes in British East Africa. PhD Thesis, Centraltryckeriet Stockholm: 101 pp.

Oestergaard, P., 1974. Abstraction of water from Lake Naivasha. Files WRH 206 vol IV/131, 5/4/74: 2 pp and WRH 206 vol III/132, 28/5/74: 5pp, Ministry of Water Development, Nairobi.

Richardson, J. L., 1966. Changes in the level of Lake Naivasha, Kenya, during postglacial times. Nature, 5620: 290–291.

Richardson, J. L. 1972. Palaeolimnological records from Rift lakes in central Kenya. In van Zinderen Bakker, E. M. (ed.), Palaeolimnology of Africa, the Surrounding Islands and Antarctica. Balkema, Cape Town: 131–136.

Richardson, J. L. & A. E. Richardson, 1972. History of an African Rift lake and its climatic implications. Ecol. Mongogr. 42: 499–534.

Sikes, H. L., 1936. Notes on the hydrology of Lake Naivasha. Journal of the East African and Uganda Natural History Society, 13: 73–89.

Stuttard, M., J. B. Hayball, G. Narciso, M. Suppo, L. Isavwa & A. Oroda 1999. Modelling lake level changes: examples from the Eastern Rift Valley, Kenya. In Harper, D. M. & A. Brown (eds), The Sustainable Management of Tropical Catchments. J. Wiley & Sons, Chichester: 337–358.

Tetley, A. E., 1948. Notes on Lake Naivasha by the Hydraulic Engineer. Directorate of Water Development, Nairobi: 9 pp.

Vincent, C. E., T. D. Davies & U. C. Beresford, 1979. Recent changes in the level of Lake Naivasha, Kenya, as an indicator of equatorial westerlies over East Africa. Climatic Change 2: 175–191.

Hydrobiologia **488:** 13–25, 2002.
D.M. Harper, R. Boar, M. Everard & P. Hickley (eds), Lake Naivasha, Kenya.

The physical attributes of the Lake Naivasha catchment rivers

Mark Everard[1,*], Jacqueline A. Vale[2], David M. Harper[3] & Håkan Tarras-Wahlberg[4]
[1]*The Natural Step U.K., 9 Imperial Square, Cheltenham, Gloucestershire GL50 1QB, U.K.*
[2]*Environment Agency, Rio House, Waterside Drive, Aztec West, Almondsbury, Bristol BS32 4UD, U.K.*
[3]*Department of Biology, Leicester University, Leicester LE1 7RH, U.K.*
[4]*Swedish Geological AB, Box 19090, 104 32 Stockholm, Sweden*
**Present address: Environment Agency, King's Meadow House, King's Meadow Road, Reading, Berkshire RG1 8DQ, U.K. E-mail: mark.everard@environment-agency.gov.uk*

Key words: RHS, river habitat, sediment, Malewa, Gilgil, Karati

Abstract

Lake Naivasha is unique in the chain of East African Rift Valley lakes in being the only freshwater ecosystem in an otherwise soda/saline lake series. Catchment rivers have never previously been surveyed. Valley topography, precipitation patterns and geology create a diversity of river systems draining into Lake Naivasha. Semi-quantitative assessment using the River Habitat Survey technique demonstrates a diversity of habitats and erosion/sedimentation patterns, reflecting the tectonic origin of the area. The survey descriptions form a basis for identification of areas where efforts will need to be directed for physical and biological protection.

Introduction

Lake Naivasha (0° 45′ S, 36° 20′ E; altitude 1890 m) lies on the floor of Africa's Eastern Rift Valley and, at approximately 100 km^2, is the second-largest freshwater lake in Kenya. It is one of a series of 23 major lakes in the East Rift Valley – eight in central Ethiopia, a further eight in Kenya and seven in Tanzania – spanning latitudes from approximately 7° N to 5° S. The overall climate of the Eastern Rift Valley is semi-arid, with the exception of regions of central Ethiopia and central Kenya. All but four of these Eastern Rift Valley lakes are alkaline or saline. Lake Naivasha is unique within the central latitudes of the valley, and indeed within the Kenyan series of lakes (from north to south these are Turkana, Baringo, Bogoria, Nakuru, Elementeita, Naivasha, Magadi and Amboseli) in being fresh, with a mean conductivity of around 250 μS cm^{-1}.

The surface inflows to the lake come via three main river systems – the Gilgil, the Malewa and the Karati – through a papyrus-dominated fringe in the northern part of the lake. The Malewa system is the largest with an estimated annual flow of 153 million m^3, the Gilgil has an estimated average annual flow of 24 million m^3, whereas the Karati only flows intermittently during the rainy season (Ase et al., 1986; Ase, 1987). The lake area fluctuates quite markedly and rapidly. For example, the heavy and prolonged precipitation caused by exceptional 'El Nino' climatic patterns between August 1997 and July 1998 resulted in a rise in lake level of 3 m, pushing inland over a kilometre in the low-gradient north. The background to the lake is described by Harper (1984) and Harper et al. (1990, 1995).

A significant body of research has contributed to the designation of Lake Naivasha as Kenya's second Ramsar site (after Lake Nakuru) in 1995. Under the Ramsar designation, a lake management plan is required to co-ordinate the sustainable development of the lake (Ramsar, 1971). (The Ramsar Commission consider sustainable development to be consistent with 'wise use' as defined by Ramsar (1996)). Uniquely for a Ramsar site, the Kenyan government has delegated responsibility for management of the site to a non-statutory body; the Lake Naivasha Riparian Association (LNRA). The plan will be required to address conflicting pressures to resolve the issue of wise use. However, wider impacts on the lake deriving from the catchment have been largely overlooked. Everard (1998) notes that the conservation and sustainable management of aquatic ecosystems – whether rivers, wetlands or stillwaters – must be considered within the context of the living landscapes of which they are a part, and of the processes that form and sustain them. Considerations must also address not merely the flow of water but also of energy and sediment (Newson, 1992) as well as of pollutants, productivity products and biological inocula, channel roughness and micro-hydrodynamics within river channels. The services provided by wetland systems within catchments, which are widely overlooked in management decisions (Everard, 1997; Cairn, 1997), deliver substantial benefits to watercourses and water bodies, and to those that use and live near them, through the various hydrological, ecological, physical and social functions they perform (Dugan, 1990). Four-fifths of Kenya is arid or semi-arid and prone to drought (Mutiso, 1989), making more urgent wise use of water resources to support an increasing population, with its resultant pollution and economic development pressures (Orie, 1996).

This study represents the first documentation of catchment morphology and land use to support both a better understanding of impacts upon the lake, and further to inform wise use decisions.

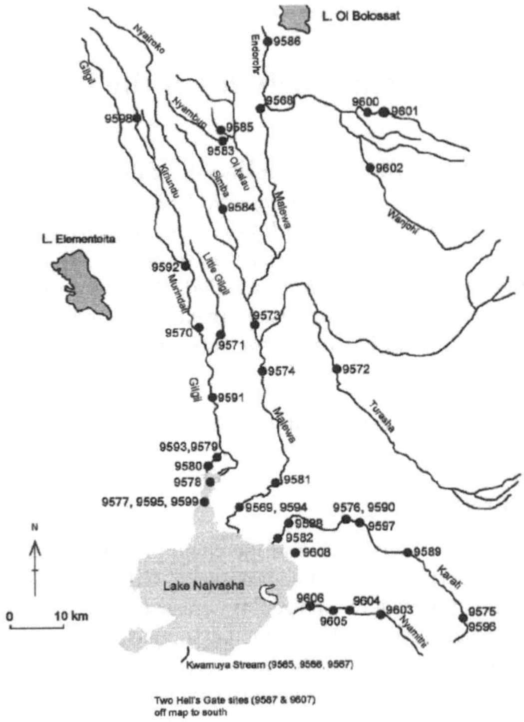

Figure 1. Lake Naivasha catchment showing the RHS sites.

Methods

Forty-four stretches of river throughout the Lake Naivasha catchment were surveyed during three sampling periods (August 1997, July 1998 and March/April 1999) using the River Habitat Survey (RHS) method. RHS is a method developed for England & Wales for the Environment Agency, based on both map-derived and field survey data, which can be interpreted statistically (Raven et al., 1998). Site selection in the Naivasha catchment was initially map-based, with the intention of establishing a representative network, but final selection was constrained significantly by accessibility and safety considerations. RHS surveys were repeated at a minority of sites between years, and two sites were evaluated outside the catchment (streams in the adjacent Hell's Gate National Park) for comparative purposes. RHS sites are listed in Table 1, their location shown in Figure 1 and RHS data and site photographs are available on CD-ROM (Environment Agency & Institute of Freshwater Ecology, 1999).

The survey was carried out in low flow conditions. Water depth and width were recorded and the percentage low/bankful flow calculated from:

$$\%\text{low/bankful} = \frac{(\text{water depth} \times \text{water width})}{[(\text{bankful height} + \text{water depth}) \times \text{bankful width}]} \times 100.$$

This calculation has no firm statistical basis since 'low flow' conditions recorded between field sampling periods may have varied, and the channel cross section calculation is simplified. However, the data are indicative of the proportion of low flow to high flow in river channels, and convey some concept of 'flashiness'.

As a surrogate of channel energy (the widespread occurrence of dry channels render flow type data unreliable) and sediment flow, the substrate distribution at 50 m transects within the RHS sites was recorded together with visible silt deposition features.

Results

The hydrology and climate of the Eastern Rift Valley, which has an unreliably low mean annual precipitation, is influenced by the topography of the escarp-

Table 1. Features of RHS sites in the Naivasha catchment (site numbers in Figure 3)

RHS site no.	RHS site (river system and site name)	Longitude/Latitude	Approximate altitude (metres)	Date
9565	KWAMUYA; BELOW ROAD	36°19′37″ E, 00°49′23″ S	1900	10/08/97
9566	KWAMUYA; 0–500 m U/S ROAD	36°19′23″ E, 00°49′47″ S	1950	10/08/97
9567	KWAMUYA; 500 m–1 km U/S ROAD	36°18′58″ E, 00°50′21″ S	2000	10/08/97
9568	MALEWA; WANJOHI ROAD	36°24′44″ E, 00°15′19″ S	2290	06/08/97
9569	MALEWA; MARULA ESTATE	36°22′03″ E, 00°40′47″ S	1895	07/08/97
9570	GILGIL; ARMY BARRACKS	36°20′41″ E, 00°29′06″ S	2000	16/08/97
9571	LITTLE GILGIL; ARMY BARRACKS	36°21′44″ E, 00°29′36″ S	2000	16/08/97
9572	TURASHA DAM	36°30′00″ E, 00°34′42″ S	2400	11/08/97
9573	MALEWA; PUMP HOUSE	36°23′41″ E, 00°29′36″ S	2050	11/08/97
9574	MALEWA; BELOW TURASHA	36°24′05″ E, 00°31′03″ S	2000	11/08/97
9575	KARATI; KINANGOP PLATEAU	36°35′16″ E, 00°47′11″ S	2610	12/08/97
9576	KARATI; GORGE	36°28′42″ E, 00°41′11″ S	2100	12/08/97
9577	GILGIL; MARULA FLOODPLAIN	36°20′55″ E, 00°40′32″ S	1890	13/08/97
9578	GILGIL; ANOSTOMOSED DELTA	36°21′10″ E, 00°38′41″ S	1890	27/08/97
9579	GILGIL; U/S NORTH LAKE ROAD	36°20′41″ E, 00°36′39″ S	1900	17/08/97
9580	GILGIL; D/S NORTH LAKE ROAD	36°20′21″ E, 00°37′23″ S	1895	17/08/97
9581	MALEWA; KARI FARM	36°25′13″ E, 00°38′26″ S	1950	26/08/97
9582	KARATI; NORTH SWAMP	36°24′44″ E, 00°42′10″ S	1890	21/08/97
9583	NYAMBUG	36°22′52″ E, 00°17′11″ S	2350	25/07/98
9584	SIMBA (OLEOLONDO)	36°22′13″ E, 00°21′24″ S	2290	25/07/98
9585	OL KALOU	36°22′47″ E, 00°16′28″ S	2350	22/07/98
9586	ENDOROHR; D/S OL BOLOSSAT	36°25′03″ E, 00°12′44″ S	2330	22/07/98
9587	HELL'S GATE; LOWER GORGE	36°19′13″ E, 00°52′47″ S	2000	21/07/98
9588	KARATI; MANERA	36°25′23″ E, 00°41′45″ S	1895	17/07/98
9589	KARATI; RIFT WALL WATERFALL	36°32′50″ E, 00°43′18″ S	2500	17/07/98
9590	KARATI; GORGE	36°28′42″ E, 00°41′11″ S	2100	17/07/98
9591	GILGIL; U/S NAKURU ROAD	36°21′44″ E, 00°32′55″ S	1950	16/07/98
9592	MORINDATI; U/S KAHUHO SCHOOL	36°20′36″ E, 00°25′08″ S	2200	19/07/98
9593	GILGIL; U/S NORTH LAKE ROAD	36°20′41″ E, 00°36′39″ S	1900	15/07/98
9594	MALEWA; MARULA	36°22′03″ E, 00°40′47″ S	1895	20/07/98
9595	GILGIL; MARULA FLOODPLAIN	36°20′55″ E, 00°40′32″ S	1890	20/07/98
9596	KARATI; KINANGOP PLATEAU	36°35′16″ E, 00°47′11″ S	2610	25/03/99
9597	KARATI; CASCADE U/S GORGE	36°29′16″ E, 00°41′02″ S	2130	30/03/99
9598	KIRIUNDU; DONDORI ROAD	36°17′41″ E, 00°14′50″ S	2560	05/04/99
9599	GILGIL; MARULA FLOODPLAIN	36°20′55″ E, 00°40′32″ S	1890	31/03/99
9600	MALEWA; 0–500m U/S KIRIMA	36°30′34″ E, 00°13′52″ S	2500	27/03/99
9601	MALEWA; 500m-1km U/S KIRIMA	36°30′54″ E, 00°14′02″ S	2700	27/03/99
9602	WANJOHI; WANJOHI	36°29′50″ E, 00°18′15″ S	2390	27/03/99
9603	NYAMITHI; D/S NEW ROAD	36°30′39″ E, 00°47′02″ S	2295	25/03/99
9604	NYAMITHI; GORGE ON SLOPE U/S OLD ROAD	36°28′13″ E, 00°46′42″ S	2060	26/03/99
9605	NYAMITHI; GORGE IN FLATLAND U/S OLD ROAD	36°27′58″ E, 00°46′42″ S	2000	26/03/99
9606	NYAMITHI; SHAMBA	36°27′00″ E, 00°46′18″ S	1990	04/04/99
9607	HELL'S GATE; CENTRAL TOWER	36°20′55″ E, 00°53′16″ S	2000	04/04/99
9608	PELICAN FARM; CONSTRUCTED WETLAND	36°25′42″ E, 00°42′24″ S	1890	06/04/99

ments. Consequently, the majority of the series of lakes are soda or alkaline due to the concentration of solutes through evapotranspiration. By contrast, Lake Naivasha is a freshwater body, apparently due to topography and orographic processes across its catchment. Lake Naivasha lies between the 'walls' of the Eastern Rift Valley: the Mau Escarpment rising to 3048 metres to the west and the Nyandarua Mountains rising to 4000 metres to the east. These ridges are among the highest ranges in the central part of the Eastern Rift Valley, and are likely to have sufficient elevation to cause climatic conditions that result in greater rainfall relative to other major lake catchments. A schematic diagram of the orographic and hydrographic regime of the catchment is provided in Figure 2 (reliable rainfall data across the catchment are not available). The presence of a number of the larger perennial rivers of the Malewa and Gilgil systems to the north and west, relative to only several temporary rivers to the south and west and few blind streams running from the Mau Escarpment to the west, provide evidence substantiating this model. The persistence of the Malewa and Gilgil systems may also suggest the existence of rainfall percolation into perched groundwater tables that feed the river system during dry periods.

Channel gradients of the major river systems are shown in Figures 3 and 4 and Table 2. Changes in channel dimensions with passage downstream are shown for the three main rivers and also for the Nyamithi (a temporary system rising on the Kinangop plateau) and the Kwamuya Stream (representative of the ephemeral dry streams from the dry southern slopes of the lakeshore) are summarised in Figures 5 and 6. As a surrogate of channel energy (the widespread occurrence of dry channels render flow type data unreliable) and sediment flow, Figures 7 and 8 shows the substrate distribution at 50 m transects within the RHS sites, respectively, in the Malewa, Gilgil, Karati and Nyamithi river systems. Silt deposition features recorded in the Malewa, Gilgil and Karati river systems are presented, respectively, in Figure 9 (no discrete depositional features were present in RHS sites on the Nyamithi system).

The Malewa river system

The Malewa river system is significantly more dendritic than other systems in the catchment, with headwaters and tributaries rising at a range of altitudes. The headwater of the main channel of the Malewa River rises in the Nyandarua Mountains at 3700 m altitude, and several small rivers feed into the Wanjohi system from the slopes of the Nyandaruas. The Endorohr system arises as a westerly outflow from Lake Ol Bolossat when lake levels are high (as observed, for example, following the 'El Nino' rains in 1997), at an altitude of 2400 m. The Endorohr here has the characteristics of a grassy swale, interpersed with *Cyperus*-dominated pools, a temporary river in a moist and cool upland setting. The Turasha sub-catchment, arising to the East of the Kipipiri mountain, is dammed to provide fresh water for the town of Nakuru, creating a fresh stillwater habitat at approximately 3000 m altitude. This feeds into an otherwise high-energy river system. Both the Malewa/Wanjohi and the Turasha sub-catchments fall rapidly from source to the altitude at which the Endorohr, Ol Kalou, Nyambug and Simba rivers rise, falling at lower gradient to approximately 40 km north of Lake Navasha where they once again fall at higher gradient cutting through the plateau north of the lake (up to 30 km distance from the river mouth).

Water width/depth data demonstrate the perennial nature of all rivers in the Malewa system, with channel sizes which tend to increase downstream with all rivers remaining relatively shallow. Channel substrates, reflective of stream energy, reflect channel topography with the steep Ol Kolau and Nyambug systems being bedrock-dominated and the Simba river being dominated by boulders. Conversely, the relatively low-gradient Endorohr and the dammed (at the RHS site) Turasha tributaries have a sediment of fines (combining soil, silt, clay and peat). Below the Turasha confluence, the substrate of the Malewa system grades from gravel/pebble to fines at the low-energy river mouth. Silt deposition features generally reflect this energy gradient, although side-bars present at Ol Kolau and Nyambug river sites suggest high sediment

Table 2. Stream orders and topography in the Naivasha catchment

River system	Stream order	Maximum horizontal channel length (km)	Maximum height drop (m)
Malewa	6/7	109	1921
Gilgil	3	60	873
Karati	1	30	760
Nyamithi	1	13	740
Kwamuya	1	2	110

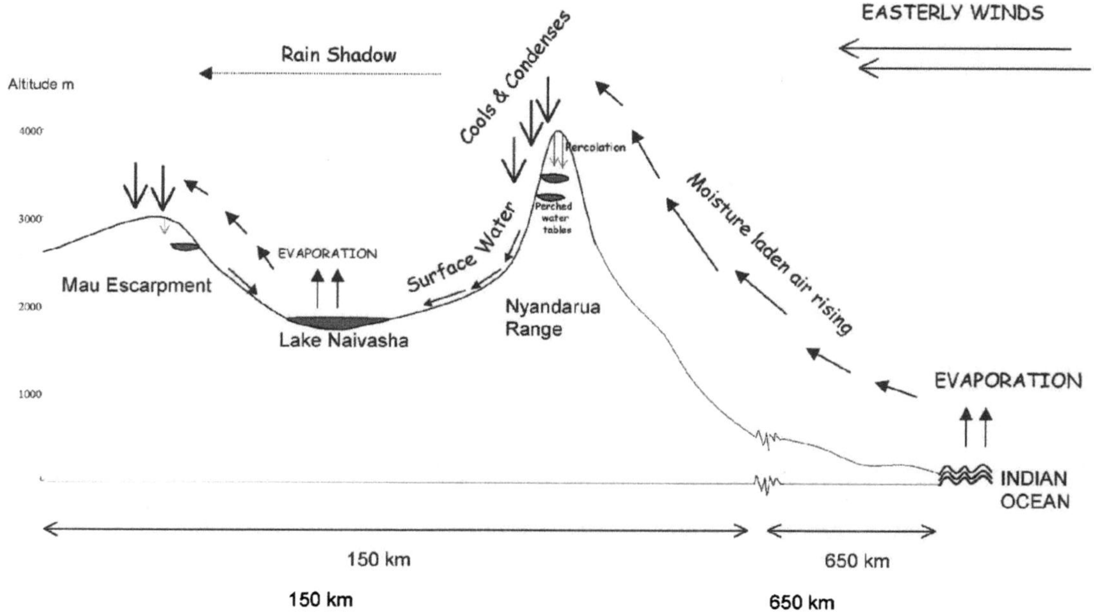

Figure 2. Profile and orographic model of the Eastern Rift Valley through the Naivasha catchment.

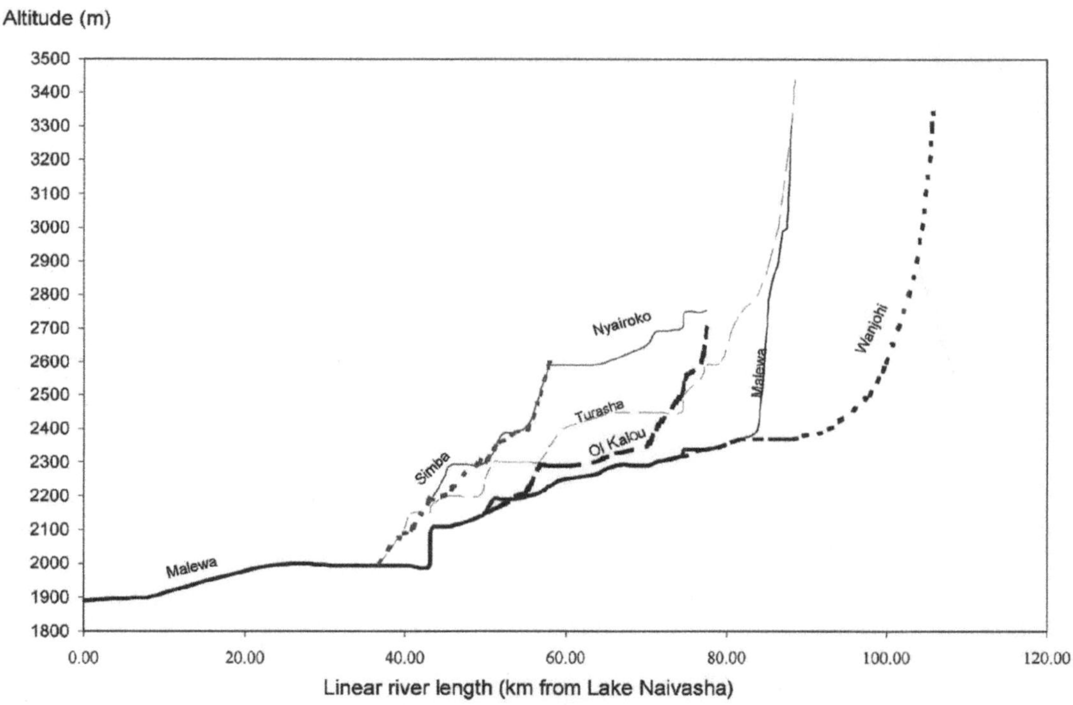

Figure 3. Channel topography of the Malewa river and its major tributaries.

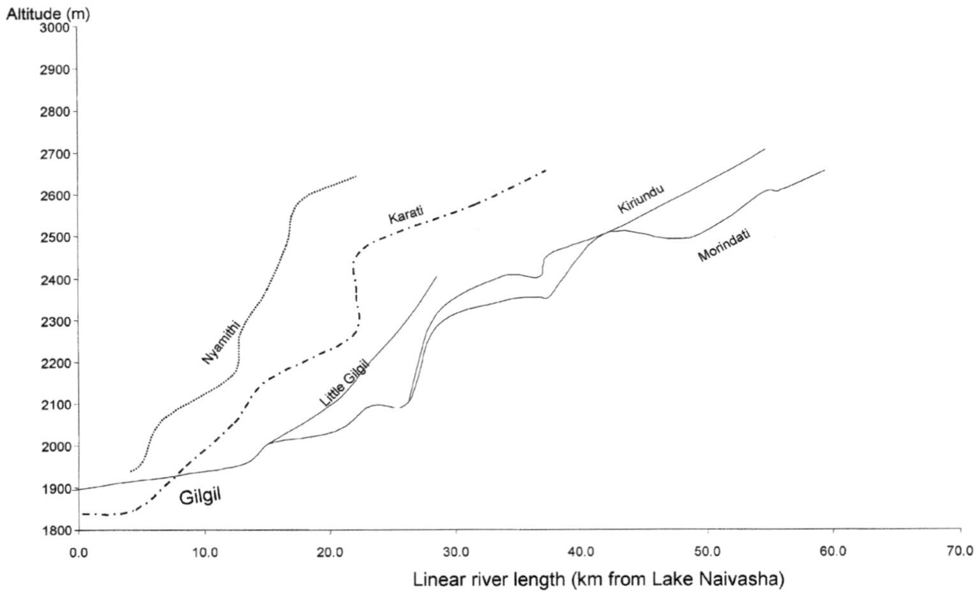

Figure 4. Channel topography of the Gilgil river and its tributaries, and the Karati and Nyamithi rivers.

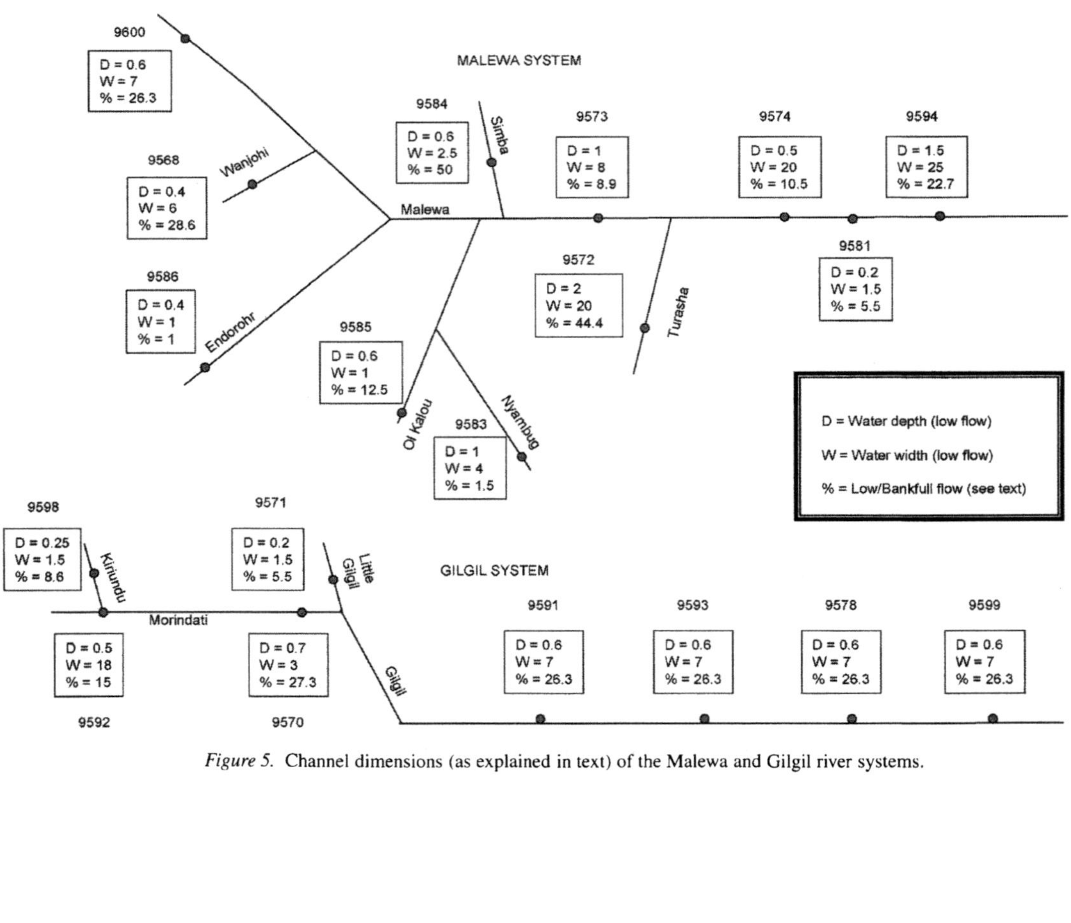

Figure 5. Channel dimensions (as explained in text) of the Malewa and Gilgil river systems.

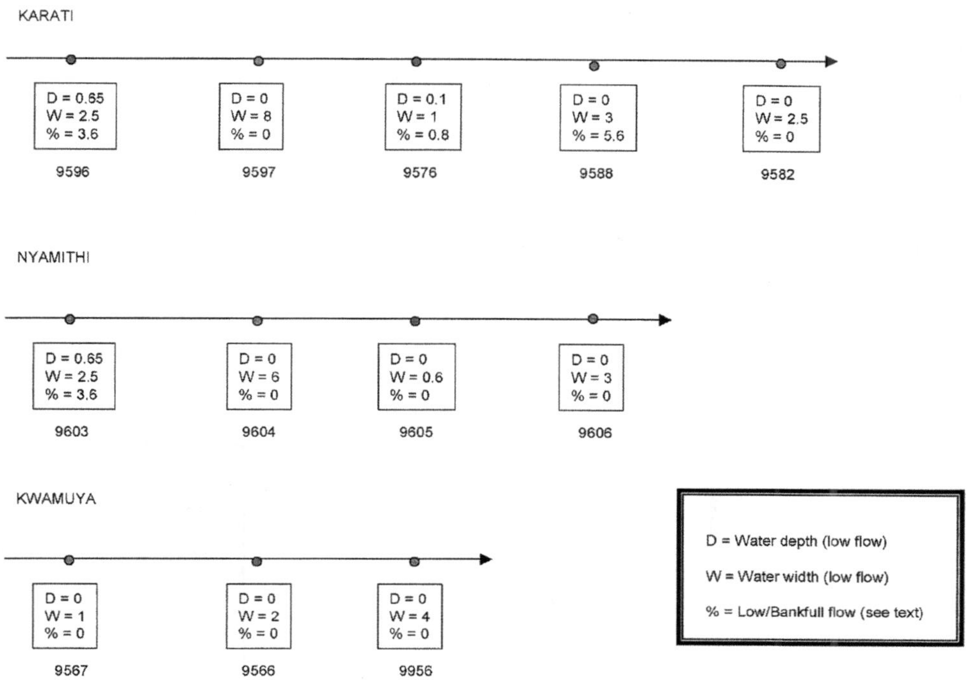

Figure 6. Channel dimensions (as explained in text) of the Karati, Nyamithi and Kwamuya river systems.

inputs upstream. Point bars present at the Turasha dam and down river at both the KARI and Marula sites suggest significant channel activity in these reaches. The Malewa discharges directly into Lake Naivasha with no dissipation into papyrus swamp as occurred in the 1970s (Gaudet, 1977). In August 1997, distinct sediment-laden plumes of Malewa river water were visible up to an estimated 500 m into the lake.

Scrub and rough pasture were the most extensive forms of riparian land use throughout the Malewa catchment although, in some areas, tillage of the land for subsistence agriculture and cash crops was evident right up to the river banks. There was also evidence of extensive erosion and poaching of the river banks in some river stretches, apparently as a result of over-grazing. Artificial features were otherwise not widespread, the most significant exception being the Turasha dam.

The Gilgil system

The Gilgil system is third-order with three headwaters: the Morindati (rising at 2700 m), the Kiriundu (2710 m) and the Little Gilgil (2400 m). The upland river sections of all three Gilgil tributaries have a relatively flat topography before cutting into the plateau approximately 40–50 km north of Lake Naivasha to the flatlands approaching the lake.

Water width/depth data indicate shallow perennial river systems, generally increasing in width downstream. Percentage of low flow/bankful appear relatively higher than the Malewa system indicating stronger base flows; there is a perennial volcanic groundwater input at Chamuka Spring near the settlement of Chokeraria. In contrast to other river systems in the catchment, the Gilgil opens into a broad floodplain immediately north of the lake, with an anostomosed section of river, now heavily modified for direct irrigation.

Bedrock channel substrate is predominant at the steep section of the Morindati River upstream of Kahuho school, and also occurs in regions of channel scour at Gilgil Weir. However, the river is otherwise gravel/pebble grading to fines reflecting topography and energy. Silt deposition features are not widespread in the river system, although they are common at the Army Barracks (Gilgil Town), Nakuru Road and Maasai Ridge sites. This is reflective of both overgraz-

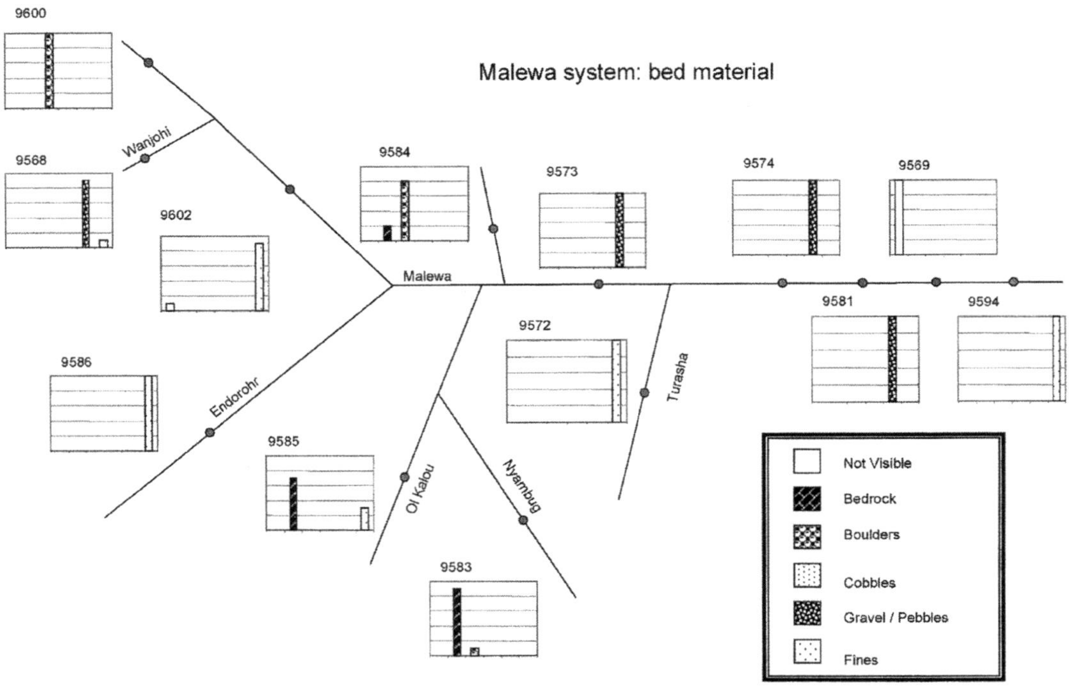

Figure 7. Substrates at ten survey transects within RHS sites on the Malewa River system.

ing/poaching and channel activity (as indicated by side bars) in this higher-gradient reach of the river.

Despite perennial flows, the broad Gilgil floodplain north of Lake Naivasha disperses river energy, and the low gradient upstream of this results in fall out of sediments (and associated substances) well short of the lake. Owing to both of these features, river energy is dispersed short of the lake, and the Gilgil River consequently lacks a clear mouth.

Scrub and rough pasture are the predominant riparian landscape types in the Gilgil catchment, although some tilled land is present near the river banks. Erosion through overgrazing and poaching, by both stock and game, are evident lower in the catchment.

The Karati system

The Karati is a relatively shorter, steeper system that has no significant tributaries and is purely first-order. The headwater rises at an altitude of 2600 m (approximately the same altitude range as the Gilgil headwaters and many of those of the Malewa) on the Kinangop plateau, but cascades steeply from the Rift Valley edge towards the flatter lands bordering the eastern lake shore.

Depth/width data indicate a predominantly dry river. Despite low % low flow/bankful, data suggest terracing, indicative of differing flow regimes over longer timescales. The flatter gradients on the Kinangop plateau result in a mid-energy gravel/pebble/cobble channel substrate, through high river energy as the Karati falls steeply through the gorge section resulting in a predominantly boulder or bedrock substrate. It rarely flows in the flatter topography near the lake itself, and the channel substrates at sites Manera and the North Swamp overwhelmingly comprise fines. Some silt deposition is evident at Kinangop, reflecting agricultural activity. Small silt deposits are also present at the Gorge site where marginal slacks in the gorge permit deposition of high sediment loads arising from grazing/poaching and riverside tillage of land. Silt deposition is also evident at the North Swamp site.

Rough pasture and scrub are common riparian landscapes throughout the catchment, however subsistence and cash crop agriculture are extensive on the Kinangop Plateau and immediately above the Karati

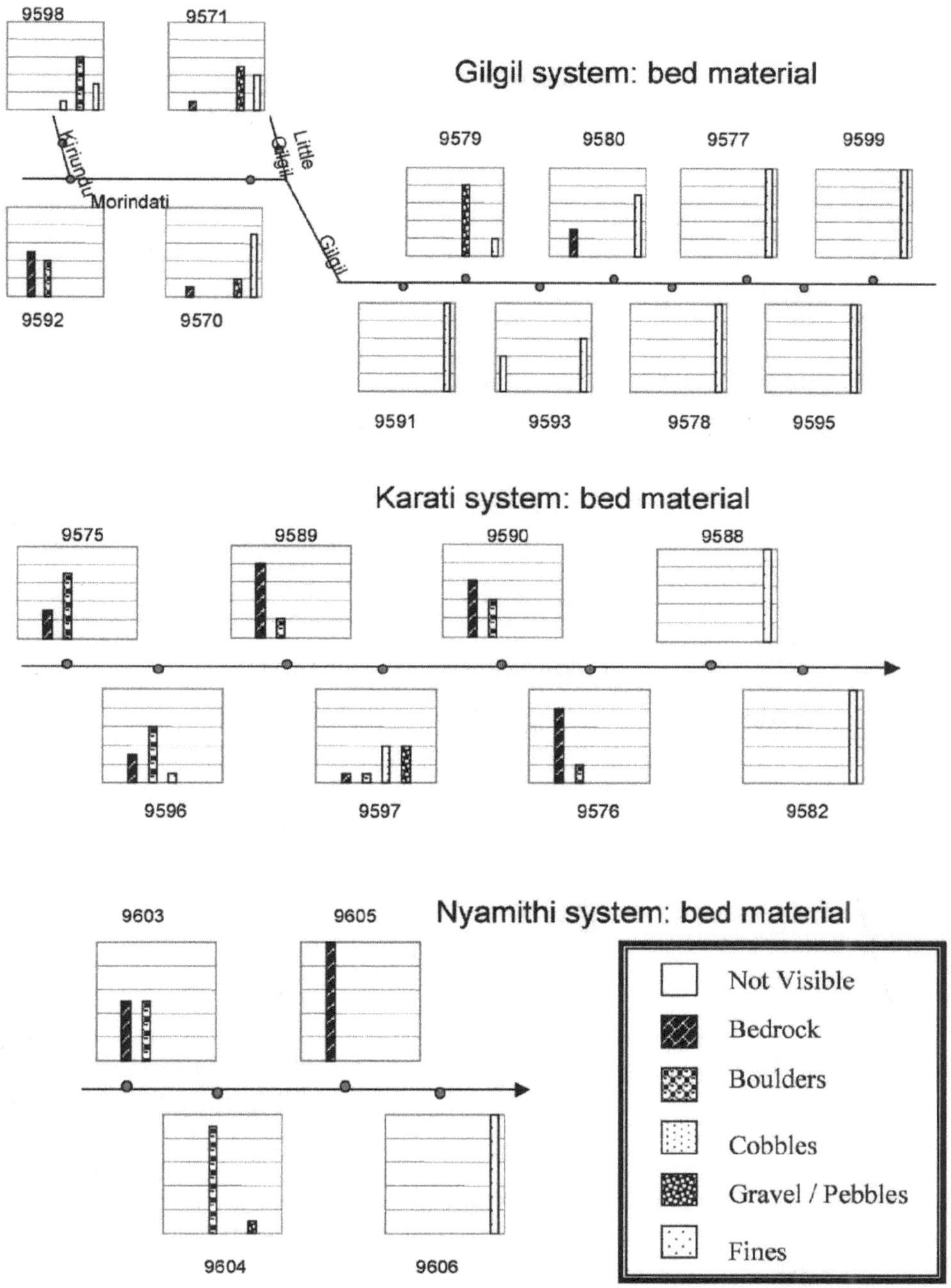

Figure 8. Substrates at ten survey transects within RHS sites on the Gilgil, Karati and Nyamithi river systems.

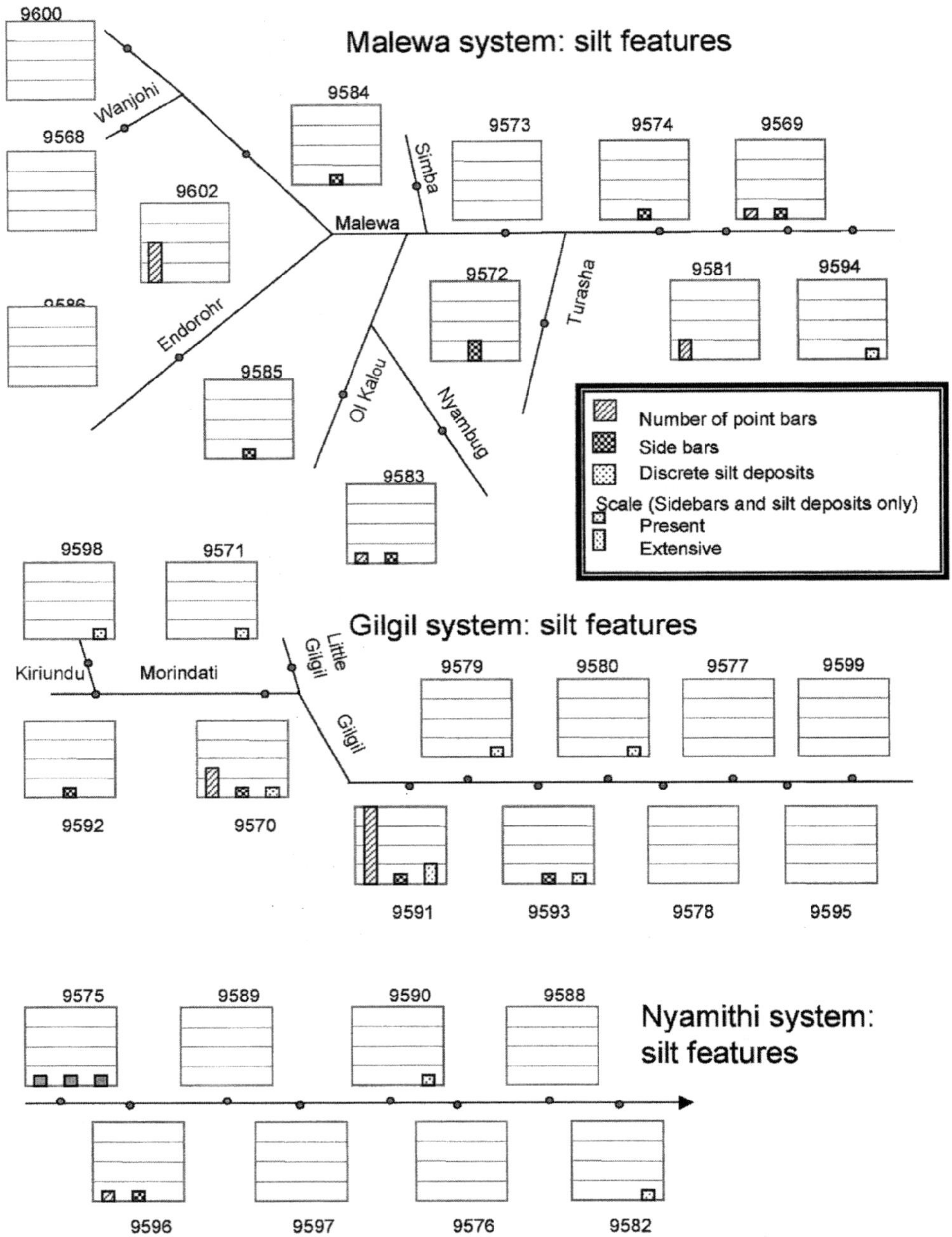

Figure 9. Silt features at ten survey transects within RHS sites on the Malewa, Gilgil and Karati river systems.

gorge. The Karati is however virtually 'blind', dispersing into the North Swamp through various distributary channels with no clear opening to the lake. Sediment flows from upstream erosion appear to contribute significantly to a delta-forming process.

The Nyamithi system

The Nyamithi is a shorter and steeper first order system even than the Karati. Despite the channel becoming indistinct some 4 km short of the lake, there is overland flow over farmland to the lake shore during heavy rainfall events. The Nyamithi is predominantly dry with a small channel width, but the bedrock/boulder substrate in all but the lowland (Shamba) site indicate substantial river energy during periods of high flow. Fines eroded from these events contribute to the silt/mud bed of the increasingly indistinct river channel at the Shamba site.

There is evidence from the terraced valleys and deeply incised/engorged channels through most of the river length that the Nyamithi had significantly greater flashing volumes over longer geological timescales than at present. These appear to have deposited substantial amounts of silt into the lake, forming a delta which current flow volumes are now incapable of penetrating.

The Kwamuya stream

The Kwamuya has been surveyed from the source to the blind end as an example of one of the several temporary rivers running from the southern slopes of the catchment. It is a steep, short first-order stream which stops some 100 m short of the lake edge. The land between the end of the river channel and the lake shore is cultivated, and the lakeside papyrus fringe is entire.

The river is dry for most of the year, reportedly running only for a few days a year following periods of heavy rainfall. Channel width increases downstream. Bedrock and boulders are predominant throughout most of the river system, although heavy overgrazing results in substantial erosion contributing to the silt deposits at the blind ending of the river. The delta-forming processes postulated for the Nyamithi system may also contribute to lowland silt deposits at a significantly smaller scale in the Kwamuya system.

Erosion, depostion and sedimentation in the river systems

The data demonstrate a largely predictable pattern predominantly of erosion (indicated by bedrock, boulder and cobble channel substrates, and gorged or deep V valley forms) in the upper catchment, and the deposition of sediments (channel substrate composed of fines, with side bars and point bars) in the lower catchment. In addition to this relationship between topography/erosion and bed type/depositional features, Tarras-Wahlberg et al. (2002) also found a distinct relationship between topography/channel pattern and the maturity, shape and size of riverine sediment particles.

Overlaid upon this pattern are poaching of river banks and erosion, sometimes severe, recorded throughout the catchment. High sediments loads from the Malewa catchment were responsible for the visible plume extending from the mouth of the Malewa river into the lake.

It is possible, based on findings of this initial study of the Naivasha catchment, to make a preliminary classification of the zones within the catchment, which has potential value in setting a geographic context for the delivery of sustainable development within the catchment (Figure 10).

Discussion

The initial purpose of the Naivasha catchment research programme was to establish a datum against which to assess future change. This is also potentially valuable in determining 'hotspots' of biological interest and problems, and supporting land use decisions. Any threat to the hydrological regime of the Naivasha catchment could have potentially substantial consequences for the lake and river ecosystems, given the apparent fine balance of inputs and outputs contributing to the unique freshwater character of Lake Naivasha.

Erosion and loss of topsoil are a significant problem worldwide and the magnitude of the problem is amplified by both the thin erosive volcanic soils and the increasing intensity and land use. Poorly-advised land use practices such as tillage to the banktop and unrestricted stock access are commonplace, and may contribute to increasing erosion. Unsympathetic riparian land management leading to vegetation loss and erosion also results in reductions in infiltration, further

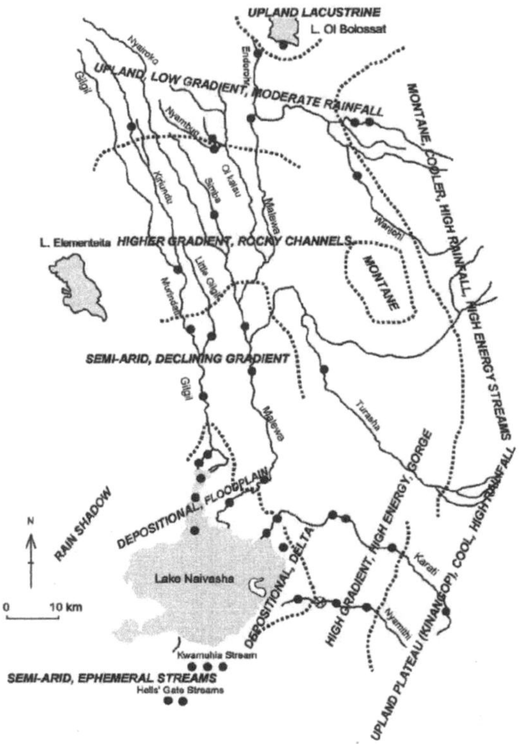

Figure 10. Preliminary classification of the Lake Naivasha catchment.

accelerating sheet and gully erosion by increasing runoff, and reducing the structure of soils as well as their capacity to retain moisture.

The former influence of the North Swamp and the fringing papyrus would have played an extremely important role in immobilising the sediment load from the river systems (Gaudet, 1979). The visible evidence of sediment plumes indicate the significance of the loss of the papyrus to the lake. Land management and erosion control in the catchment therefore appear to be a significant issue potentially deteriorating the quality of Lake Naivasha.

The proposal to dam the main stem of the River Malewa to supply growing demands for fresh water at Nakuru Town is also of concern. Hydrological, geomorphological, chemical and ecological consequences in both the Naivasha and Nakuru catchments, together with the social and economic needs sustained by these ecosystem services, need to be considered in detail – informed by sound science – to evaluate the sustainability of any proposal to impound rivers within the catchment. These theoretical problems are more than just speculation, based on lessons being learned elsewhere in the Rift Valley and across the tropics.

Adverse development in catchments, often remote from the receiving water body itself, is widely recognised as a major contributor to degradation in lake quality. However, recognition of river conservation in the developing world as a means for stemming ecological damage, and consequent adverse social and economic harm, is a more recent phenomenon (Barel et al., 1985). Sustainable development must be addressed at catchment scale if it is to be effective, taking an integrated approach to protection of both lake and catchment ecosystems and the livelihoods they sustain.

To protect the unique characteristics of Lake Naivasha, and the social and economic development that depends upon it, considerable further study of the river catchments is required. An improved understanding of interactions between the lake and influent rivers, and changing patterns of land use and habitation across the catchment, are essential to support sustainable policies. Rivers also integrate the influences of changing land use within the catchment (Newson, 1992) and so river monitoring can prove a cost-effective method for monitoring whether sustainable catchment management is being achieved (Ormerod & Juttner, 1998). Once appropriate indicators are identified, routine river monitoring will become a priority to determine, and develop strategies for managing, impacts on Lake Naivasha deriving from its catchment.

Acknowledgements

This study formed part of the work of the University of Leicester research project at Lake Naivasha, which since 1987 has been authorised by the Office of the President, Government of Kenya under research permit to Dr D. M. Harper no. OP 13/001/12C 46. The project was funded by the Earthwatch Institute, Boston USA and Oxford, England. The data collection would not have been possible without the assistance of numerous Earthwatch volunteers or the facilities at the erstwhile Elsamere Conservation Centre. Our sincere thanks go to the numerous colleagues in Kenya for logistical help, in particular Jill and Angus Simpson and Velia Carn, and Chris Campbell Clause. The authors benefited from dialogue with R. Stadtmuller. Peter Scarlett and Hugh Dawson were extremely helpful in managing input of RHS data to the CD-

ROM/central RHS database, and extracting relevant pieces of data for subsequent analysis.

References

Ase, L.-E., 1987. A note on the water budget of Lake Naivasha, Kenya. Geografiska Annaler: 69: 415–429.

Ase, L.-E., K. Sernbo & P. Syren, 1986. Studies of Lake Naivasha, Kenya and its drainage area. Stockholms Universitet Naturgeografiska Institutionen Stockholm. 63: 77 pp.

Barel, C. N. D., R. Dorit & D. H. Greenwood, 1985. Destruction of fisheries in Africa's lakes. Nature 315: 19–20.

Cairn, J. Jr., 1997. Defining goals and conditions for a sustainable world. Environmental Health Perspectives 105: 1164–1170.

Denny, P. (ed.), 1985. The Ecology and Mangement of African Wetland Vegetation. Dr W. Junk Publishers, Dordrecht.

Dugan, P. J., 1990. Wetland Conservation: A Review of Current Issues and Required Action. IUCN, Gland, Switzerland 96 pp.

Environment Agency & Institute of Freshwater Ecology, 1999. River Habitat Survey Database 1999 (CD-ROM). Environment Agency, Rio House, Aztec West, Bristol, U.K.

Everard, M., 1997. Development of a British wetland strategy. Aquat. Conserv. 7: 223–238.

Everard, M., 1998. Application of The Natural Step to water management. J. Inst. Environ. Sci. 7: 10–13.

Gaudet, J. J., 1979. Seasonal changes in the nutrients in a tropical swamp: North Swamp, Lake Naivasha. J. Ecol. 67: 953–981.

Harper, D. M., 1984. Recent changes in the ecology of Lake Naivasha, Kenya. Verh. int. Ver. Limnol. 22: 1193–1197.

Harper, D. M., C. Adams & K. Mavuti, 1995. The aquatic plant communities of the Lake Naivasha wetland, Kenya: pattern, dynamics and conservation. Wetlands Ecology and Management 3: 111–123.

Harper, D. M., K. M. Mavuti & S. M. Muchiri, 1990. Ecology and management of Lake Naivasha, Kenya in relation to climate change, alien species introductions and agricultural development. Environ. Conserv. 17: 328–336.

Mutiso, G. C., 1989. Managing Kenya's Arid and Semi-Arid Areas. In Kiriro, A. & C. Juma (eds), Gaining Ground: Institutional Innovations in Land Use Management in Kenya. Nairobi, African Centre for Technology Studies: 76 pp.

Newson, M., 1992. Land, Water and Development: River Basin Systems and Their Sustainable Management. Routledge, London.

Orie, K. K., 1996. Proposed amendment to Kenya's water legislation: some missing sustainable water resources management principles. Wat. Res. Dev. 12: 79–88.

O'Riordan, T., 1993. The politics of sustainability. In Turner, K. (ed.), Sustainable Environmental Economics and Management: Principles and Practice. Belhaven, London: 37–69.

Ormerod, S. & I. Juttner, 1998. Catchment sustainability and river biodiversity in Asia: a case study from Nepal. In Harper, D. M. & A. Brown (eds), John Wiley and Sons Ltd: 187–207.

Ramsar, 1971. Convention on Wetlands of International Importance Especially as Waterfowl Habitat. 2/2/71, Ramsar, Iran.

Ramsar, 1996. The Ramsar 25th Anniversary Statement. Resolution VI.14, 6th Meeting of the Conference of the Contracting Parties, Brisbane, March 1996.

Raven, P. J., N. T. H. Holmes, F. H. Dawson, P. J. A. Fox, M. Everard, I. R. Fozzard & K. J. Rouen, 1998. River Habitat Quality: The Physical Character of Rivers and Streams in the U.K. and the Isle of Man. Environment Agency, Bristol.

Tarras-Wahlberg, H., M. Everard & D. M. Harper, 2002. Geochemical and physical characteristics of river and lake sediments at Naivasha, Kenya. Hydrobiologia 488 (Dev. Hydrobiol. 168): 27–41.

Hydrobiologia **488:** 27–41, 2002.
D.M. Harper, R. Boar, M. Everard & P. Hickley (eds), Lake Naivasha, Kenya.

Geochemical and physical characteristics of river and lake sediments at Naivasha, Kenya

Håkan Tarras-Wahlberg[1,*], Mark Everard[2] & David M. Harper[3]
[1]*Swedish Geological AB, Box 19090, 104 32 Stockholm, Sweden*
[2]*The Natural Step UK, 9 Imperial Square, Cheltenham, Gloucestershire GL50 1QB, U.K.*
[3]*Department of Biology, Leicester University, Leicester LE1 7RH, U.K.*
**Author for correspondence. E-mail: hakan.tarras-wahlberg@hifab.se*

Key words: Lake Naivasha, sediment stratigraphy, sediment geochemistry, metals

Abstract

Prior studies on Lake Naivasha relevant to understanding sediment dynamics include a bathymetric map, a paleolimnological study of fossil invertebrate assemblages in lake sediment, an overview of lake level fluctuations throughout the 20th century, and identification of a dynamic assemblage of macrophyte zones that has responded both to these changes in lake level and to more recent, alien species. Sediment samples collected from the rivers systems and the lake were examined physically and chemically. River sediment characteristics reflect geology and geomorphological processes in the catchment, whereas lake sediment stratigraphy has responded to past lake level changes. Such changes have caused significant changes in aquatic vegetation assemblages. Present day sediment dynamics in the lake are governed by the presence of river point sources in the north and wave-induced re-suspension, such that sediments introduced by rivers are transported in easterly and southerly directions, and are eventually deposited in the eastern, central and southern parts of the lake. Sedimentary deposition is also occurring in northern areas that once were protected by papyrus swamp vegetation but now only have a narrow fringe, highlighting the important role of swamp vegetation in filtering out suspended particulates and thereby controlling water quality in the lake. Geochemical analyses of river and lake sediments indicate that they represent fairly undisturbed background conditions. Higher-than-expected concentrations of cadmium, iron, nickel and zinc found in both river and lake sediment are likely to derive from volcanic rocks and/or lateritic soils found in the lake catchment.

Introduction

Lakes are strongly influenced by the natural and human characteristics of their catchments. River sediment in low order channels derive from a plethora of hillslope and channel sources in the upstream catchment (Bowie & Mutchler, 1986). During downstream transport, sediment from different sources will mix and, therefore, investigations of the geochemical and physical characteristics of river sediment may provide evidence of processes occurring in the entire catchment. As a consequence, stream sediment sampling has long been used as a tool to prospect for metalliferous mineral deposits, and it is generally accepted that such surveys provide more representative and cost effective results than surveys of soils or vegetation (Levinson, 1980).

Geochemical surveys of river sediment are increasingly being used also in environmental investigations, and applications include:

(i) assessments of environmental impacts, as sediments' contaminant concentrations downstream of industrial, agricultural or other human activities often are elevated (Darnley et al., 1995);
(ii) human and animal health studies as geologically controlled deficiencies/excesses of certain nutrients/toxins in water and/or food products may

cause health problems (Fordyce et al., 1996); and

(iii) to provide a geochemical baseline against which any future pertubations may be appraised (Williams et al., 2000).

Similarly, investigations of lake sediments can also be used in environmental surveys. When rivers enter a lake, the decrease of velocity results in sediment deposition. Investigations of geochemical changes with sediment depth may, thus, provide indications of changes in the characteristics of sediment input with time. Such investigations may therefore be used to determine both the character of contamination and, if sedimentation rates are known, changes in contaminant loading with time (Nicholson, 1996).

Environmental applications of geochemistry often investigate metal contamination, and it is generally acknowledged that metals in their dissolved ionic state control direct toxic effects (Förstner, 1983; 1988; Stumm & Morgan, 1996). However, metals in aquatic systems are usually predominantly associated with sediment (Förstner, 1983; Windom et al., 1991). Therefore, the concentration of metals in sediment is generally a more sensitive and accurate indicator of contamination of an aquatic system, compared to the concentration of dissolved metals (Carelli et al., 1995). It is also recognised that sediment-bound metals have the potential of changing to the ionic state (Willford et al., 1987). Investigations of sediment geochemistry is, thus, directly relevant to assessing the possibility of toxic effects of metal contamination. Furthermore, as metals are present in much higher concentrations in sediment than in water, the analysis of sediment is comparatively easy to perform and, therefore, less costly, enabling more samples to be analysed, and a greater geographical coverage.

The interpretation of sediment geochemistry data is, however, not without its problems. In rivers, different morphological units often have quite different geochemical characteristics, resulting from physical sorting processes within the stream bed (Ladd et al., 1998). Similarily, metals are often found preferentially concentrated in the finer fraction of sediment (Förstner, 1983). Consequently, care needs to be taken to ensure that the same or at least similar morphological units and size fractions are sampled in order for results from different rivers or even reaches of the same river to be comparable (Ladd et al., 1998). Further, whereas environmental criteria for water have been established for a varity of purposes, including the protection of aquatic life, attempts to elaborate similar criteria for sediment have been less succesful. It is difficult to determine to what extent potentially harmful elements in sediment are, or may become, available for biological uptake. Consequently, environmental criteria for sediment are generally seen to be of limited value due to variations through time and space in the conditions that determine bioavailability (Darnley et al., 1995; Maher et al., 1999). Non-statutory systems for sediment quality guidelines have, nevertheless, been developed in a number of countries, including Canada and Sweden.

Studies of the input of nutrients and other material from the Naivasha catchment to lake have been undertaken by Gaudet & Melack (1981), Harper et al. (1993), Harper et al. (1995), Everard et al. (2002), and Kitaka et al. (2002). None have yet investigated lake and river sediments. The aims of the study were: (i) to investigate the geochemical and physical characteristics of river and lake sediments, (ii) to investigate the sediment stratigraphy of the lake, and (iii) to assess the implications of (i) and (ii) in informing of how to best manage Lake Naivasha and its catchment.

Study site description

Water level and vegetation changes

The water level and aquatic vegetation of Lake Naivasha changes continuously (Becht & Harper, 2002), and such fluctuations will have significant effects upon sediment characteristics. The lake was first described in the writing of European explorers in the 1880s but its earliest record is that according to local tradition, there was a period immediately before the arrival of the first Europeans when the lake was almost completely dry (Åse et al., 1986). Towards the end of the 19th century, writings by European travellers indicate that the lake's level was continuously rising until 1895, when a maximum lake level of 1896 m above sea level was reached. This historic high water level is some 4–5 m higher than any level reached since that time. Levels then declined around the turn of the century, so that when continuous recordings commenced in 1908 the level was 1890 m (Åse, 1987). Thereafter, the records show a progressive decline in lake levels during the 1920s and 1930s, culminating in a historic (recorded) low of 1882 m in 1946. The lake level remained low until about 1956, whereafter it rose to an intermediate level of 1887 m in 1980 (Åse et al., 1986). Lake level declined progressively during the following decade, rising rapidly by 1 m in May 1988, declining for a decade then rising by 3 m in 1997. Both these latter events were following heavy rains associated with the ENSO climatic phenomenon (see Becht & Harper (2002) for details of lake level changes).

The vegetation of Lake Naivasha is, in turn, influenced by changing lake levels. Beadle (1932) provided the first scientific description of the lake's vegetation. At that time, during a period of declining lake levels, papyrus (*Cyperus papyrus*) formed an extensive floating swamp in the whole northern part of the lake, north of an east–west horizontal line along the northern edge of Crescent Island (known colloquially as the North Swamp). Elsewhere varying widths of it fringed the lake and formed a multitude of small, mobile islands. Various submerged macrophytes and the blue water lily (*Nymphaea nouchalii* var. *caerulea*) grew in shallow lagoons inside these papyrus islands. To the lakeward side, as water deepened, the floating-leaved followed by dense beds of submerged macrophytes formed clear zonation. This pattern was consistent throughout the water level changes up to the mid-1970s. Thereafter, it underwent dramatic and continuous changes. The driving forces behind these changes have been various anthropogenic influences, including the introduction of alien species of animals and plants (Harper, 1984; Tarras-Wahlberg, 1986; Gouder et al., 1998), and the clearance of lakeside lands for intensive agriculture, (Harper et al., 1990).

Lake bathymetry

Åse et al. (1986) provide a bathymetric map of the lake referring to a lake-surface of 1889 masl (reproduced as Fig. 1). The map shows that the lake may be divided into three semi-separate basins: the main lake basin, the Crescent island crater in the east and Lake Oloidien in the south. The main lake is characterized by a smooth flat relief with a maximum depth just off Hippo Point in the southwest. The flat bottom topography contrasts sharply with the hilly topography of the surrounding land, indicating that the lake basin is filled with large amounts of sediments (Åse et al., 1986). A shallower area in the northern end of the lake is interpreted as a deltaic sediment accumulation deriving from the inflows of the Gilgil, Karati and Malewa rivers. At the time of the survey by Åse et al. (1986), the deltaic feature coincided closely with the papyrus-dominated North Swamp. A feedback process is, thus, indicated in which the shallow delta area provides a preferred habitat for papyrus whilst at the same time the papyrus swamp functions as a sediment trap and protects the delta from wave erosion and, in turn, promotes delta development.

Lake sedimentology

Verschuren (1996) conducted stratigraphic and geochronological analyses of three sediment cores taken from the main lake. He noted that sedimentation patterns in the lake vary due to the effects of wind-induced re-suspension and re-deposition of sediments, and the presence of river point sources of sediment in the north. As a result, two of the cores sampled did not represent longer periods of undisturbed sedimentation. In the central part of the lake, however, he sampled what was interpreted as a complete and uninterrupted sequence of sediments deposited during the 20th century, non-conformably overlying sediments dating back as far as the 16th–17th centuries. In this core, he identified four main units: Unit 0, Unit I, Unit II and Unit III.

- **Unit 0**, found at the base of the core at 114–105 cm depths, was a peat horizon with abundant fragments of *Cyperus papyrus* and *Ceratophyllum*. Verschuren interpreted it as representing a period when a low lake level transformed the lake into a shallow, fragmented wetland overgrown by a papyrus swamp. A radiocarbon age of 290 ± 50 years of sediments directly overlying this horizon indicates that Unit 0 was laid down, not during the mid 19th century lowstand but during an earlier event in the 16th–17th century.
- The clayey muds of **Unit I**, between 105 and 82 cm, was interpreted as being composed of sediments laid down during low to intermediate water levels between the 16th–17th century and the end of the 19th century.
- **Unit II** sediments, between 82 and 44 cm, were interpreted as being laid down during high lake levels in the late 19th and early 20th century. At 48–58 cm there was a black layer with a significant admixture of coarse partially decomposed papyrus material. The unit's black color was imparted by the decomposed papyrus fragments, and the unit was interpreted as representing sedimentation in shallow areas near papyrus stands, which grew on exposed mudflats, created as lake levels decreased in the 1920s and 1930s.
- Lake levels declined in the 1920s, 1930s and 1940s and reached a lowstand of 1881.5 masl in 1946.

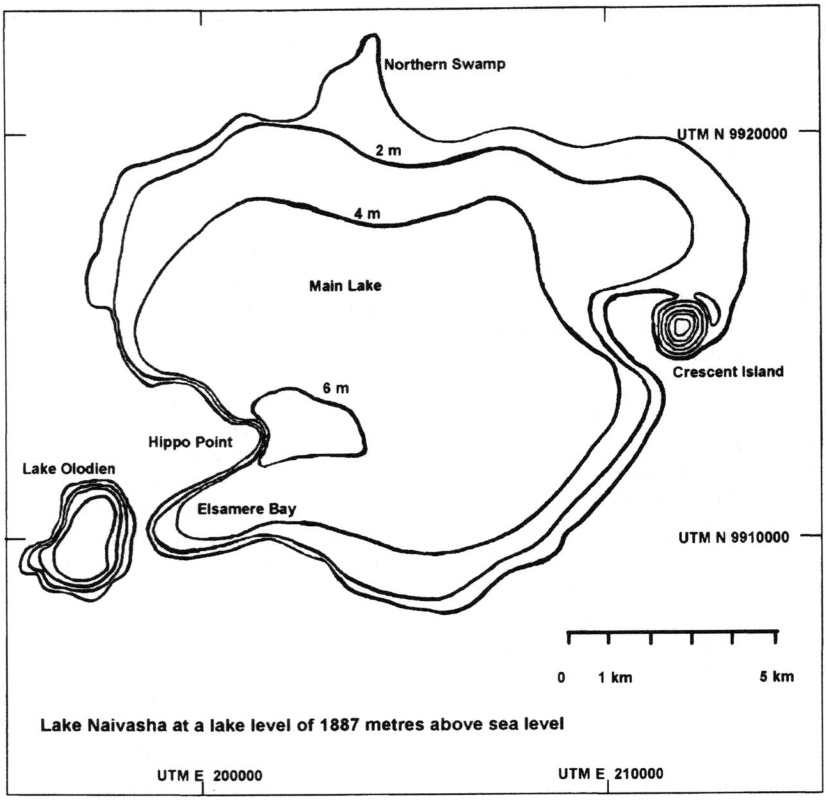

Figure 1. Bathymetric map of Lake Naivasha. Water depth contours are drawn at 2-m intervals. The lake contour represents a water level of approximately 1887 m above sea level, (adapted from Åse et al., 1986).

The transition to **Unit III** sediments at 44 cm were interpreted as marking this return to lower lake levels. Near the transition of Unit II and III, Verschuren noted the presence of a thin, light felt horizon, dated at 1941 ±7 years, composed of numerous needle shaped diatoms. The mechanism of formation is unclear, though Verschuren speculates that its origin could be either a result of mass sedimentation of diatoms triggered by unfavorable conditions during low lake levels, or the sedimentary redistribution and associated aggregation of previously deposited diatoms. Above this level, Unit III consisted of a fairly homogenous organic, clayey mud indicating that the depositional environment has remained similar over the past 40 years. The sedimentation rate in the main lake during this time was estimated at about 1 cm per year.

Methods

Sampling was undertaken during the dry season in July/August 1997 and during October 2000.

Sediment samples from rivers were collected from sixteen river sites during the dry season, between 19th July and 7th August 1997. Sample sites were selected in low (first to third) order reaches to represent the overall characteristics of river sediments in the entire lake catchment (Fig. 2). During this time, the Karati river contained water in isolated pools only. The samples were taken with an Ekman grab from active, oxygenated, mid-river sediments at sites where the channel is straight and the flow fairly rapid (with the exception of the Karati samples, which were taken from standing pools of water and sample S10, taken from the Turasha dam in the Malewa catchment). At each site, about 1 kg of sediment was collected in two to three grabs and these were subsequently mixed. A subsample of approximately 150 g (wet weight) was retrieved from the master sample. This subsample was subsequently wet sieved through a 2-mm mesh (discarding the +2 mm fraction), and air dried at approximately 40 °C.

The sample were then divided in two, and one subsample was investigated under the microscope (×10 magnification) in order to determine grain size, angularity and sorting in accordance with tables provided in Pettijohn (1975). The other sub-sample was sent for analysis of bulk and trace geochemistry by X–ray Fluorescence spectrometry (XRF) at the Anglo American research Laboratories in Johannesburg, South Africa.

Care was taken to avoid cross sample contamination and all equipment used were thoroughly cleaned, first in river water and then in de-ionised water. The accuracy and repeatability of the sampling procedure and the XRF analysis was checked by sampling and analysing the sediment from one site in duplicate.

Sampling of lake sediment was performed by using a sediment corer in 1997, and by using an Ekman grab in 2000.

Eighteen lake sites were sampled during 5–18 August 1997. Sediment cores were collected in 110-cm long Perspex tubes, attached to a single-drive piston corer, which was deployed from a small boat. The location of sample sites was fixed by using a handheld Garmin GPS 12, Global Position System. The exact sampling locations are shown in Figure 3. The cores were extruded and described with regards to their physical and biological character and appearance. Each stratigraphic unit identified was described in detail.

Two cores were selected for geochemical analysis, one from the central lake (no. 12) and one in the Northern swamp (no. 9), close to the outflow of the Malewa river. One sample of 50–100 g (wet weight) was collected from each of the stratigraphic units identified in the two cores, making a total of 10 samples. Care was taken to take these samples from the centre of the core, so as to avoid cross-sample contamination. The samples were kept cool in a refrigerator before being sent for geochemical analysis at the Anglo-American research laboratories in Johannesburg, South Africa. The analysis was performed by plasma emission spectroscopy (ICP-AES) following digestion in aqua regia. The accuracy and repeatability of the sampling and analysis was checked by sampling and analysing one stratigraphic unit twice. Seven lake sediment samples were retrieved during the period 20–24 October 2000. At each site, about one kilo of sediment was collected in three grabs, which were subsequently mixed. A subsample of approximately 150 grams (wet weight) was retrieved from the master sample. This subsample was air dried at approximately 40 °C, and thereafter kept cool in a refrigerator before being sent for analysis of Loss on Ignition (LOI) and metal content at the Leeds University School of Geography's laboratories in the UK. The metal analysis was performed by plasma emission spectroscopy (ICP-AES) following digestion in aqua regia. Again, the accuracy and repeatability of the sampling and analytical procedure was checked by sampling and analysing one sample twice.

The evaluation of sediment analysis was carried out by comparison of the observed major and minor element sediment concentrations with global averages for shale (Turekian & Wedepohl, 1961; Mason & More, 1982), and Environment Canada's provisional sediment quality guidelines for the protection of aquatic life (Environment Canada, 1995). The latter assume 100% bioavailablity for sediment-bound metals which, in turn, make them possibly over protective (Williams et al., 2000). Further, caution must attend comparisons between river and lake sediments as the analytical procedures differ.

Results

River sediment description

Site descriptions are provided in Table 1, together with a sediment description and a linkage to RHS sites examined by Everard et al. (2002). Sediment characteristics differ markedly between the three rivers sampled, reflecting the hydrological, topographic, and geological differences in their catchments.

Sediment in the Karati is generally immature (i.e., poorly sorted and angular in nature), reflecting the fact that the river's gradient beneath the Kinangop Plateau is steep and that it flows only during the rainy season, providing little time for sorting and abrasion. As a consequence of the episodic nature of the river's flow, Karati sediment contain both fine particles (clay and silt) and coarse material (gravel and coarse sand).

Sediment in the Gilgil system is different from that of the Karati. Everard et al. (2002) demonstrate that the topography of the Gilgil system is more gentle than the Karati, with few rapids. The flow of water in the Gilgil system are also perennial. Consequently, the sediments found at Gilgil sampling sites are mod-

Table 1. Descriptions of the sampling locations, the physical characteristics of the sampled sediment and the analysis performed

Sample (RHS number*)	Sampling date	Location description and land use	River morphology	Sample description	XRF analysis
s1 (9576)	19 Jul 97	River pool near bridge over the Karati. Near small village (10 huts). Mix of natural vegetation and small scale agriculture (maize). Pool used for washing	No continuous water flow. Relatively steep gradient with gravelly riverbed. Channel width of 2–4 m and depth of 0.3 m (in pool)	Sandy, clayey silt suba – subr m-p sorted	
s2 (no RHS)	19 Jul 97	Pool at upper part of a deep gorge near bridge over the Karati. Near small village (30 huts). Pastures and small scale farming (beans).	No continuous water flow. Steep gradient with gravelly riverbed. Channel width of 2–3 m, and depth of 0.8 m (in pool)	Sandy, clayey silt subs – subr m-p sorted	
s3 (9575)	19 Jul 97	At bridge over the upper Karati on the Njabeni – south Kinangop road. Near two farmhouses. Pastures, small-scale farming	No continuous water flow. Steep gradient with gravelly riverbed. Channel width of 2 m and depth of of 0.4 m (in pool)	Gravelly, clayey sand a m-p sorted	Y
s4 (No RHS[†])	19 Jul 97	Upstream of bridge over Karati on the Naivasha–Nakuru highway. Pastures and extensive watering of cattle	No continuous water flow Gentle gradient and smooth riverbed. Channel width of 3 m, and depth of 0.15 m (in pool)	silty sand subr – suba m sorted	Y
s5 (9579)	20 Jul 97	Downstream of bridge over the Gilgil on the northern Lake road. Dense acacia vegetation, 20–100 m wide, pastures.	Small rapid at bridge, otherwise gentle gradient and smooth riverbed. Channel width of 5 m, and depth of 0.5 m	silty fine sand suba – subr w-m sorted	
s6 (No RHS[†])	20 Jul 97	Downstream of bridge over the Malewa on the Naivasha-Nakuru highway. Near Naivasha town, extensive settlements. Washing and bathing downstream	Gentle gradient. Slow flowing water. Channel width of 10–12 m, and depth of 0.5 m	slightly silty sand subr – r w sorted	
s7 (9571)	20 Jul 97	At bridge over Little Gilgil at the Pembroke school. Euphorbia and eucalyptus woodland. Acacia woodland downstream. Small village nearby (50 houses).	Smooth and gentle gradient. Grassed banks. Channel width of 1 m, and depth of 0.1 m	fine sand subr m sorted	Y
s8 (9574)	21 Jul 97	Upstream of bridge over Malewa. Dense acacia woodland. Watering place for cattle.	Mainly smooth with gentle gradient. Small rapid below sampling area. Channel width of 20 m, and depth of 0.5 m	fine sand suba – subr m sorted	Y

Table 1. Continued

Sample (RHS number*)	Sampling date	Location description and land use	River morphology	Sample description	XRF analysis
s9 (9573)	21 Jul 97	At pumpsation just upstream of Malewa's confluence with the Turasha. Small scale farming (maize) and acacia woodland	Fairly smooth gradient but with intermittent small rapids. Channel width of 16 m, and depth of 0.4 m	Fine sand suba – subr m sorted	
s10 (9572)	24 Jul 97	At small bridge over the Turasha dam reservoir. Small scale farming (maize) and acacia woodland	Reservoir dam with submerged macrophytes and floating vegetation	Sandy silt subr – suba w sorted	Y
s11 (9570)	24 Jul 97	At bridge over Gilgil, 3 km east of Gilgil township, Pastures	Smooth and gentle gradient. Grassed banks. Channel width of 2–4 m, and depth of 0.4 m	Sandy silt subr – suba w-m sorted	Y
s12 (9568)	24 Jul 97	At small bridge downstream of bigger bridge over the Malewa on the Ol Kalou road, east of Ol Kalou, Pastures	Smooth and gentle gradient. Grassed banks. Channel width of 5–6 m, and depth of 0.7 m	fine sand suba – subr m sorted	
s14 (9582)	25 Jul 97	On the Karati at the Delamere farm, next to fence by the northern swamp. Pastures and near Buffalo/cattle watering place	No continuous water flow. Gentle gradient with grassed banks. Channel width of 1–2 m, and depth of of 0.1 m (in pool).	silty clay r w sorted	Y
s15 (No RHS)[<]	25 Jul 97	Gilgil river on the farm of Mr A. Simpson. Farmland and pastures, near cattle watering location	Smooth and gentle gradient. Channel width of 1–2 m, and depth of 0.1 m (in pool).	clayey, sandy silt s – mr m sorted	Y
m1 (9577)	7 Aug 97	Eastern fringe of Gilgil papyrus swamp, on Merula farm, swamp/wetland vegetation for > 50 m each side	River fringed by shallow pools of water with abundant papyrus and salvinia. Channel width of 5m, and depth of 1 m	silty clay r – sr w sorted	Y
m3 (9569)	7 Aug 97	Southern part of Merula farm, at cattle watering place near small farm village (~10 houses)	Gentle gradient, steep grassed bands. Channel width of 15 m, and depth of 1.5 m	silty sand subr m-w sorted	Y

Notes: *RHS site numbers and background data are referred to in Everard et al. (2002).
[†] Neither of these sites are an RHS site, but both are close to Malewa RHS site 9581 (KARI farm). [<] This is not an RHS site, but is upstream of RHS site 9579 and formed the upper limit of bird survey stretch in Everard et al. (2002).
Sample description: a – angular; r – rounded; w – well; m – moderately; p – poorly
Note: XRF–X-ray fluorescence – major and trace geochemistry.

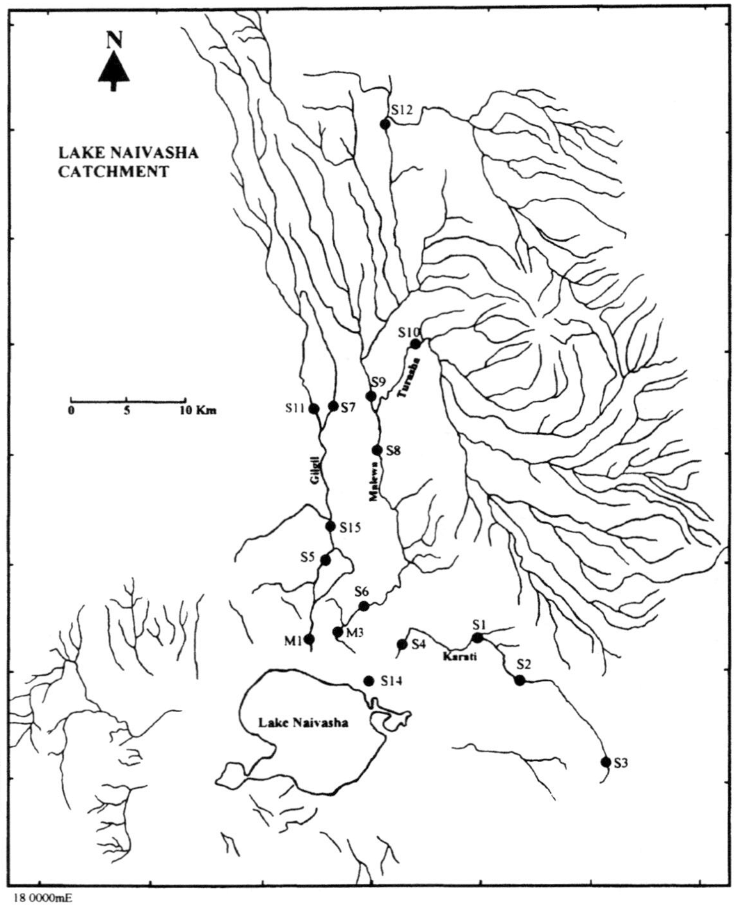

Figure 2. Sample site locations in the river sediment survey.

erately to well sorted, and the individual grains are sub-rounded to rounded. Since the flow of the Gilgil is relatively slower and more gentle, the sediment is generally fine (fine silt or clayey silt).

Sediments in the Malewa system represent an intermediary stage between the poorly sorted and angular sediment of the Karati and the fine, well-sorted sediment of the Gilgil. This follows from the upper reaches of the Malewa being steep and similar to those of the Karati (although these upland reaches were not sampled in this study), whereas the lower reaches are similar to the Gilgil. Compared to the Karati and the Gilgil, the Malewa is, furthermore, a more forceful river. Consequently, the sediment in the lower reaches is, though generally well-sorted, coarser-grained than those of the Gilgil.

Lake sediment description

The lake stratigraphy is conveniently divided into six sub-areas: central lake, delta front, northeast, northwest, southeast, and southwest. These divisions are not absolute, as differences between areas are gradational. The division reflects differences in water depth and long term sediment dynamics, and serves to make both analysis and presentation of results more effective.

The *Central Lake* sub-area occurs in the centre of Lake Naivasha with depths of 5–6 m. The general stratigraphy of the four cores corresponded to that suggested by Verschuren (1996), and is summarised in Table 2 . A thickness of 10–20 cm of sediments in the central lake sub-area, interpreted as being laid down during the past 10–15 years, gives a current sediment-

Table 2. Idealised stratigraphic section of the central Lake Naivasha basin

Depth of interface	Type of sediment and interpretation
0 cm	*Brown, organic mud* Trace macrophytes and papyrus, and trace diatoms. Interpreted as sediment laid down during the last 10–15 years during which time papyrus rafts have been more or less completely absent from this part of the lake. **Distinct contact**
10–20 cm	*Light brown, organic mud* Abundant macrophytes and papyrus and trace diatoms. Interpreted as sediments laid down during progressive increasing water levels from about 1950 to 1980. A few cm thick layer of *yellowish brown diatomaceous felt* may in places be present at the base of this unit. Together with the unit above they seem to correspond to Unit III of Verschuren (1996). **Gradational contact**
40–50 cm	*Black rooty and diatomaceous mud* Abundant macrophytes, papyrus and abundant diatoms. Interpreted as sediments laid down during a progressive lowering of water levels in the early 20th century when papyrus stands were continuously germinating on exposed mudflats. Probably corresponding to Unit II of Verschuren (1996). **Distinct non-conformity**
70–75 cm	*Dark brown peaty mud* Abundant macrophytes, papyrus and diatoms. Interpreted as sediment laid down during low lake level stand of the mid 19th century. Probably truncated by a period when the lake may have been completely dry. Probably corresponding to Unit I of Verschuren (1996). **Distinct non-conformity**
80–90 cm	*Black Peat* Abundant papyrus and trace macrophytes. Hard and massive peat laid down when the whole lake must have been a shallow, fragmented wetland overgrown with papyrus. Probably corresponding to Unit 0 of Verschuren (1996) and thus laid down during 16th to 17th century.

ation rate of about 1 cm per year. The upper 40–50 cm appears to have been laid down conformably. Below this level, two distinct non-conformities, caused by erosion during times when lake levels were low, were identified. A distinct horizon of yellowish brown, diatomaceous felt was identified at the base of the light brown, organic mud in cores 10 and 15 only. The fact that this unit was not noted in all cores indicates that the unit is not continuous over the whole central basin. An explanation for the unit's formation must therefore be sought in a process that was not basin-wide. A system of smaller sub-basins, each with its specific salinity and water quality conditions, situated within a larger wetland, may have created conditions for the mass mortality and subsequent sedimentation of diatoms in some basins but not others. A similar situation, albeit at a larger scale, occurs in the present day where water quality, and hence diatom assemblages, vary somewhat between the three sub-basins of the main lake, that is the Main Lake, Oloidien Bay and Crescent Island crater lake (Verschuren, 1996).

The *Delta front* sub-area occurs along the lakeward fringe of the North Swamp, at depths up to about 1 m. The deltafront sub-area is dynamic where sedimentation patterns are sensitive to changes in water level, changes in location of river inflows and the extent of swamp vegetation. Consequently, sediments in the area are an heterogeneous mix of massive brown clays, sandy mud, organic mud and peat horizons.

The *Northeast* sub-area, at depths of about 1–3 m, contains sandy muds derived from the river inlets. These sediments are continuously being reworked and re-suspended by wave action. Below these active sediments are stiff grey clays, interpreted as having formed from the previous aerial exposure and mineralization of lacustrine sediments, and brown clays which are interpreted as representing deltaic mudflat deposits.

The *Northwest* sub-area, depths of 2–3 m, contains sediments which appears to correspond to the upper three units of the central lake stratigraphy with two brown organic mud units overlying a black rooted mud unit.

Core 11 was situated only about 3 km west of the Malewa outflow. Nevertheless, there was no evidence of material directly derived from the river in this core. It is therefore indicated that river sediment input is, subsequent to being introduced to the lake, transported in an easterly direction only.

The *Southeast* sub-area, together with one core taken about 4 km further north, 2.5 km west of Crescent Island (core 5), but included in this group because of its stratigraphic similarity, appear to correspond to the upper four units of the central lake stratigraphy with brown organic mud units above a black or dark brown rooted mud unit over a dark brown peaty mud.

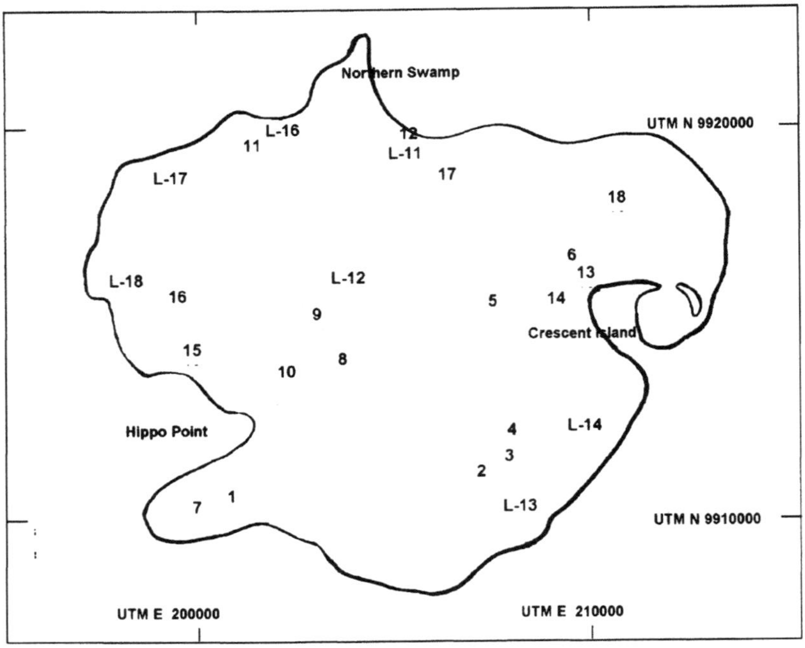

Figure 3. Sample site locations in the two lake sediment surveys. Sites in the 2000 survey are identified by the prescript L.

In the *Southwest Bay* sub-area, at depths of 3–4 m, a far greater rate of deposition was evident. The stratigraphies was somewhat different from the central lake stratigraphy. The two cores consisted of organic mud units, interpreted as corresponding to the upper two units of the central lake stratigraphy. However, though both cores were nearly 100 cm in length, neither intersected the organic rich, black mud units laid down in other parts of the lake in early to middle part of this century. Consequently, the sedimentation rate in this relatively sheltered part of the lake appears to be more rapid than in the central part of the lake. A 10-cm thick brown organic mud unit with abundant *Salvinia* and papyrus fragments at a depth of about 40 cm is interpreted as a unit laid from the 10–15 years before the 1987 low level. At this time, both papyrus rafts and *Salvinia* mats were common in this part of the lake (Tarras-Wahlberg, 1986). Such an interpretation leads to a sedimentation rate of about 3 cm per year in this part of the lake, which is considerably more than estimated sedimentation rates in other parts of the lake.

Present-day sediment dynamics

Comparisons of the sediment composition at the sediment-water interface of the 18 cores enables an analysis of present day sedimentation patterns.

Near the outflow of the Malewa the surface sediments are either massive brown clay or sandy mud units. The clay is interpreted as a deltaic deposit (a mud flat) which is now actively being eroded by wave action, whereas the sandy mud is material directly derived from the Malewa and Gilgil Rivers. Consequently, the sedimentary regime of the pro-delta area is dominated by erosion and sedimentary transport, rather than sedimentation. Presumably, sedimentation and associated deltaic growth occurs in the reduced areas now protected by swamp vegetation.

The surface sediments in the bay in the northeastern part of the lake and areas directly west and north west of Crescent Island are, like the pro-delta area, dominated by sandy muds and clay units. Consequently, the sedimentary regime here is also dominated by transport and erosion. However, the stiff clays here are not only brown clays (interpreted as deltaic deposits) but also stiff clays that are grey or dark grey in colour. These stiff clays are interpreted as having formed from lacustrine sediments that were exposed to air, such that they became mineralised and oxidised. This may be a consequence of historic low lake levels, and is consistent with the inferences about delta-forming processes based on RHS and other geomorphological observations by Everard et al. (2002). Organic muds dominate the surface sediments collected elsewhere across the lake in this study and, hence, these areas represent depositional sedimentary environments.

Geochemistry

Tables 3 and 4 show the results of the geochemical analyses. Lake and river sediments are geochemically similar, and the contents of major and minor elements are mostly comparable to global averages (Table 3). The only noticeable differences between river and lake sediments are that river sediment contain somewhat more P_2O_5 and Pb, somewhat less Cu and Ni, and significantly less SO_3, than lake sediment. These differences are, with the exception of SO_3, less than one order of magnitude and may arise from the different analytical methods used. The higher content of SO_3 in lake sediment compared to river sediment is, however, to be expected due to the tendency of animals and plants, more common in the lake compared to the rivers, to concentrate sulphur. This hypothesis is corroborated by the fact that the only two lake sediment samples to have low SO_3 concentrations are 12 A and 12 D, both of which represent inorganic clay units.

River sample S10, taken from the Turasha dam in the Malewa catchment, differs from the other samples in that it contains more organic and volatile materials (as indicated by the low total percentage). This reflects, the low energy environment of the dam where fine organic material can settle out to a greater extent than in-stream.

All sediment samples contain elevated levels of Fe_2O_3, and the concentrations of 6–13% are considerably higher than the 4% expected in an average shale. The high iron content could either be caused by the bed rock (i.e., the volcanic rocks of the area) being naturally enriched in iron, or caused by the erosion of iron rich lateritic soils, which are common throughout the area.

The content of minor and potentially toxic metals and metalloids are, in comparison to global averages, mostly unremarkable. Exceptions are elevated concentrations of Zn in all samples, Ni in lake sediment and in the Turasha dam sediment, and Cd in two lake sediment samples.

Zinc concentrations are similar in all samples, including samples taken from remote parts of the cathment (e.g., S12 and S3), which strongly indicates that relatively high concentrations are natural in origin. Zinc concentrations vary from 88 to 190 mg kg^{-1}, which can be compared to the 'threshold effect level' of the sediment quality criteria, set at 123 mg kg^{-1}. The latter assumes however, 100% bioavailability, which is unlikely to be the case, so the Zn levels are not likely to be of concern.

Nickel levels in sediment are elevated in comparison to the sediment quality guidelines, although considerably lower than the global average shale. Further, the Ni concentrations are considered as only 'somewhat elevated levels', as defined by the Swedish Environmental Protetion Agency (SEPA, 1998). Thus, it may be the case that the Canadian criteria are overly conservative. Nickel is, nevertheless, enriched in basic rocks such as the basalts of the Nyandarua range, which may explain the levels encountered. Further, Ni concentrations are somewhat higher in lake sediment and in the Turasha dam sediment compared to river sediments, indicating that this metal, for reasons unknown, is preferentially enriched in still water sediment compared to active river sediment. Possible anthropogenic sources of Ni include batteries, electroplating and paints. None of these sources appear overly likely in the Lake Naivasha area, which harbours little or no such industry.

The lake sediment samples 9C and 12B contain 7 and 6 mg kg^{-1} Cd, respectively, which is significantly above global average levels and also above the 'probable effect level' of the Canadian sediment quality criteria. There are a number of explanations for these seemingly high concentrations. Firstly, the concentrations are fairly close to the low limit of detection of the analytical method used (2 mg kg^{-1}) and therefore analytical error cannot be ruled out. Secondly, cadmium is a common contaminant of rock phosphorus, the major ingredient of chemical fertilisers (Driver et al., 1999). Excessive inputs of Cd into the lake may therefore result from the use and release of chemical fertilisers. Both units represent sediments at depth below the sediment–water interface (39–49 and 8–44 cm, respectively). The high Cd concentrations may thus be a result of Cd being mobilised from the upper oxidising sediments and re-precipitating out at depths where conditions are reducing. Thus a horizon enriched in Cd could be generated from the concentra-

Table 3. Concentration of selected major and minor elements in river sediment samples from Lake Naivasha's catchment. Analyses were performed by X-Ray Fluorescence (XRF). Lower totals are caused by organic matter and other volatile material

Sample	Total %	SiO_2 %	Al_2O_3 %	Fe_2O_3 %	K_2O %	Na_2O %	MgO %	P_2O_5 %	SO_3 %	CaO %	MnO %	As mg/kg	Co mg/kg	Cr mg/kg	Cu mg/kg	Ni mg/kg	Pb mg/kg	Zn mg/kg
S3	92	57	12	13	3.2	2.2	0.53	0.31	0.04	0.91	0.22	<10	<6	<100	2	2	14	150
S4	91	57	13	11	4.2	2.7	0.73	0.29	0.04	1.2	0.41	<10	<6	<100	3	7	17	150
S7	91	57	13	12	3.6	2.0	0.81	0.20	0.03	1.7	0.38	<10	<6	<100	4	10	19	150
S8	90	55	14	9.9	3.6	2.3	0.85	0.26	0.05	1.8	0.23	<10	8	<100	8	11	16	140
S10	83	49	14	11	2.0	1.1	0.84	0.54	0.08	1.6	0.31	<10	14	<100	22	20	12	170
S11	90	58	13	10	3.1	1.7	0.72	0.41	0.03	1.0	0.38	<10	<6	<100	5	11	16	150
S14	93	61	14	9.3	2.6	1.4	1.0	0.31	0.03	1.6	0.23	<10	<6	<100	20	16	12	170
S15	90	61	13	8.0	3.7	2.2	0.66	0.24	0.04	1.0	0.20	<10	<6	<100	6	6	14	140
M1	89	58	15	9.4	2.0	0.90	0.88	0.49	0.04	0.99	0.12	<10	<6	<100	19	15	11	190
M1-repeat	87	56	15	9.3	2.0	0.90	0.86	0.28	0.04	1.0	0.14	<10	8	<100	17	14	11	190
M3	91	55	14	10	3.6	2.4	0.89	0.37	0.03	2.1	0.17	<10	11	<100	7	10	13	130
Average shale[1]		58.1	15.4	4.02	3.24	1.30	2.44	0.17	0.64	3.1	0.10	–	19	90	45	68	20	95
Sediment quality criteria for protection of aquatic life[2]								Threshold effect level:				5.9	–	37	36	18	35	123
								Probable effect level:				17	–	90	197	36	91	315

[1]Major elements taken from Mason & More (1982) and minor elements from Turekian & Wedepohl (1961). [2]Environment Canada (1996)

Table 4. Concentration of selected major and minor elements in lake sediment samples from Lake Naivasha and its catchment. Analyses of lake samples were performed by ICP-AES after digestion in aqua regia

	LOI %	Al_2O_3 %	Fe_2O_3 %	MgO %	P_2O_5 %	SO_3 %	CaO %	MnO %	As mg/kg	Cd mg/kg	Co mg/kg	Cr mg/kg	Cu mg/kg	Ni mg/kg	Pb mg/kg	Zn mg/kg
Core samples (1997)																
9A (0–11cm)	–	17	11	0.94	0.18	0.63	0.94	0.21	2	<2	9	9	23	28	<5	160
9B (11–39 cm)	–	14	9.0	0.65	0.16	0.70	1.1	0.22	3	<2	8	18	29	41	<5	130
9C (39–49 cm)	–	13	7.3	0.56	0.16	0.58	2.0	0.15	5	7	7	6	9	18	<5	110
9D (49–73 cm)	–	9.9	9.5	0.48	0.16	0.68	1.3	0.28	<2	<2	11	36	17	38	<5	88
9E (73–85 cm)	–	12	7.9	0.55	0.16	0.60	0.95	0.15	2	<2	9	30	13	35	<5	140
9F (85–101cm)	–	13	7.9	0.55	0.09	0.50	0.97	0.12	2	<2	6	22	13	22	<5	150
12A (0–8 cm)	–	11	7.3	0.48	0.11	0.05	0.67	0.15	<2	<2	10	15	10	19	5	170
12A-repeat	–	11	7.5	0.50	0.09	0.05	0.67	0.15	<2	<2	11	11	10	18	5	190
12B (8–44 cm)	–	10	6.0	0.51	0.09	0.30	1.1	0.22	4	6	10	3	16	21	6	160
12C (44–60 cm)	–	16	7.2	0.63	0.07	0.25	0.83	0.08	4	<2	10	7	18	24	7	180
12D (60–73 cm)	–	19	8.6	0.70	0.07	0.08	0.60	0.06	5	<2	11	14	17	22	7	190
Grab samples (2000)																
L-11	4.8	–	–	–	–	–	–	–	–	–	–	–	–	–	–	–
L-12	24	–	–	–	–	–	–	–	0.98	<0.17	–	–	15	17	<0.3	160
L-13	31	–	–	–	–	–	–	–	0.84	<0.17	–	–	11	13	<0.3	140
L-14	20	–	–	–	–	–	–	–	0.55	<0.17	–	–	7.4	6.4	3.7	100
L-14-repeat	20	–	–	–	–	–	–	–	0.52	<0.17	–	–	7.6	5.7	3.7	100
L-16	26	–	–	–	–	–	–	–	0.99	<0.17	–	–	17	15	<0.3	180
L-17	33	–	–	–	–	–	–	–	–	–	–	–	–	–	–	–
L-18	39	–	–	–	–	–	–	–	0.87	<0.17	–	–	13	19	3.6	112
Average shale[1]		15.4	4.02	3.24	0.17	0.64	3.11	0.10	–	0.3	19	90	45	68	20	95
Sediment criteria for protection of aquatic life[2]						Threshold effect level:			5.9	0.6	–	37	36	18	35	123
						Probable effect level:			17	3.5	–	90	197	36	91	315

[1]Major elements taken from Mason & More (1982) and minor elements from Turekian & Wedepohl (1961). [2]Environment Canada (1996).

tion of existing natural Cd or, alternatively, through the similar concentration of contaminant Cd. The analysis of surface sediments performed in 2000 does not indicate any significant supply of contaminant Cd from fertilisers. This follows since Cd levels are equally low at sites situated near the big flower farms on the eastern shore of the lake (L-13 and L-14), as at sites near the much less developed western shore (L-16 and L-18). Hence, the relatively high Cd levels encountered are unlikely to be caused by contamination, but rather due to geochemical redistribution of naturally existing Cd. The content of other minor and potentially toxic metals and metalloids is low.

Conclusions

The description and analyses of Lake Naivasha sediment have provided extensive data on both present and past sedimentation patterns. Importantly, the findings show that natural lake level changes have caused dramatic changes to the lake ecosystem in the past, and that the instability of the lake's ecosystem is in part natural.

At present, sediment dynamics in the lake are governed by the presence of river point sources in the north, the deposition of silt and sand sized material near the river outlets; and wave-induced re-suspension of fine and organic rich sediment in the north, which are transported in easterly and southerly directions and deposited in the lake's central, southern and eastern parts. Deposition of fine, organic rich sediment is also occurring in northern areas that are protected by papyrus swamp vegetation.

The sedimentation rate in the central lake is found to be about 1 cm per year, which is in agreements with a previous estimate (Verschuren, 1996). However, the sedimentation rate in Elsamere Bay in the southwest appears to be higher, at up to 3 cm per year.

In the north-western part of the lake, historic lows in lake levels has lead to the formation of a possibly continuous layer of stiff grey, mineralised clay. This layer will act as an effective water barrier and its existence indicates that ground water recharge in this part of the lake, suggested by among others Åse (1987) may be impossible.

The geochemical analysis of a selected number of sediment samples from the lake and its catchment indicate that the concentrations of Fe, Zn, Cd and Ni sediment are elevated compared to global averages and/or sediment quality criteria. The reasons for the relatively high concentrations are likely to be found in the sediment source materials (volcanic rocks or laterite soils), which are naturally rich in these elements. The concentrations of other potentially toxic elements are low.

Lake Naivasha area is experiencing a period of rapid development with ever-increasing use of land and irrigation water for flower farming and the expansion of the Olkaria hydrothermal plants (Becht & Harper, 2002). In addition to these intensive uses, Everard et al. (2002) note an increasing intensity of subsistence agriculture in the catchment, including widespread poor land management practices. Consequently, it is not unlikely that the lake's catchment may soon be severely affected by impacts related to these developments as well as the pressures of a rising population. In addition to these immediate pressures, the planned construction of a water supply reservoir for the nearby town of Nakuru, that will export a substantial proportion of the river Malewa's water, will have an immediate and large scale affect both on the Malewa River and the lake itself (Harper et al., 1995).

The present study indicates that the river and lake sediment of the Lake Naivasha catchment are still rather pristine, and that anthropogenic impacts are not yet significant. The data obtained may therefore be used as reference values for assessing the possible environmental impacts of future developments around the lake.

Acknowledgements

This study formed past of the work of the University of Leicester research project at Lake Naivasha, which since 1987 has been authorised by the Office of the President, Government of Kenya under research permit to Dr D. M. Harper no. OP 13/001/12C 46. The project was funded by the Earthwatch Institute, Boston, USA, and Oxford, England. The data collection would not have been possible without the assistance of numerous Earthwatch volunteers. Our sincere thanks go to the numerous colleagues in Kenya in particular Jill & Angus Simpson and Velia Carn.

References

Åse, L. E., K. Sernbo & P. Syren, 1986. Studies of Lake Naivasha, Kenya and its Drainage Area. Forkningsrapporter 63, Stockholms Universitet Naturgeografiska Institutionen, Stockholm: 75 pp.

Åse, L. E., 1987. A note on the waterbudget of Lake Naivasha, Kenya. Geografiska Annaler, 69A: 415–429.

Beadle, L. C., 1932. Scientific results of the Cambridge Expedition to East Africa. Linn. Soc. J. Zool. 38: 137–141.

Becht, R. & D. M. Harper, 2002. Towards an understanding of human impact upon the hydrology of Lake Naivasha, Kenya. Hydrobiologia 488 (Dev. Hydrobiol. 168): 1–11.

Bowie, A. J. & C. K. Mutchler, 1986. Sediment sources and yields from complex watersheds. In: Wang, S. Y., Shen, H. W. & Ding, L. Z. (eds), Proceedings of the Third International Symposium on River Sedimentation. pp 1224–1232. University of Miss., Jackson.

Carelli, G., N. Sannolo, G. De Lorenzo & N. Castellino 1995. Ecosystems. Chapter 3 In Inorganic Lead Exposure – Metabolism and Intoxication, Lewis Publishers, Boca Raton, FL: 15–51.

Darnley, A. G., A. Björklund, B. Bolviken, N. Gustavsson & P. Koval, 1995. A Global Geochemical Database: Recommendations for International Geochemical Mapping. Final Report of IGCP Project 259. UNESCO, Paris.

Driver, J., P. Lijmbach and I. Steen, 1999. Why recover phosphorus for recycling, and how? Environ. Tech. 20: 651–662.

Environment Canada, 1995. Interim Sediment Quality Guidelines, Draft. Prepared by Environment Canada, Ecosystem Conservation Directorate, Evaluation and Interpretation Branch, Guidelines Division, Ottawa.

Everard, M., J. A. Vale, D. M. Harper & H. Tarras-Wahlberg, 2002. The physical attributes of the Lake Naivasha catchment rivers. Hydrobiologia 488 (Dev. Hydrobiol. 168): 13–25.

Fordyce, F. M., D. Masara & J. D. Appleton, 1996. Stream sediment, soil and forage chemistry as indicators of cattle mineral status in Northeast Zimbawe. In Appleton J. D., R. Fuge & G. J. H. McCall (eds), Environmental Geochemistry and Health, with Special Reference to Developing Countries. Geological Society Special Publication No. 113, The Geological Society, London: 23–37.

Förstner, U., 1983. Assessment of metal pollution in rivers and estuaries. In Thornton, I. (ed.), Applied Environmental Geochemistry. Academic Press, London.

Förstner, U. & G. T. W. Wittman, 1983. Metal Pollution and the Aquatic Environment. Springer Verlag, Amsterdam.

Gaudet, J. J. & J. M. Melack, 1981. Major ion chemistry in a tropical African lake basin. Freshwat. Biol. 11: 309–333.

Gouder de Beauregard, A.-C., D. M. Harper, F. Malaisse & J.-J. Symoens, 1998. Dynamique récente et cartographie de la végétation aquatique (1960–1996) du lac Naivasha (Rift Valley, Kenya). Bull. Séanc. Acad. r. Sci. Outre-Mer 44 (1998–3): 373–389.

Harper, D. M., 1984. Recent changes in the ecology of Lake Naivasha. Verh. int. Ver. theor. angew. Limnol. 22: 1192–1197.

Harper, D. M, K. M. Mavuti & S. M. Muchiri, 1990. Ecology and management of Lake Naivasha, Kenya, in relation to climate change, alien species' introductions, and agricultural development. Environ. Conserv. 17: 328–336.

Harper, D. M., G. Phillips, A. Chilvers, N. Kitaka & K. M. Mavuti, 1993. Eutrophication prognosis for Lake Naivasha, Kenya. Verh. int. Ver. theor. Angew. Limnol. 25: 861–865.

Harper, D. M., C. Adams & K. M. Mavuti, 1995. The aquatic plant communities of the Lake Naivasha wetland, Kenya: pattern, dynamics and conservation. Wetlands Ecology and Manage. 3: 111–123.

Kitaka, N., D. M. Harper, N. Pacini & M. Kenneth Mavuti, 2002. Chemical characteristics, with particular reference to phosphorus, of the rivers draining into Lake Naivasha, Kenya. Hydrobiologia 448 (Dev. Hydrobiol. 168): 55–70.

Ladd, S. C., W. A. Marcus & S. Cherry, 1998. Differences in trace metal concentrations among fluvial morphological units and implications for sampling. Environ. Geol. 3: 259–270.

Levinson, A. A., 1980. Introduction to Exploration Geochemistry. Applied Publishing Ltd., Calgary.

Mason, B. & C. B. More, 1982. Principles of Geochemistry, 4th edn. John Wiley and Sons, New York.

Maher, W., G. E. Batley & I. Lawrence, 1999. Assessing the health of sediment ecosystems: use of chemical measurements. Freshwat. Biol. 41: 361–372.

Nicholson, K., 1996. Lacustrine sediment geochemistry as a tool in retrospective environmental impact assessment of mining and urban development in tropical environments: examples from Papua New Guinea. In Appleton, J. D., R. Fuge & G. J. H. McCall (eds), Environmental Geochemistry and Health, with special reference to developing countries. Geological Society Special Publication No. 113. The Geological Society, London: 195–200.

Pettijohn, F. J., 1975. Sedimentary Rocks. 3rd edn. Harper and Row, New York.

Stumm, W. & J. J. Morgan, 1996. Aquatic Chemistry. Chemical Equilibria and Rates in Natural Waters, 3rd edn. Wiley-Interscience, New York.

Swedish Environmental Protection Agency, 1998. Bakgrundsdokument till bedömningsgrunder för sjöar och vattendrag, SNV, Stockholm (in Swedish).

Tarras-Wahlberg, N., 1986. Observations on Salvinia and its environment at Lake Naivasha (Kenya). J. East Africa nat. Hist. Soc. nat. Mus. 76: 189.

Turekian, K. K. & K. H. Wedepohl, 1961. Distribution of the elements in some major units of the earth's crust. Bull. geol. Soc. Am. 72: 175–192.

Verschuren, D., 1996. Recent and late-Holocene palaeolimnology of Lake Naivasha and Sonachi, Kenya. PhD thesis, University of Minnesota.

Windom, H. L., J. T. Byrd, R. G. Smith & F. Huan, 1991. Inadequacy of NASQAN data for assessing metal trends in the nation's rivers. Environ. Sci. Technol. 25: 1137–1142.

Willford, W. A., M. J. Mac & R. J. Hesselberg, 1987. Assessing the bioaccumulation of contaminants from sediments by fish and other aquatic organisms. Hydrobiologia 149: 107–111.

Williams, T. M., P. N. Dunkley, E. Cruz, V. Acitimbay, A. Gaibor, E. Lopez, N. Baez & J. A. Aspden, 2000. Regional geochemical reconnaissance of the Cordillera Occidental of Ecuador: economic and environmental applications. Appl. Geochem. 15: 531–550.

Hydrobiologia **488:** 43–55, 2002.
D.M. Harper, R. Boar, M. Everard & P. Hickley (eds), Lake Naivasha, Kenya.

Aspects of the biodiversity of the rivers in the Lake Naivasha catchment

Mark Everard[1,*], Anthony Kuria[2], Michael Macharia[2], Jacqueline A. Vale[4] & David M. Harper[3]
[1]*The Natural Step UK, 9 Imperial Square, Cheltenham, Gloucestershire GL50 1QB, U.K.*
[2]*Department of Ornithology, National Museums of Kenya, P.O. Box 40658, 00100-GPO, Nairobi, Kenya*
[3]*Department of Biology, University of Leicester, Leicester LE1 7RH, U.K.*
[4]*Environment Agency, Rio House, Waterside Drive, Aztec West, Almondsbury, Bristol BS32 4UD, U.K.*
**Present address: Environment Agency, King's Meadow House, King's Meadow Road, Reading, Berkshire RG1 8DQ, U.K. E-mail: mark.everard@environment-agency.gov.uk*

Key words: tropical river, RHS, birds, plants, feeding guilds

Abstract

The diverse river systems of the Lake Naivasha catchment provide a wide range of habitats. Distribution of a number of flowering plant species along river corridors appears most strongly linked with altitude and topography, although some species have an affinity with river edges. A distinct dry land/savannah flora is represented close to river channels in the lower, temporary rivers, while other plant species are ubiquitous. More physically diverse river habitats support more diverse plant assemblages. Bird biodiversity and abundance appears to be strongly linked with habitat diversity and feeding guild data suggest that this is related to the availability of a range of food items. Diversity is not linked with rarity. Changes in land use patterns and intensity in the catchment can affect its capacity to support rich biological communities.

Introduction

The shores of Lake Naivasha, Kenya (0° 45′ S, 36° 20′ E), offer fertile soils, a relatively constant climate, equatorial intensity of solar radiation over a near-constant day-length, ready access to fresh water, and an inexpensive labour force with which to pursue intensive agriculture. This has attracted a large number of horticultural interprises to the flat lands around the lake shore. Production has been predominantly of luxury products – cut flowers, fruit and out-of-season vegetables – for European markets, to where they are daily flown from Nairobi. This form of development has proven extremely profitable. Together with geothermal energy production, tourism, and commercial fisheries, it comprises a major element of the economic development of the Naivasha region (Anon, 1999) and horticulture is now the largest industry in Kenya.

The lake, which is at the center of this activity, was designated as a Ramsar site in 1995 (Ramsar, 1971, 1996; Anon, 1999). By contrast, the wider catchment comprises predominantly land tilled for subsistence and cash crops with some rough pasture in the river valleys. Its environmental protection is non-existent and knowledge of its value is confined to rare bird distribution (Bennun, 2001).

Many tropical wetland areas are now sites for development due to their high productivity (Dugan, 1990). Many wetland systems have suffered in the past from unsympathetic intensive development, which has led to their destruction or degradation with the loss of wetland 'services' and biodiversity (Dugan, 1990; Denny, 1991, 1994). A Lake Naivasha management plan has begun to address lakeside issues through a process of consensus-building (Enniskillen, this volume) but there is no equivalent impetus in the catchment.

Tropical catchments are infrequently inventoried to gain an understanding of their total contribution to biodiversity (Bennun, 2001). Those that are studied often use the rivers as indicators of the overall catchment health (Ormerod & Jüttner, 1999). Many catchments include small wetlands: swamps and marshes, occupying an intermediate position between fully aquatic and fully dry habitat types, which perform many physico-chemical, hydrological and ecological functions, valuable to ecosystems and local human populations (Dugan, 1990; Thenya, 2001). Wetlands are a precious wildlife resource, threatened at the global scale yet crucially important for both their wildlife and the resources they provide for human society. They are also highly productive, resulting in their all too frequent over-exploitation for agriculture and other human uses which, if too intensive and poorly planned, generally destroy the wetland habitat and with it the 'services' that it provides (Denny, 1995).

This paper describes a preliminary inventory of the plant and bird biodiversity in the Naivasha catchment as the first step in understanding the conservation needs of a predominantly subsistence rural economy and as a precursor to a full management strategy for the protection of the Lake Naivasha Ramsar site.

Methods

The catchment (see Fig. 3) was biologically surveyed at the same time as the River Habitat Survey described in Everard et al. (2002). A complete inventory of flowering plant and fern species was undertaken in the river channel and up to one metre from the landward edge of the bank top at each 500 m river stretch. Abundance data were not generally recorded.

A series of transects were taken across the eastern Gilgil floodplain on the Marula Estate immediately upstream of the lake's North swamp. Four replicate quadrats of one square metre, each separated by one metre in a series parallel to the river channel, were taken at 25 m intervals from the *Cyperus papryrus* swamp edge to the *Acacia xanthaphloea* woodland at the floodplain fringe. Taxa were assessed in each quadrat using the DAFOR scale.

Bird data were also collected at RHS sites on the river systems (visited once), sections of the lake shore (visited twice) and some sites outside the catchment (for comparative purposes). Cumulative bird sampling along the whole Kwamuya River system (i.e. three contiguous RHS sites) resulted in a survey length of 1500 m, and the two contiguous Gilgil sites at Maasai Ridge were recorded with a contiguous bird survey length of 1000 m. Shannon species diversity indices were calculated from abundance in timed counts and additional indices calculated separately for foraging guilds (Owiunji & Plumptre, Unpubl. report). A coefficient of species overlap was used to compare sites, taken from the coefficient of community overlap in Whittaker (1975).

Results

Plant taxon richness

Gross numbers of taxa (species and families) of flowering plants are reported by site along river corridors in Figures 1 and 2. Species richness ranged from 11 (at the stressed Kwamuya site downstream of the South Lake Road) to 102 (at the boulder-strewn dry Karati River upstream of the gorge). Diversity was particularly high (greater than 60) at eight sites:

Simba River (73 species)
Ol Kalau River (80 species)
Nyambug River (67 species)
Kiriundu River (83 species)
Morindati River at Kahuho School (67 species)
Karati River upstream of the gorge (102 species)
Karati River at the gorge (62 species)
Nyamithi River gorge (60 species)

All these sites are energetic (boulder/bedrock dominated) with surrounding trees, with the exception of the Kiriundu River site on the headwaters of the Gilgil system. This contains diverse habitat however, comprising a rocky gorge section enclosed in shrubs downstream, open pasture on the mid section, and mature woodland at the upstream end. This habitat diversity appears also to enhance plant diversity, as the site also contains many unique taxa: 13 species from 12 families occur only here, including unique types such as epiphytic ferns and an epiphytic orchid. It is also the only site at which the families Balsaminaceae, Orchidaceae and Papaveraceae are represented.

Number of species per family varies from 1.2 (0–500 m upstream of the road on the Kwamuya stream) to 2.5 (Nyambug river, Morindati at the Army Barracks, Gilgil at Nakuru Road, and the Karati Gorge). Of the four sites with a species per family ratio of 1.5 or less, three were observed to be subject to severe erosion or silt input (1.2 at Kwamuya stream, 0–500 m upstream of the South Lake Road, 1.3 on the same stream downstream of the road, and 1.5 at the Malewa River at KARI). The value of 1.5 recorded at the Malewa downstream of the Turasha confluence may therefore suggest a similar stress.

The Karati site immediately above the Karati gorge is exceptional in that it contains both the most species

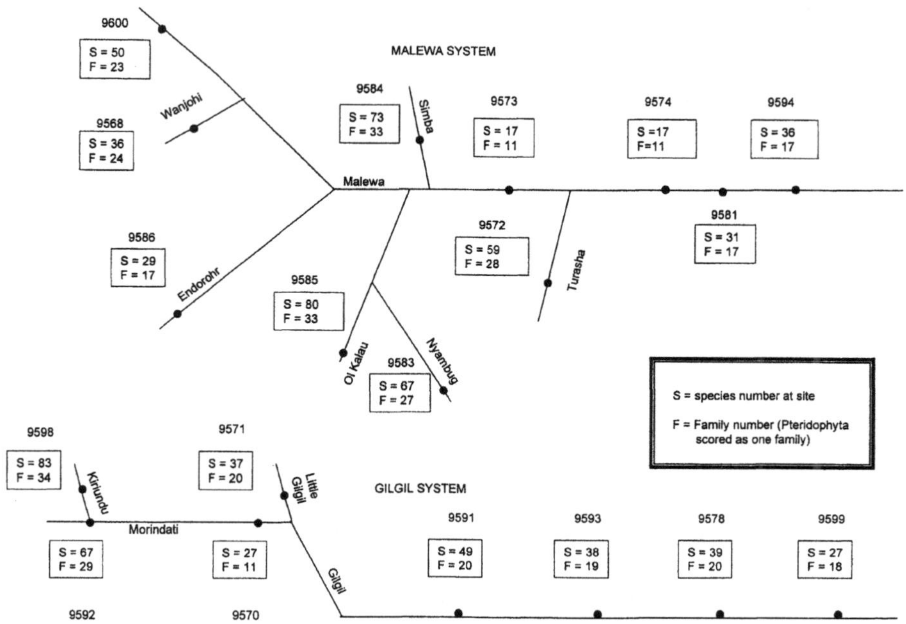

Figure 1. Numbers of plant species and families at RHS sites in the Malewa and Gilgil river systems.

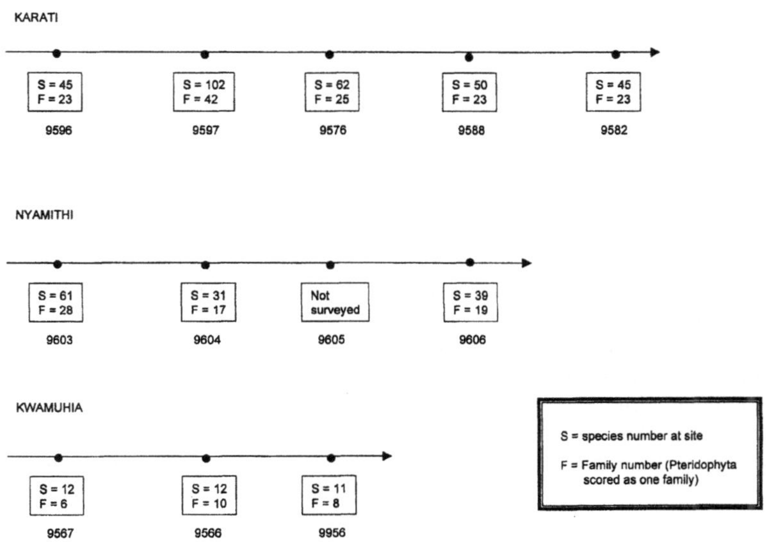

Figure 2. Numbers of plant species and families at RHS sites in the Karati and Nyamithi river systems and the Kwamuya stream.

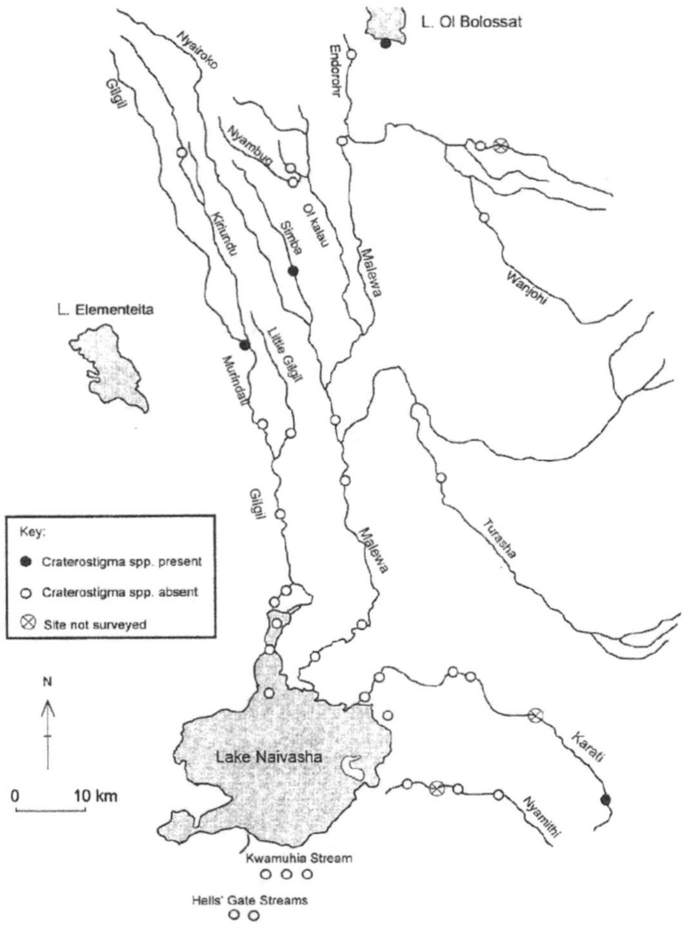

Figure 3. Distribution of *Craterostigma* spp. at RHS sites in the Naivasha catchment, and on the riparian zone of Ol Bolossat and the North Swamp.

(102) and families (43), as well as one of the highest species to family ratios (2.4), recorded at any site. As well as being a rocky gorge site, it is also an interface reach between a plateau immediately upstream and the rock gorge below, and it has semi-continuous tree cover on each bank with attendant shading of the channel, overhanging boughs and exposed bankside roots. Energetic rivers at mid-altitudes (2400–2700 m) therefore appear generally to be host to high plant richness.

Flowering plant assemblages are more closely related to altitude than to the river system in or near which they occur. The catchment may be divided into three altitude classes plus the wetlands formed at the river inlets by low lake levels. These are:

Montane rivers (>2500 m). Only one montane river is represented, the Malewa at Kirima (altitude 2500–2700 m). Species richness was nevertheless higher than average (50 species, with a species to family ratio of 2.2) and three species of plants occur only at one other site: *Brachylaena huillensis* (Compositae) which is also at the higher altitude and nearby Wanjohi River site (RHS 9602); *Artemesia afra* (Compositae) is also present in a disturbed temporary stream in Hell's Gate (outside of the catchment RHS 9607); and *Plantago palmata* (Plantaginaceae) which is also present at the diverse Kiriundu site (RHS 9598).

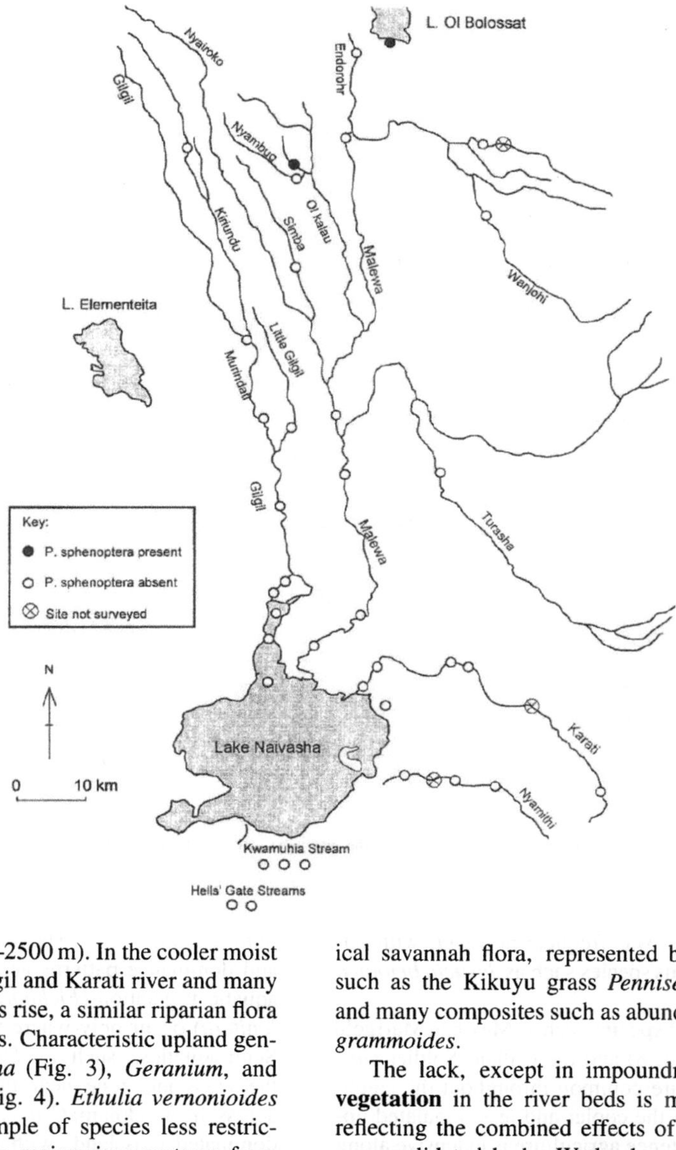

Figure 4. Distribution of *Poly* … olossat and the North Swamp.

Upland rivers (2000–2500 m). In the cooler moist uplands in which the Gilgil and Karati river and many of the Malewa headwaters rise, a similar riparian flora occurs in all river systems. Characteristic upland genera include *Craterostigma* (Fig. 3), *Geranium*, and *Polygala sphenoptera* (Fig. 4). *Ethulia vernonioides* (Compositae) is an example of species less restricted, occuring in all three major river systems from montane through to medium altitudes (Fig. 5). At altitudes below 2400 m, a wider mix of herbs and grasses are present, with the virtual absence of some of the characteristic upland herbs.

Lower catchment rivers (1890–2000 m). The flatter lands near the lake shore contain a more typical savannah flora, represented by dryland species such as the Kikuyu grass *Pennisetum clandestinum* and many composites such as abundant *Senecio mesogrammoides*.

The lack, except in impoundments, of **aquatic vegetation** in the river beds is marked, apparently reflecting the combined effects of turbid waters and unconsolidated beds. Wetland species are relatively rare in the catchment and occur predominantly near the water's edge in low-gradient lowland rivers, as well as opportunistically in stillwaters (e.g. the Turasha Dam) and pools in rock gorges (e.g. in the Karati gorge). Examples include *Ludwigia stolonifera* (Fig. 6) and *Sphaeranthus suaveolens* (Fig. 7). Marginal

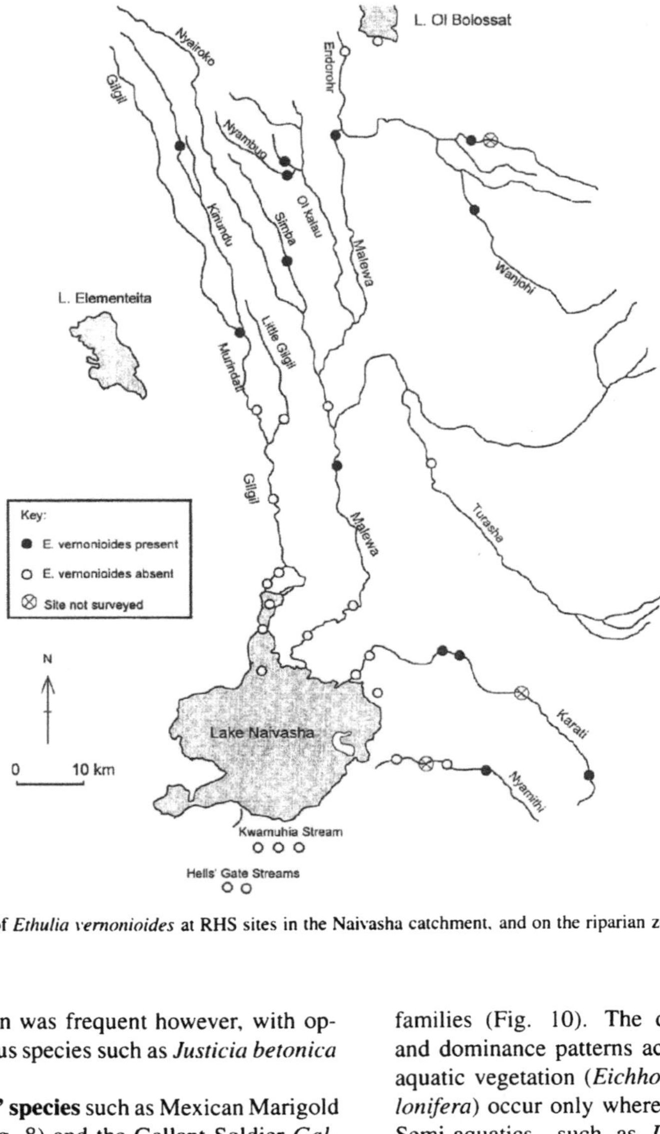

Figure 5. Distribution of *Ethulia vernonioides* at RHS sites in the Naivasha catchment, and on the riparian zone of 01 Bolossat and the North Swamp.

emergent vegetation was frequent however, with opportunist amphibious species such as *Justicia betonica* or *Commelina* spp.

Ruderal 'weed' species such as Mexican Marigold *Tagetes minuta* (Fig. 8) and the Gallant Soldier *Galinsoga parviflora* are common throughout the catchment, right up into the cooler and less populated uplands where subsistence agriculture is extensive along river valleys. A few species such as the star grass, *Cynodon dactylon* (Fig. 9), occur ubiquitously across the catchment.

In the floodplain of the Gilgil on former lake bed (Fig. 3 shows the extent of the lake at its highest water levels) there were 27 species representing 18 families (Fig. 10). The data illustrate distribution and dominance patterns across the floodplain. Fully aquatic vegetation (*Eichhornia crassipes* and *L. stolonifera*) occur only where water is present or close. Semi-aquatics, such as *Diplachne fusca* occur in the grassland zone near the water's edge. Further across the floodplain, *C. dactylon* and *P. clandestinum*-dominated grassland, with opportunist species such as the crucifer *Lepidium bonariense*, were extensive, grading into *Achyranthes aspera/Solanum incarnum* scrub before reaching mature *A. xanthophloea* woodland. The relatively low flowering plant species diversity therefore masks the structural diversity.

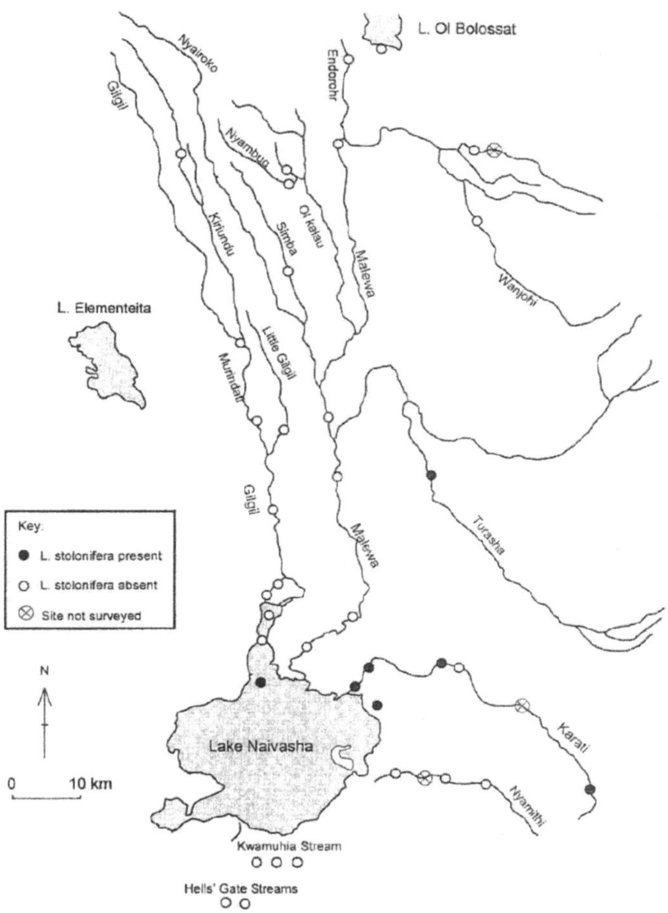

Figure 6. Distribution of *Ludwigia stolonifera* at RHS sites in the Naivasha catchment, and on the riparian zone of Ol Bolossat and the North Swamp.

Bird richess and diversity

One hundred and thirty species of birds were recorded, belonging to 17 orders and 45 families (Table 1). These included the Jackson's Widowbird (*Euplectes jacksoni*), a near-threatened species in Kenya, recorded at the Turasha Dam site. Feral lovebirds, which originally occurred only around the lake, were found at many sites along the Gilgil and Karati rivers far from the lake edge. Long-toed Lapwing (*Vanellus crassirostris*), once described as common (Henderson & Harper, 1992), were quite rare, recorded only twice along the lake edge.

Little grebe (*Podiceps ruficollis*), moorhen (*Gallinula chloropus*), African black duck (*Anas sparsa*), giant kingfisher (*Cerlye maxima*) and black crake (*Limnocorax flavirostra*) represent some of the water species recorded at the Turasha Dam. Other important species included Hautlaub's Turaco (*Tauraco hartilaubi*) recorded along the Gilgil riverine sites where fruiting trees were common and Grey Crowned Cranes (*Balearica regulorum*), a wetland indicator species, recorded at the Marula site on the Gilgil floodplain and Lake Ol Bolossat.

The KARI site on the Malewa system had the highest richness (34 species) followed by the Marula Floodplain site on the Gilgil River (29). The Gilgil

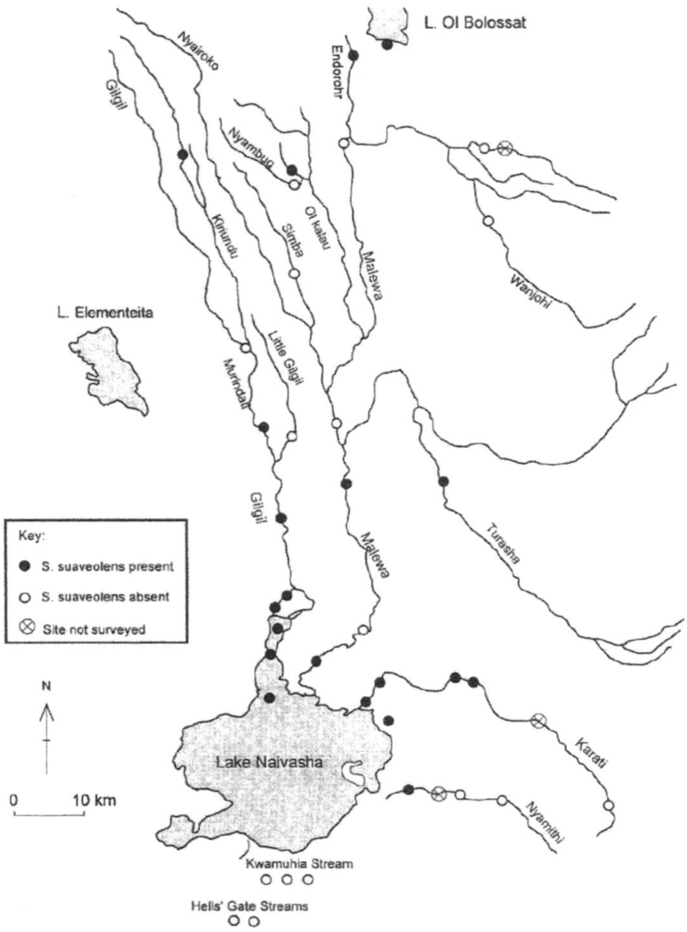

Figure 7. Distribution of *Sphaeranthus suaveolens* at RHS sites in the Naivasha catchment, and on the riparian zone of Ol Bolossat and the North Swamp.

site at Maasai Ridge and the Kwamuya river each had a total of 25 recorded species. The Gilgil river site at the Army Barracks was species-poor (only 8 species) whereas the poorest score of all was at the Karati site at Kinangop (4). The Kwamuya River had the highest bird species diversity (2.9) followed by the Karati North Swamp (2.4) then Malewa Pumphouse site (2.3). The Gilgil floodplain at the Marula Estate was the least diverse (0.31) despite having 29 species. No diversity index was calculated for the KARI site on the Malewa as numbers observed were not accurately recorded.

Grouping sites along rivers, the Kwamuya River had the highest diversity index (2.9), but on a mean basis (excluding the Kwamuya channel), Malewa sites were the most diverse (2.1) while the lake was the least (1.1). However, there was no significant difference in the mean species diversity between the river locations (Kruskal–Wallis test: $H(4, N = 15) = 8.02667$, $P > 0.05$).

The species data converted into feeding category, yielded 8 'forage' guilds (Table 2). The mean forage guild diversity index (FGDI) for all sites combined was 0.53, with a range of 0.44–0.67 at the Malewa mouth and the Turasha Dam respectively (Table 1). Karati River sites grouped together had the highest mean forage guild diversity (0.56) while the lake (open water) had lowest mean (0.46). The Kwamuya sys-

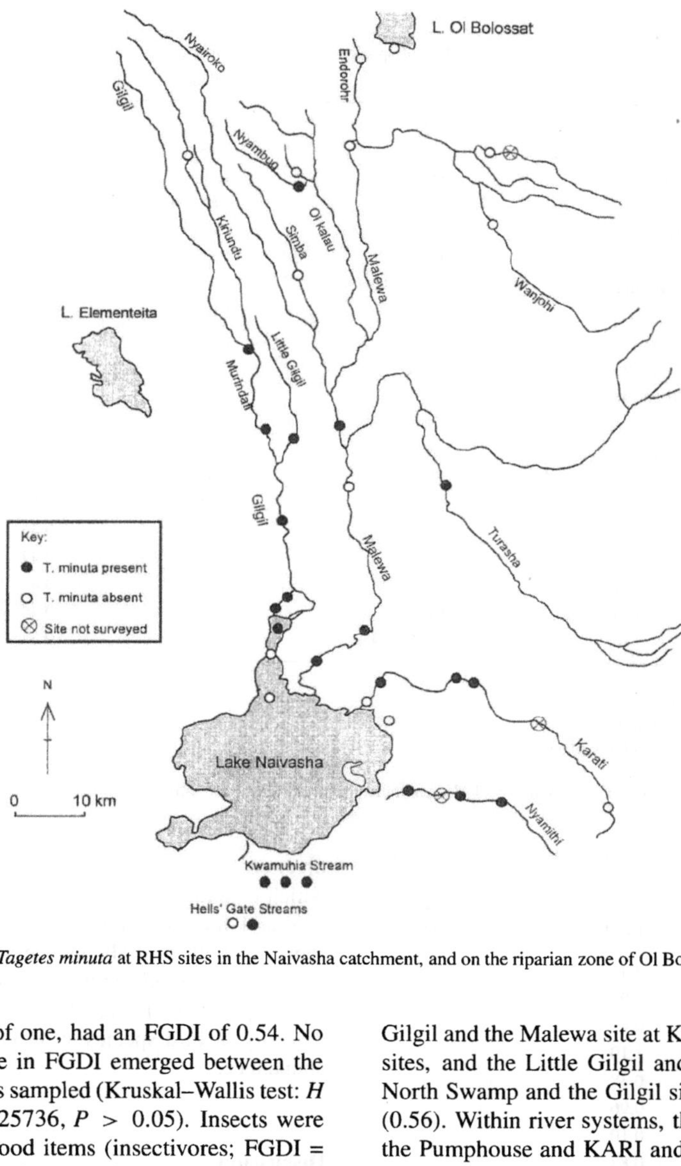

Figure 8. Distribution of *Tagetes minuta* at RHS sites in the Naivasha catchment, and on the riparian zone of Ol Bolossat and the North Swamp.

tem, with an effort of one, had an FGDI of 0.54. No significant difference in FGDI emerged between the five general locations sampled (Kruskal–Wallis test: H $(4, N = 16) - 5.325736, P > 0.05$). Insects were the most abundant food items (insectivores; FGDI = 2.42) followed by seeds (gramnivores; FGDI = 2.06) over the whole study area. Amongst the least abundant foragers were 'scavengers' (0.10) and 'carnivores' (0.49).

The data above, recalculated as Coefficients of Species Overlap (CSOs) (Table 3), showed strong overlap (0.77) between the Gilgil Mouth and the Malewa mouth sites but only weak link with the other lake edge site (the KWS Fishery Annex site). Relatively strong overlap (0.64) occurred between Little Gilgil and the Malewa site at KARI and the Kwamuya sites, and the Little Gilgil and the Karati site at the North Swamp and the Gilgil sites the Army Barracks (0.56). Within river systems, the two Malewa sites at the Pumphouse and KARI and the two Gilgil system sites at Little Gilgil and the Army Barracks had relatively strong species overlap coefficient (each 0.56) between them. Mean coefficient of species overlap from a combination of sites on discrete lake/river system locations revealed fairly strong overlap within the lake (0.47) followed by the Malewa (0.37). Overlap along the Karati River was poorest (0.17), perhaps reflecting the diverse topography of that river system. The Pumphouse and below the Turasha confluence

Table 1. Bird species richness and indices

Location and site number	No. of species	Richness index	FGDI	CSO
Malewa River	(96)	2.03	0.54	0.37
Turasha Dam (1)	23	2.19	0.67	
Malewa Pumphouse (2)	24	2.28	0.46	
Malewa Below Turasha (3)	15	1.63	0.53	
Malewa at KARI (4)	34	–	0.51	
Gilgil River	(99)	1.13	0.55	0.33
Gilgil Floodplain at Marula (5)	29, ((9.7))	0.31	0.46	
Little Gilgil (6)	17	1.06	0.52	
Gilgil at Army Barracks (7)	8	1.57	0.62	
Gilgil Maasai Ridge (2 sites) (8)	25, ((12.5))	0.64	0.51	
Gilgil below Lake Road (9)	20	2.1	0.61	
Karati River	(52)	1.95	0.56	0.17
Karati at Kinangop (10)	4	1.37	0.58	
Karati gorge (11)	17	2.08	0.53	
Karati North Swamp (12)	21	2.41	0.56	
Lake Open Water	(57)	1.12	0.46	0.47
KWS Fishery Annex (13)	23	1.56	0.47	
North Swamp, Gilgil Mouth (14)	15	1.13	0.47	
North Swamp, Malewa Mouth (15)	19	0.67	0.44	
Kwamuya (16)	(25) ((11.4))	2.89	0.54	

Notes:
All are mean figures by location.
() total number of birds recorded in that river.
(()) extended figure for extent of sample distance.
FGDI = Forage Guild Diversity Index.
CSO = coefficient of species overlap (from Whitaker, 1975, coefficient of community overlap).

Table 2. Foraging guilds and indices for birds recorded in the Lake Naivasha catchment

Foraging guild	FGDI (Guild Diversity Index)
INS (Insectivores)	2.42
GRA (Gramnivores)	2.06
CAR (Carnivores)	0.49
NEC (Nectarinivores)	1.19
HER (Herbivores)	0.38
FRU (Frugivores)	0.96
SCA (Scavengers)	0.10
PSI (Piscivores)	0.87

sites on the Malewa system, the Karati gorge, and the Gilgil site at the Army Barracks had no overlap.

The data demonstrate considerable richness of bird species within the catchment, and between habitat types within the catchment. Hotspots of bird diversity do not necessarily overlap strongly with those of plant diversity. For example the Karati site above the gorge, which contained the maximum diversity of both species and families of plants, had the second-lowest score for both bird abundance and species diversity. Conversely, the Kiriundu site, which had extraordinary proportions of rare plants, also held the highest diversity of bird species as well as the third highest abundance.

Discussion

The initial purpose of the Naivasha catchment research programme was to establish a baseline, identify 'hotspots' of biological interest and problems, thus supporting land use decisions. Diverse river flow types and flow regimes stem from the local Rift Valley topography and microclimate, ranging from perennial to ephemeral channels, open confluences to the lake to blind channel endings, erosive rocky gorges and cascades to slack depositional channels, and mont-

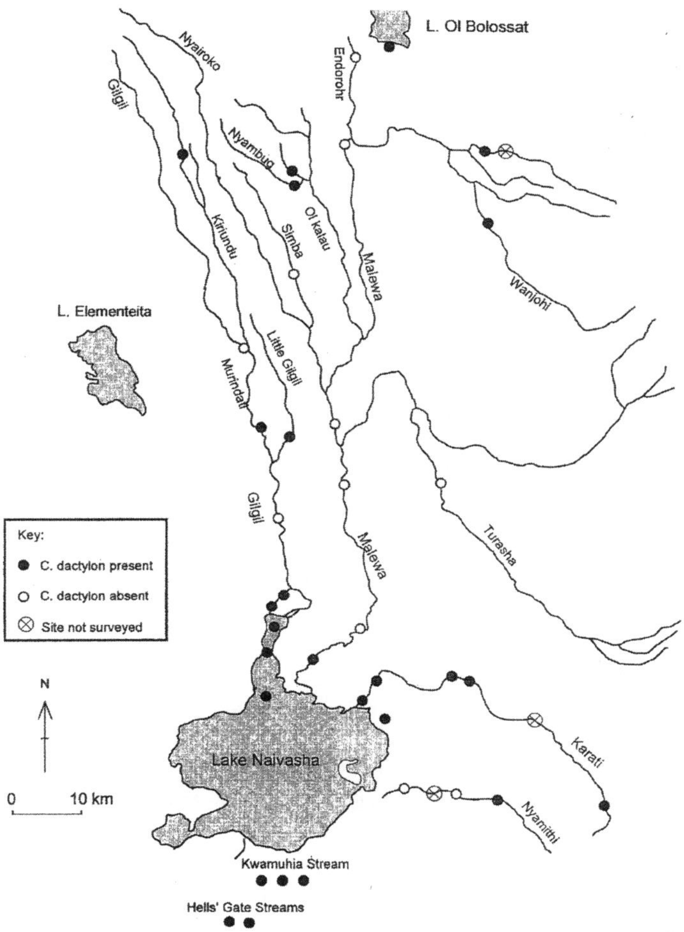

Figure 9. Distribution of *Cynodon dactylon* at RHS sites in the Naivasha catchment, and on the riparian zone of Ol Bolossat and the North Swamp.

ane forest to papyrus-dominated swampland. Biological diversity in the river corridors is related to this considerable diversity of morphology and hydrology.

Data on physical habitat diversity, plants and birds serve to form an initial description of elements of the biodiversity of the Lake Naivasha catchment. Extension of this work is important to add further substance to insights into ecological, physical and biogeochemical processes of the river systems. Expansion of research into other elements of biodiversity (other taxa, genetic diversity, habitat and land use away from the river channels) would add depth to this understanding. An understanding of biodiversity and ecosystem processes is in turn fundamental to achieving sustainable development by the human population.

River conservation has only relatively recently been perceived as an important issue in the developing world as a means for stemming ecological damage, and consequent adverse social and economic harm (Barel et al., 1985). It is well known that adverse development in lake catchments, often remote from the lake itself, can contribute to lake degradation, promoting the need to address integrated sustainable development at the catchment scale rather than merely the local lake ecosystem (Everard, 1999). Through this integrated approach to protection of the wider lake and catchment ecosystems, the natural resources

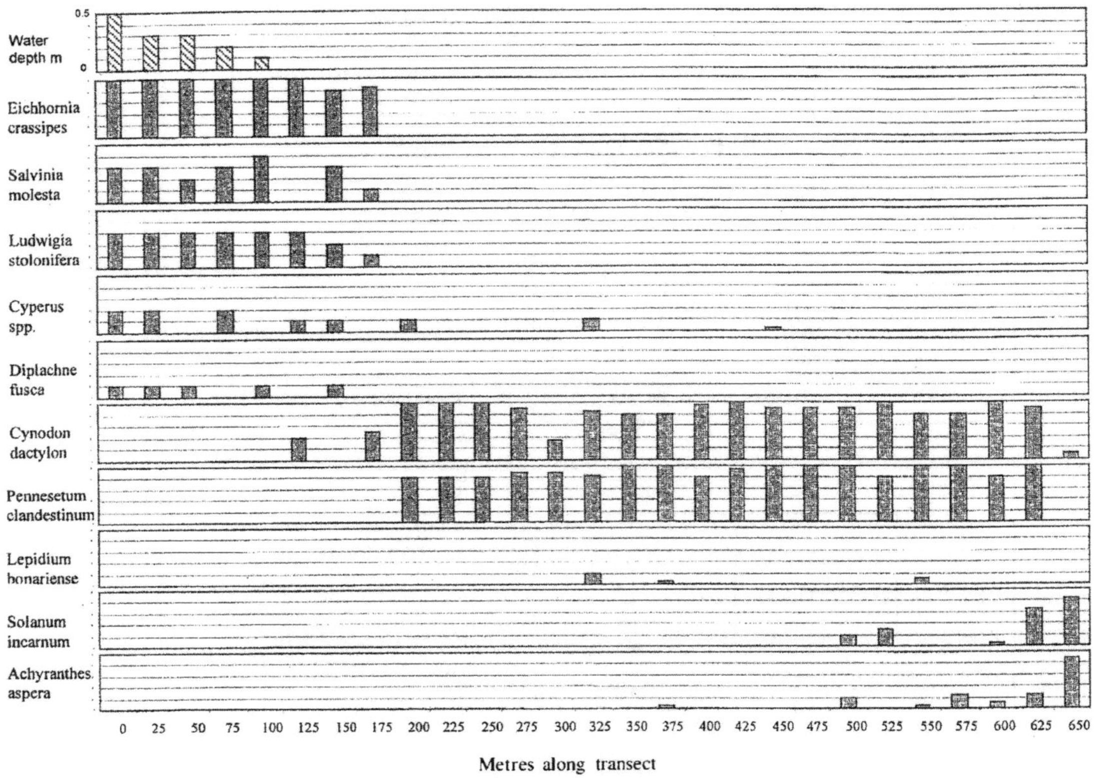

Figure 10. Distribution of macrophytes in a transect across the Gilgil floodplain at the Marula Estate, based on DAFOR measurements in replicate quadrats.

and services supporting social and economic needs can be sustained. Although this reasoning may appear self-evident, the attainment of sustainability is beset with many obstacles ranging from vested local interests, myopia about the distant effects of unsympathetic wetland development within catchments and socio-political barriers (O'Riordon, 1993). Tropical wetlands in particular have all too often been degraded or destroyed through unsympathetic development for short-term gain, resulting in catastrophic loss of wildlife and ecosystem services (Denny, 1995).

The river catchments draining into Lake Naivasha require considerable further study to determine a datum, the mechanics sustaining the unique properties of the lake, biodiversity, and the social and economic needs that they support. A proposal exists, although currently 'on ice', to dam the main stem of the River Malewa to supply growing demands for fresh water at the town of Nakuru. The consequences in both the Naivasha and Nakuru catchments, together with the social and economic needs sustained by the ecosystems, could be severe. It is necessary to distinguish impacts on the lake deriving from the catchment, as distinct from those arising from intensive lake edge agriculture, species introductions to the lake, and urban and industrial development (principally in the town of Naivasha on the eastern shore of the lake). Rivers also integrate the influences of changing land use within the catchment (Newson, 1992) and so river monitoring can prove a cost-effective method for monitoring whether sustainable catchment management is being achieved (Ormerod & Jüttner, 1998).

Acknowledgements

This study formed part of the work of the University of Leicester research project at Lake Naivasha, which since 1987 has been authorized by the Office of the President, Government of Kenya under research per-

Table 3. Coefficient of Species Overlap (CSO) between sites

0	1	2	3	4	5	6	7	8	9	10	11	12	13	14	15	16	Mean CSO by location
1	23																
2	0.298	24															
3	0.298	0.41	15														*0.368*
4	0.085	0.564	0.552	34													Malewa
5	0.255	0.308	0.138	0.348	29												
6	0.255	0.462	0.414	0.64	0.13	17											
7	0.17	0.308	0.414	0.606	0.174	0.56	8										
8	0.255	0.41	0.345	0.333	0.304	0.48	0.242	25									*0.284*
9	0.34	0.359	0.138	0.333	0.435	0.4	0.364	0.208	20								Gilgil
10	0.17	0.359	0.207	0.387	0.258	0.129	0.129	0.194	0.258	14							
11	0.255	0.513	0.45	0.435	0.087	0.261	0.217	0.174	0.174	0.323	17						*0.165*
12	0.085	0.359	0.207	0.278	0.391	0.56	0.182	0.333	0.5	0.129	0.043	21					Karati
13	0.17	0	0	0.091	0.13	0	0	0.042	0.045	0.065	0	0.273	23				
14	0.17	0	0	0.059	0.174	0.08	0	0	0.235	0.129	0	0.222	0.318	15			*0.467*
15	0.128	0	0	0.038	0.13	0.08	0	0	0.113	0.065	0	0.167	0.318	0.765	19		Open Lake
16	0.213	0.513	0.483	0.148	0.13	0.64	0.303	0.292	0.133	0.323	0.348	0.222	0	0	0	25	

Note: Numbering scheme is according to the numbers in Table 1.

mit to Dr D. M. Harper no. OP 13/001/12C 46. The project was funded by the Earthwatch Institute, Boston, U.S.A. and Oxford, England. The data collection would not have been possible without the assistance of numerous Earthwatch volunteers. Assistance with taxonomy was given by Twala (KWS Training Institute, Naivasha) , several staff of the National Museums of Kenya Herbarium, Nairobi, Dennis Otieno (University of Nairobi) and Henk Beentje (Royal Botanic Gardens, Kew). Our sincere thanks go to the numerous colleagues in Kenya for logistical support, in particular Jill and Angus Simpson and Velia Carn.

References

Anon, 1999. The Lake Naivasha Management Plan. Lake Naivasha Riparian Association, Naivasha.

Barel, C. N. D., R. Dorit & D. H. Greenwood, 1985. Destruction of fisheries in Africa's lakes. Nature, 315: 19–20.

Bennun, L., 2001. Long-term monitoring and the conservation of tropical wetlands: high ideals and harsh realities. Hydrobiologia 458: 9–19.

Denny, P., 1985. Submerged and floating-leaved aquatic macrophytes (euhydrophytes). In Denny, P. (ed.), The Ecology and Mangement of African Wetland Vegetation. Dr W. Junk Publishers, Dordrecht: 19–42.

Denny, P., 1991. African Wetlands. In Finlayson, M. & M. Moser (eds), Wetlands. International Wetlands and Wildfowl Research Bureau, Slimbridge: 115–148.

Denny, P., 1994. Biodiversity and wetlands. Wetlands Ecology and Management 3: 55–61.

Denny, P., 1995. Benefits and priorities for wetland conservation; the case for national wetland conservation strategies. In Cox, M., V. Straker & D. Taylor (eds), Proceedings of the International Conference on Wetland Archaeology and Nature Conservation. HMSO, London.

Dugan, P. J., 1990. Wetland Conservation: A Review of Current Issues and Required Action. IUCN, Gland, Switzerland: 96 pp.

Enniskillen, A., 2002. Introduction: The Lake Naivasha Management Plan – Consensus building to conserve an international gem. Hydrobiologia 488 (Dev. Hydrobiol. 168): ix–xii.

Everard, M., 1999. Towards sustainable development of still water resources. Hydrobiologia 395/396: 29–38.

Everard, M., J. A. Vale, D. M. Harper & H. Tarras-Wahlberg, 2002. The physical attributes of the Lake Naivasha catchment rivers. Hydrobiologia 488 (Dev. Hydrobiol. 168): 13–25.

Henderson, I. G. & D. M. Harper, 1992. Bird distribution and habitat structure on Lake Naivasha. Afr. J. Ecol 30: 223–232.

Newson, M., 1992. Land, Water and Development: River Basin Systems and Their Sustainable Management. Routledge, London.

O'Riordan, T., 1993. The politics of sustainability. In Turner, K. (ed.), Sustainable Environmental Economics and Management: Principles and Practice. Belhaven, London: 37–69.

Ormerod, S. & I. Jüttner, 1998 Catchment sustainability and river biodiversity in Asia: a case study from Nepal. In Harper, D. M. & T. Brown (eds), The Sustainable Management of Tropical Catchments. John Wiley and Sons: 187–207.

Owiunji, I. & A. Plumptre, Unpubl. report on foraging guilds of forest birds. Department of Ornithology, National Museums of Kenya, Nairobi.

Ramsar, 1971. Convention on Wetlands of International Importance Especially as Waterfowl Habitat. 2/2/71, Ramsar, Iran.

Ramsar, 1996. The Ramsar 25th Anniversary Statement. Resolution VI.14, 6th Meeting of the Conference of the Contracting Parties, Brisbane, March 1996.

Thenya, T., 2001. Challenges of conservation of dryland shallow waters: Ewaso Narok swamp, Laikipia District, Kenya. Hydrobiologia 458: 107–119.

Whittaker, R. H., 1975. Communities and Ecosystems (2nd edn). Macmillan, London.

Hydrobiologia **488:** 57–71, 2002.
D.M. Harper, R. Boar, M. Everard & P. Hickley (eds), Lake Naivasha, Kenya.

Chemical characteristics, with particular reference to phosphorus, of the rivers draining into Lake Naivasha, Kenya

Nzula Kitaka[1,2], David M. Harper[1,2,*], Kenneth M. Mavuti[3] & Nic Pacini[1,4]
[1]*Department of Biology, University of Leicester, LE1 7RH, U.K.*
[2]*Department of Zoology, Egerton University, Box 536, Njoro, Kenya*
[3]*Department of Zoology, Nairobi University, Box 30197, Nairobi, Kenya*
[4]*Present address: Ministry of the Environment, Rome, Italy*
**Author for correspondence. E-mail: dmh@le.ac.uk*

Key words: phosphorus, suspended solids, conductivity, seasonality, El Niño

Abstract

The loss of phosphorus from the catchment of Lake Naivasha, Kenya, was 0.2 kg ha^{-1} ann^{-1}, 76% of it particulate in a 'normal' year of wet and dry periods. It rose to a mean of 1.8 kg ha^{-1} ann^{-1}, 90% of which was particulate, in the months following the extreme rainfall which followed the 1997–1998 El Niño event in the Southern Atlantic. Total and particulate phosphorus were positively correlated with suspended solids and with discharge, and conductivity was negatively correlated with discharge. The magnitude of losses pose both threats to the water quality of Lake Naivasha and to the sustainable soils resources of the catchment.

Introduction

The extent of soil degradation, the losses of potential agricultural production (Dregne, 1990) and the decrease in water quality (Calamari, 1987; Dejoux, 1988) are relatively poorly known for Africa. Yet its human population growth makes the necessity for an agricultural production increase essential (Fardeau & Frossard, 1991). Knowledge of the negative consequences of this, then becomes paramount.

In many regions of the tropics, a sustainable level of food production is possible only by achieving better agricultural yields with the current land resources and not by extending cultivation on marginal lands (Dregne, loc.cit.; Downing et al., 1990). This means that for a large part of Africa, an increase in agricultural output is possible only if pathways of phosphorus cycling are changed by a greater P uptake by crops (Fardeau & Frossard, loc.cit). Phosphorus is low in Kenyan soils and is commonly the primary limiting nutrient (Hinga, 1973; Nyandat, 1981). It is lost from cultivated land mainly in particulate form, where the phosphate adsorption capacity is dependent on particles' surface area and on their content of hydrated metal compounds (Fe, Al, Mn); most particles with these characteristics are clays (Viner, 1987). Their iron content depends on particle size and a strong relationship exists between available P, Fe content and the size of different clay fractions (Viner, loc. cit.). Organic matter also plays a significant role in the transport of particulate phosphorus. Organic particulates transport a larger fraction of adsorbed P than inorganic particulates (Sharpley & Smith, 1990). Conditions of intensive weathering in the tropics are associated with the leaching of Ca (Uriyo & Singh, 1978) and correlates with high abundance of amorphous Fe and Al (Wolf et al., 1985).

Total phosphorus and suspended solids (SS) generally increase in parallel during rain events when erosional processes in the catchment and in the channel are active. Discharge is initially positively correlated to TP and SS concentrations. Most of the particulate matter is mobilised during the rising limb of the hydrograph. Its concentration may decrease before the peak in discharge is reached. Alternatively peaks of SS and TP concentration can be observed during the falling limb of the hydrograph when the mobilisation of sediments continues after the peak in discharge. The concentration of TP generally correlates with the concentration of sediments (Pacini & Harper, 1999).

P export is maximal in storm events as the P carried by high discharge during periods of increased surface runoff represents the highest losses of total and dissolved phosphorus (TP and DP). Once in the river, dissolved and adsorbed P forms behave differently in relation to discharge. The concentration of DP is often reported to be higher in baseflow and to decrease during storms as a consequence of dilution by high water levels. Some studies have shown the opposite: higher DP export during surface runoff has been observed and explained as a reduction in contact time between the percolating water and the soil constituents which results in less P adsorption of the DP carried by surface waters (Sharpley & Syers, 1976).

The irregular flow regimes of tropical rivers are related to the particular rainfall pattern characterized by short rainy seasons with intense peaks, which generate a very high response in runoff (Jones, 1979; Viner, 1981). In the Orange river (South Africa) as much as 90% of the suspended and dissolved yearly load is mobilized during one rainy season. In Morocco, short-term monitoring revealed that four hours of flood could yield 98% and 74% of the annual TP and TN loads, respectively (Viner, 1981). These examples point out the challenge represented by attempts to monitor sediment and TP in tropical catchments. Roseboom & Lotriet (1993) found, after a detailed analysis of sediment export studies in South Africa, that a minimum of 8 years of continuous monitoring would be necessary to obtain reasonable estimates of suspended sediment export from a single catchment. A recent sediment yield map for South Africa was still based on simplifying assumptions defining broad regions of similar SS export potential, however, despite several decades of sediment monitoring and a detailed survey of reservoir sedimentation across the country, (Roseboom & Lotriet, 1993).

In Kenya, information on the physical and biological processes characterizing the cycle of phosphorus is scarce and fragmentary. Data are periodically collected by soil scientists (Hinga, 1973; Nyandat, 1981), by limnologists (Talling & Talling, 1965; Kilham, 1972; Gaudet, 1976, 1977, 1979; Melack, 1976, Peters & McIntyre, 1976; Melack et al., 1982; Njuguna, 1988; Melack & McIntyre, 1993; Hecky, 1993) and by agriculturists (Smaling & Bouma, 1992). There have been no surveys, however that would integrate different compartments of the P cycle and no quantitative estimates of annual P losses through runoff. Melack & McIntyre (1991) reviewed the available data on P in surface waters in tropical Africa and its potential for the limitation of aquatic ecosystems. A large part of their data concerns lakes and rivers in Kenya. In their analysis, they stressed the diversity of chemical composition of surface waters in the country and the lack of data on P in rivers during a full hydrological cycle.

A study of the major sources and sinks of P and other nutrients in different agro-ecological zones in Kenya (Smaling, 1993) suggested that human intervention in nutrient cycles in modern tropical Africa may be a cause of excessive enrichment in surface waters. This is supported by reports of on-going eutrophication in Lake Victoria (Hecky, 1993) and by concern expressed about the impact of increasing intensive agricultural land use around Lake Naivasha (Harper et al., 1993).

This study examines the sources and quantity of phosphorus in the rivers of the Lake Naivasha catchment, particularly the river which carries the main input, the Malewa. It was carried out from September 1997 to March 1998, when heavy rains fell over the country, believed to be a consequence of 'El Niño' in the southern Atlantic, and this period was then compared with a more 'normal' period, between 1998 and 1999.

Methods

Sampling stations were established on a range of headwater rivers and streams and the three major rivers flowing into lake Naivasha; the Malewa, Gilgil and Karati (Fig. 1). The positions of the sampling stations were on road access to the river, in most cases, below bridges.

Instantaneous river discharge at the time of sampling was estimated using a mechanical flow meter of General Oceanics, model 2030R to give mean column velocity at 60% water depth, at 1 m interval in stations with more than 10 m width and 0.5 m intervals in stations with less than 10 m width. Discharge was calculated by the velocity area method. In the lower reaches and in rainy seasons, the meter could not be safely used; water velocity was estimated by timing an orange flowing through a known distance from a bridge or vantage point. The values obtained during low flows were calibrated using graduated gauges positioned at the riverbank and in the middle of the river

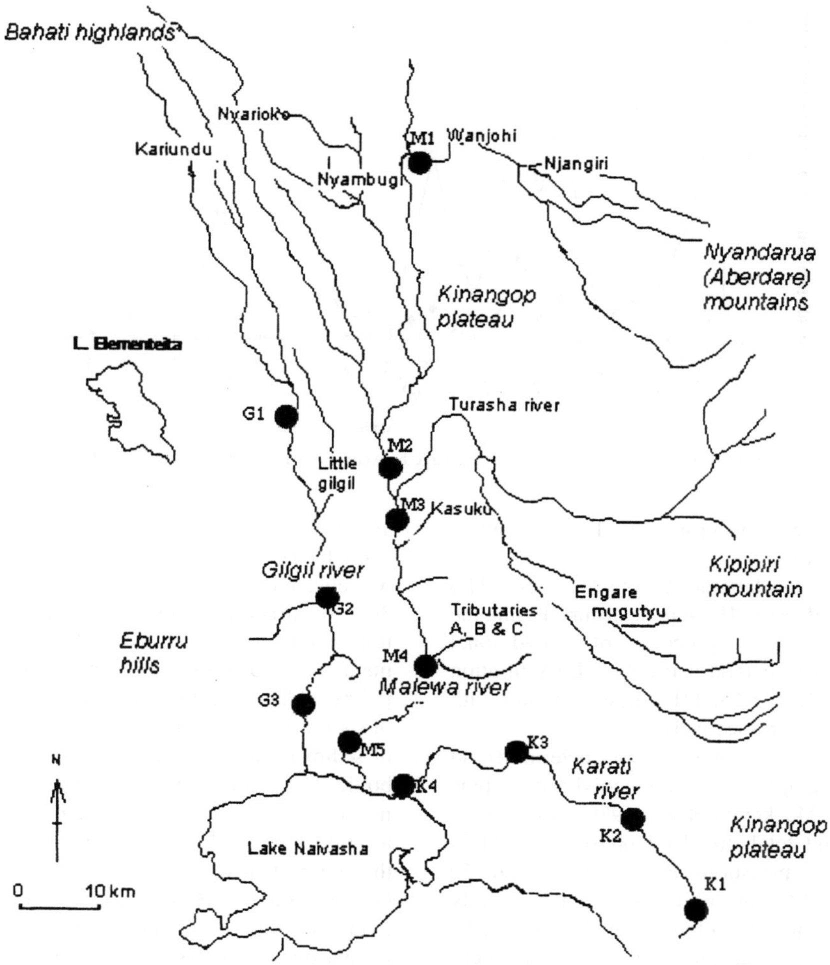

Figure 1. The lake Naivasha catchment showing sample stations.

channel by the Kenyan Ministry of Water at the lower, M_4 and G_2 stations. The gauge height values (GHT) were calibrated using Åse et al. (1986).

Water samples were collected by dipping an acid-washed polyethylene bottle a few centimetres below the water surface in the middle of the river channel. During high water level, samples were collected by lowering down an acid-washed plastic bucket tied on a rope from the bridge to the middle of the river. The water was then transferred into an acid-washed polyethylene sample bottle. The samples for dissolved nutrients were filtered immediately after collection in the field using an acid-washed 60 ml syringe and a Swinnex filter holder with a GF/C glass microfibre filter. The samples were transported in a cool box to the laboratory for analysis. The water samples were analysed the same day of collection or, were stored in the refrigerator for analyses the following day. During the sample visit, conductivity, total dissolved solids (TDS) and temperature were measured by probe using a portable Hach Conductivity/TDS meter whilst pH measurement was determined using an Orion SA 250 model pH meter.

The alkalinity was determined by titrating a 50 ml sample in the field using a Hach Digital Titrator model 16900, with a 0.16N H_2SO_4 titration cartridge

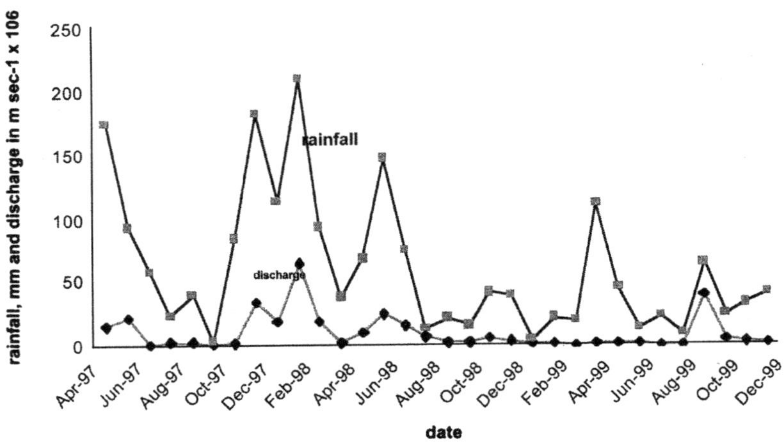

Figure 2. The average rainfall and Malewa river discharge between 1997 and 1999. Source: R. Becht, pers. comm.

and phenolphthalein and bromocresol green powder pillows as indicators.

On arrival in the laboratory, total hardness (TH) and calcium hardness (CH) were determined using the Hach calorimetric version procedure of the traditional Solochrome Black method using the EDTA titration technique with Manver 2 and Hardness solution 1 with Calver 2 indicators, respectively.

Soluble reactive phosphorus was determined in filtered water samples using the molybdenum blue technique after Mackereth et al. (1979). Total phosphorus (TP) and total dissolved phosphorus (TDP) were obtained by measuring P after digestion of 20 ml unfiltered and filtered water samples respectively with 0.5 g potassium persulphate and 1.2 ml of 10N sulphuric acid. Particulate phosphorus (PP) was calculated as the difference between TP and TDP; dissolved organic phosphorus as the difference between TDP and SRP (Lennox et al., 1997).

Total suspended solids (TSS) were determined by filtering a known volume of water sample on arrival in the laboratory through a pre-weighed, oven dried GF/C glass microfibre filter. The amount of water filtered depended on turbidity of the water samples. The filters were then dried in the oven at 100 °C for 24 h before re-weighing them. The re-weighed filter papers were then ignited at 500 °C in a muffle furnace for 2 h to determine the loss on ignition (Dean, 1974).

River sediment was analysed in the same way after oven drying for 24 h. Total phosphorus (TP) in the sediment was analysed by heating 10 gm of sediment after LOI determination in a volumetric flask with 10 ml of 1N HCl to dryness in a hot plate at 100 °C. After cooling the flask, 20 ml of 50% HCl was added slowly until the residue dissolved. The flask was covered with a self-sealing Nescofilm and left for 36 h to extract. After extraction, phosphorus concentration was determined using the molybdenum blue method as above. Iron-bound phosphorus was extracted by adding 25 ml of bicarbonate buffered solution of sodium dithionite (BD-reagents) to a known weight of sediment sample in a screw capped polypropylene bottle. The bottles were then incubated at 40 °C for 30 min in a water bath with a shaker. After centrifuging at 5000 rpm for 10 min, the phosphorus content of the supernatant was analysed using the molybdenum blue method after 30 min aeration of each sample to oxidise the dithionite, which is known to interfere with the phosphorus analysis (Nürnberg, 1988).

Data were analysed using Microsft Excel and Minitab.

Results

The patterns of rainfall and river discharge in the major river, the Malewa, are shown in Figure 2. They do not indicate a true relationship, because the rainfall gauging is predominantly at the lower altitudes of the catchment, below 3000 m, whereas the disharge is generated by the high-altitude rainfall, which is not gauged reliably. For this reason, the rains in early 1999 did not produce a corresponding increase in discharge because they reflected the more local situation on the floor of the Rift Valley, whereas those in 1997–1998 did, because they were a reflection of the region-wide

Table 1. The characteristics of the Naivasha headwaters

Name of tributary	Drainage region	Altitude m.a.s.l	Catchment area km^2	Stream order	Temp °C	Condy $\mu s\ cm^{-1}$	pH	Total hardness $mg\ l^{-1}$ as $CaCO_3$	Ca hardness $mg;^{-1}$ as $CaCO_3$	Alkalinity $ms\ l^{-1}$ as $CaCO_3$
Nyambugi	Bahati hills	2320–260	17	1	20.9	81	7.8	47	19	11
Nyairoko	Bahati hills	2160–2760	35	2	22.2	92	7.7	40	20	12
Njangiri	Nyandarua ranges	2360–3620	10.5	2	16.6	72	7.4	33	28	22
Wanjohi	Nyandarua ranges	2340–3800	137.5	3	13.2	76	8.2	58	30	22
Engare Mugutyu	Kipipiri	2469–2620	8	2	20.5	103	7.6	57	40	28
Turasha	Kipipiri	2000–2906	133.5	4	20.4	80	7.5	56	33	23
Kasuku	Kinangop plateau	2012–2316	<1	1	18.2	372	7.6	189	58	43
Unnamed A	Kinangop plateau	2012–2316	<1	1	20.1	384	8.0	213	40	36
Unnamed B	Kinangop plateau	2012–2316	<1	1	19.9	500	8.2	288	44	40
Unnamed C	Kinangop plateau	2012–2316	<1	1	19.7	462	8.1	218	54	48

heavy rains following the 1997 El Niño event in the Southern Atlantic.

The chemical characteristics of the headwaters streams are shown in Table 1. The streams on the north eastern part of the catchment draining the Bahati highlands and the Nyandarua mountain ranges were the most dilute, with conductivity below 100, slightly alkaline with low alkalinities. The small streams draining the southern part – the Kinangop plateau – were approximately 4 times richer and harder. This is an area of older underlying rocks, with regosol, calcareous, deep loam soils.

Total phosphorus was generally below 100 $\mu g\ l^{-1}$ (Fig. 3) with the three exceptions, which were either influenced by cattle (Njangiri, evidenced by high ammonium, Table 2 and Kitaka, pers. obs.) or at lower altitude and so more agricultural (Little gilgil) or urban (Gilgil, below the town of the same name). Nitrate was generally low. The general chemical characteristics of the main rivers were otherwise similar (Tables 3–5).

Table 2. The nutrient chemistry of the Naivasha headwaters

Tributary	SRP/TP ratio	PP% of TP	Mean NH_4-N $\mu g\ l^{-1}$	Mean NO_3-N $\mu g\ l^{-1}$
Nyambugi	0.4	59.2	82.9 ± 35	0.1 ± 0.03
Nyairoko	0.2	75.7	8.6 ± 3	0.04 ± 0.01
Njangiri	0.1	89.6	604.3 ± 34	0.04 ± 0.01
Wanjohi	0.5	47.5	50 ± 2	0.11 ± 0.03
Engare Mugutyu	0.4	59.6	68.1 ± 9	0.56 ± 0.3
Turasha	0.6	50.1	40.8 ± 7	0.58 ± 0.1
Unnamed A	0.5	54.9	11.9 ± 3	0.13 ± 0.02
Unnamed B	0.5	49.7	36.6 ± 7	0.11 ± 0.03
Unnamed C	0.7	34.6	21.2 ± 4	0.17 ± 0.01
Kariundu	0.2	80.3	< detection	0.63 ± 0.2
Little Gilgil	0.6	63.7	26.5 ± 14	0.44 ± 0.07

The means hid considerable seasonal changes. The TP of three headwater streams show (Fig. 4) that two of them – the Little gilgil and the Turasha, had elevated levels three times higher than the means at the beginning of the heavy rainy season (see Fig. 1) but less as the season progressed. The wet season brought a higher proportion of particulate phosphorus, as a consequence of soil erosion and sediment transport. The regressions of TP and PP against TSS were strong for the Little gilgil (TP R^2 0.97, $p < 0.0001$; PP R^2 0.98, $p < 0.001$), not significant at the Turasha (TP R^2 0.15, n.s.; PP R^2 0.27, n.s.), and non-existent for the Nyambugi (Fig. 5), most likely because of high values of phosphorus at low discharges.

The effect of seasonal fluctuation of discharge on chemical composition is more clearly seen in the main rivers. In the Malewa, suspended sediment increased with discharge (Fig. 6). Most phosphorus was bound into the TSS (Fig. 7), because most of the phosphorus is particulate (Fig. 8).

The TSS was mostly organic in all the three of the main rivers, with higher values at the start of the rainy season than at its middle to end (Figs 9–11), There was an indication of higher organic content percentages in the upper and the middle-course stations in all the three rivers. (The Karati in September and October 1997 and the Gilgil in October were dry with pools in places).

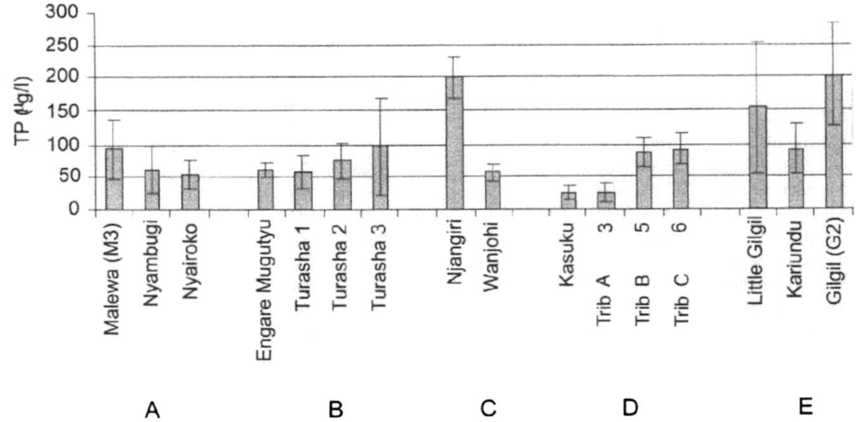

KEY:

Drainage areas (see Fig. 1)

A = Bahati Hills

B = Kipipiri mountain

C = Nyandarua mountain ranges

D = Kinangop plateau

E = Gilgil

Figure 3. Mean total phosphorus (TP) concentrations in the Naivasha headwater streams, together with the Malewa (M3) and Gilgil (G2) for comparison.

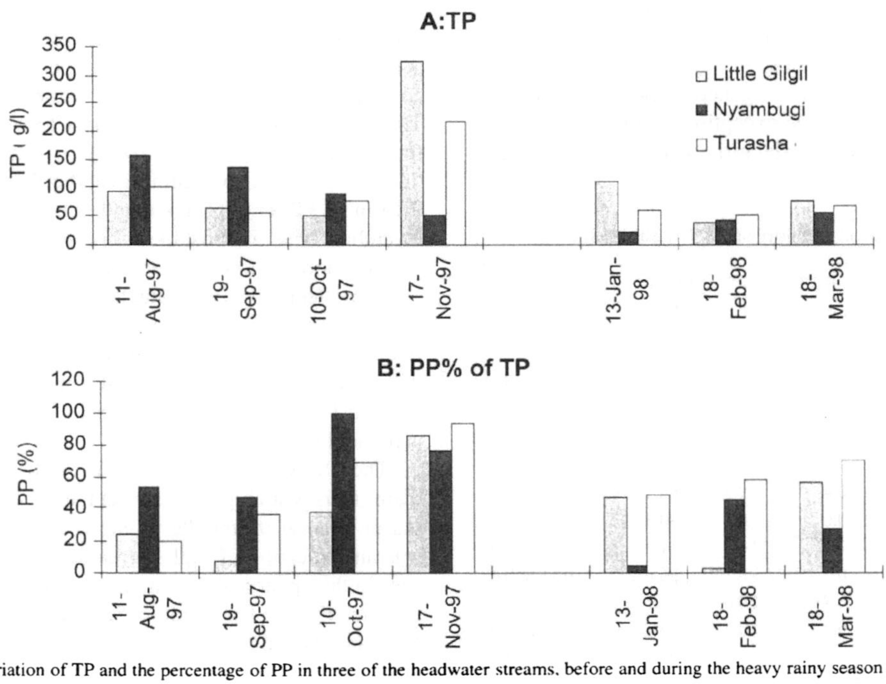

Figure 4. Variation of TP and the percentage of PP in three of the headwater streams, before and during the heavy rainy season of 1997–1998.

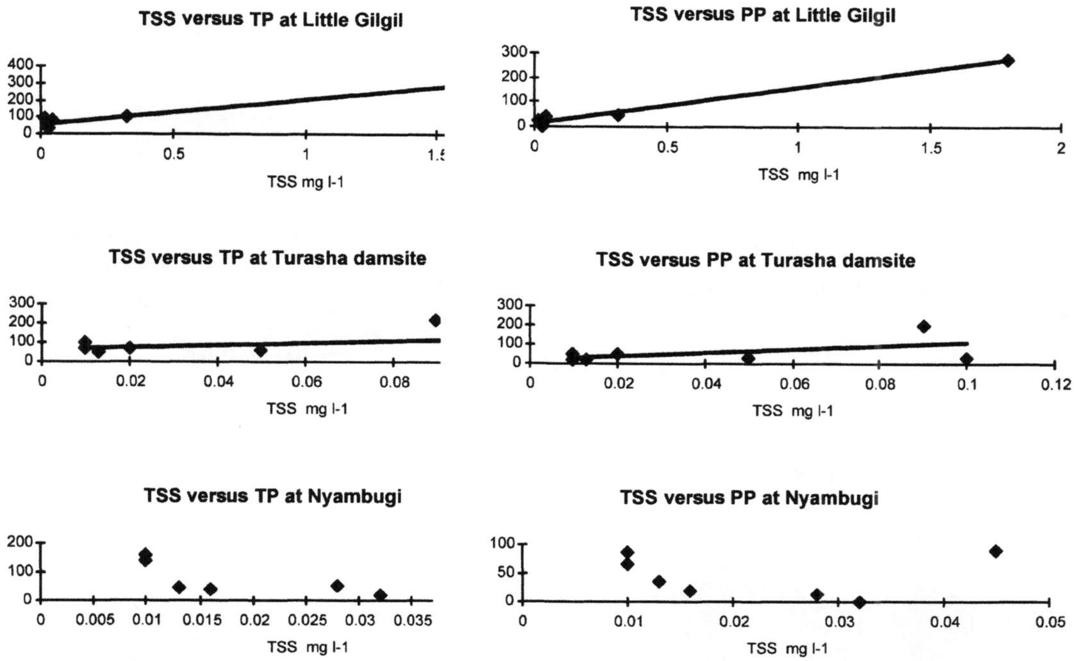

Figure 5. Regressions between forms of phosphorus and suspended solids in three upper catchment streams.

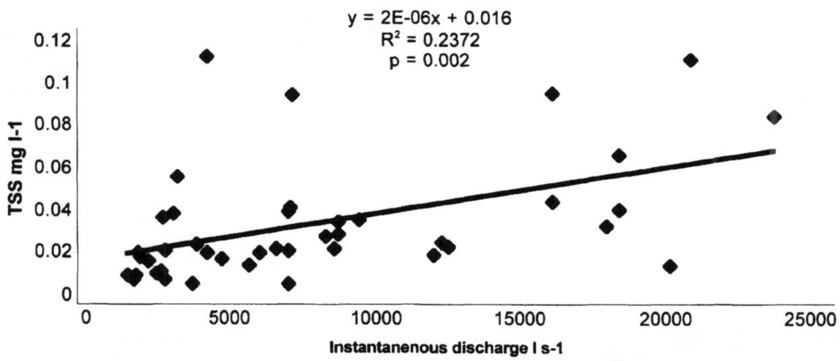

Figure 6. The relationship between instantaneous discharge and TSS in the river Malewa.

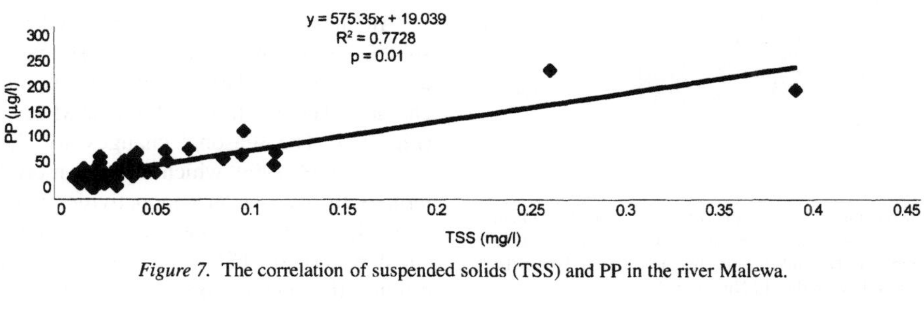

Figure 7. The correlation of suspended solids (TSS) and PP in the river Malewa.

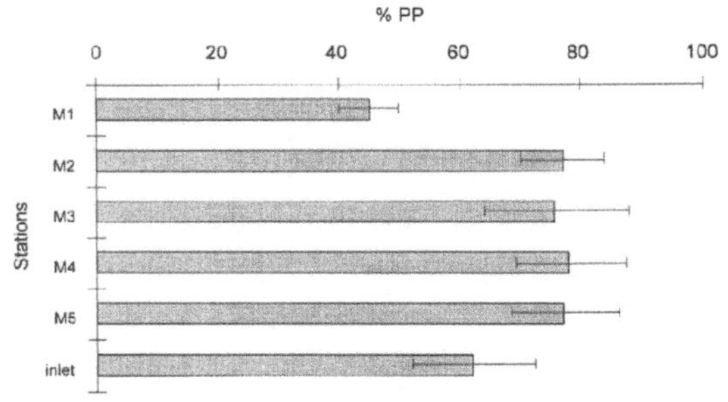

Figure 8. Mean particulate phosphorus (PP) content (%) along the river Malewa.

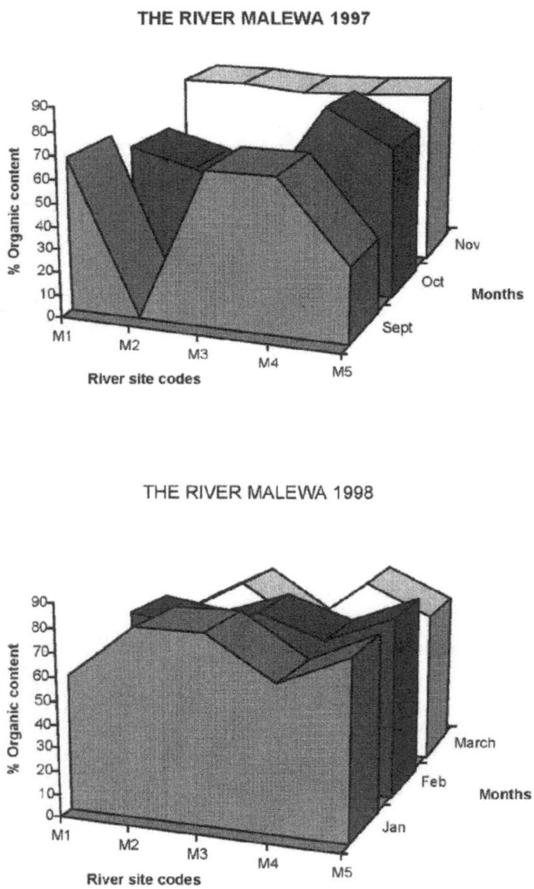

Figure 9. The percentage organic matter of the suspended sediment in the river Malewa just before (September & October 1997) at the start of (November 1997) and during (January–March 1998) the heavy wet season following the 'El Nino rains'.

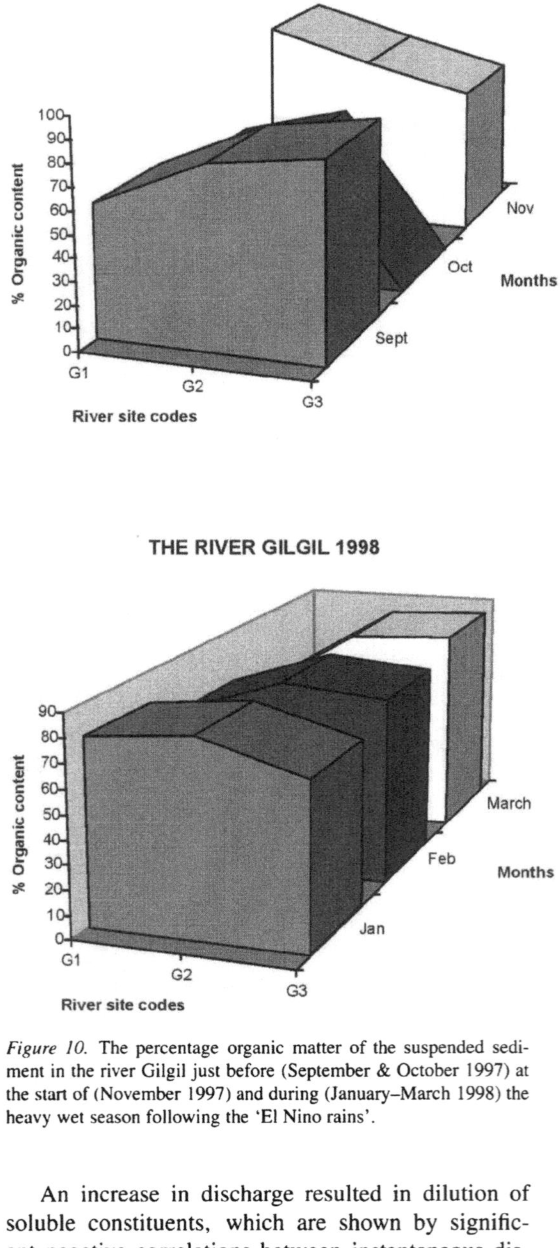

Figure 10. The percentage organic matter of the suspended sediment in the river Gilgil just before (September & October 1997) at the start of (November 1997) and during (January–March 1998) the heavy wet season following the 'El Nino rains'.

An increase in discharge resulted in dilution of soluble constituents, which are shown by significant negative correlations between instantaneous discharge and conductivity in both the Malewa and Gilgil (Fig. 12). The seasonal changes are shown for 13 months 1998–1999, which reflect a more 'usual' pattern, with dry season conductivities which are over twice the wet season ones, a mirror image of the seasonal pattern of TSS (Fig. 13). At high flow conditions, the rivers approached the proportional con-

Table 3. Chemical characteristics of the Malewa river

Station code	Location	Distance km from lake	Altitude (m.a.s.l.)	pH	Alkalinity mg l^{-1} as $CaCO_3$	Total hardness mg l^{-1} as $CaCO_3$	Ca hardness mg l^{-1} as $CaCO_3$
M1	Upstream Ol Kalou	68	2360	7.98 ± 0.4	46 ± 10	26.2 ± 3.99	17.7 ± 2.7
M2	Gilgil pump house	35.5	2222	7.67 ± 0.3	49.8 ± 6.5	26.8 ± 2.6	18.1 ± 1.8
M3	Turasha ridge settlement	30.5	2012	7.69 ± 0.3	48.4 ± 4.7	27.4 ± 2.4	17.4 ± 1.9
M4	Naivasha-Nakuru road bge	10.5	1903	7.44 ± 0.2	57.8 ± 9.9	27.6 ± 2.1	19.0 ± 2.5
M5	Marula	2.5	1890	7.69 ± 0.2	50.6 ± 6.2	27.4 ± 2.0	19.2 ± 2.5

Mean ± standard error ($n = 16$).

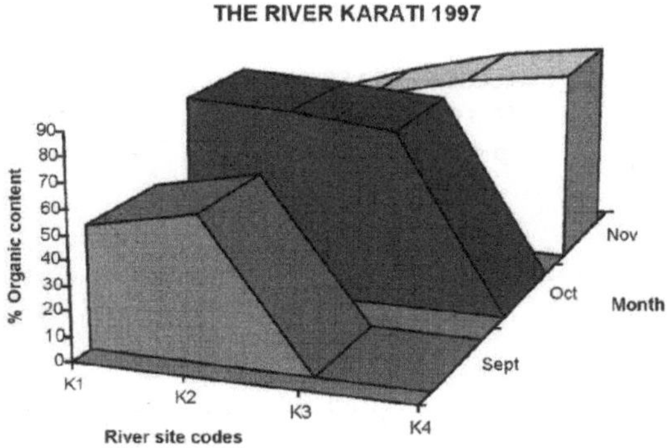

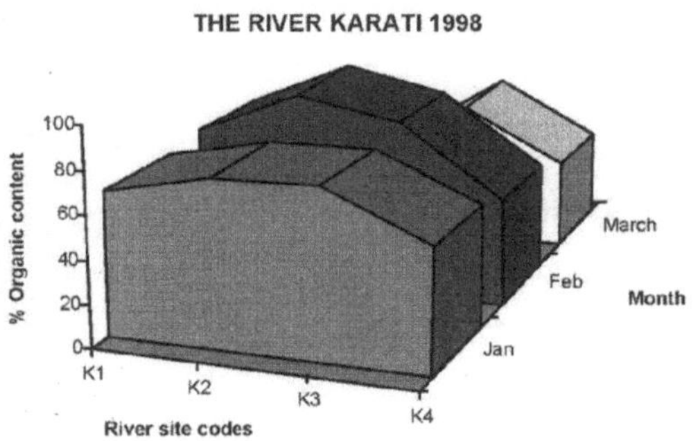

Figure 11. The percentage organic matter of the suspended sediment in the river Karati just before (September & October 1997) at the start of (November 1997) and during (January–March 1998) the heavy wet season following the 'El Nino rains'.

centration of the upper catchment streams; about 4–5 times the concentration of rainwater in the catchment (mean conductivity 10.6 μS cm^{-1}, $n = 8$).

The patterns of the phosphorus forms in the two rivers were dissimilar (Fig. 14). Total phosphorus more closely followed the rainfall pattern (see Fig. 1) than discharge, with dissolved phosphorus more often the predominant form in low flows and particulate in high flows.

The daily loss of TP from the catchment reached over 4×10^3 kg day^{-1} at the beginning of the heavy rainy season in late 1997 (Fig. 15) and, although strongly correlated with discharge (Fig. 16), fell tenfold in the later part of the season in early 1998. At its height in 1997, this phosphorus loss was 75–95% particulate, but in the more normal discharges of 1998–1999 the loss of soluble and particulate was almost equal (mean 41 and 42 kg day^{-1}), respectively (Fig. 17). These gave an annual phosphorus loss of 1.8 kg ha^{-1} ann^{-1}, 90% particulate, in the 'El Niño' year and 0.2 kg ha^{-1} ann^{-1}, 76% particulate in the 'normal' year.

Sediment phosphorus, from 8 headwater streams, averaged 0.47 TP mg g^{-1} dry wt sediment and iron content 51.3 mg g^{-1} dry wt., giving a mean ratio of 109 Fe:P.

Discussion

The calculated catchment losses are derived from regular samplings. They spanned a heavy rainfall period by chance, but the sample frequency must be borne in mind when considering the annual loss estimate. Pacini (1994) found, from intensive study of a catchment draining eastwards from the Nyandarua, that a single day exported 10% of the annual TP loss and about 75% of the TP and SS losses occurred during the rainy events.

The background composition of the river water matches the scenario described for rivers draining mature, leached tropical soils (Viner, 1975; Faniran

Table 4. Chemical characteristics of the Gilgil river

Station code	Location	Distance km from lake	Altitude (m.a.s.l.)	pH	Alkalinity mg l^{-1} as $CaCO_3$	Total hardness mg l^{-1} as $CaCO_3$	Ca hardness mg l^{-1} as $CaCO_3$
G1	Gilgil	3.0 (0)	2012	7.96 ± 0.2	38.7 ± 7.2	17.68 ± 2.0	11.05 ± 1.4
G2	North lake road	15.0 (15.0)	1950	7.54 ± 0.2	47.3 ± 5.4	21.85 ± 3.3	15.35 ± 3.4
G3	Marula	2.0 (28.0)	1890	7.48 ± 0.2	55.1 ± 3.3	24.44 ± 2.6	14.97 ± 1.4

Mean $\pm$ standard error ($n = 16$).

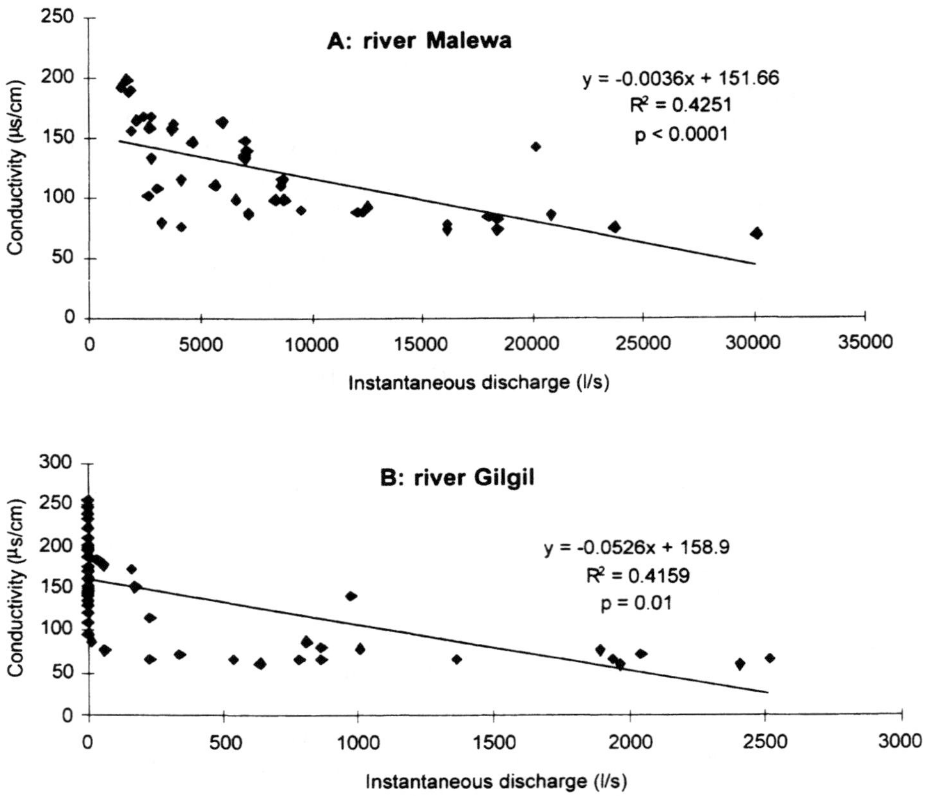

Figure 12. The relationship between instantaneous discharge and conductivity in the rivers Malewa (A) and Gilgil (B).

& Areola, 1978). The catchment lies in a volcanic area consisting of trachytes, alkaline basalt and phonolites with acid to alkaline soils (Odingo, 1971) and the drainage chemistry similar to Kilham's 'common' African rivers (Kilham, 1972) in contrast with more 'rain-dominated' catchments such as found in the Amazon basin (Gibbs, 1970).

The Naivasha rivers and streams have more phosphorus concentration than the 10 $\mu g\ l^{-1}$ (SRP) or 25 $\mu g\ l^{-1}$ (TDP) calculated by Meybeck (1982) from unpolluted rivers world-wide. Phosphorus in the

Table 5. Chemical characteristics of the Karati river

Station code	Location	Distance km from lake	Altitude (m a.s.l.)	pH	Alkalinity mg l^{-1} as $CaCO_3$	Total hardness mg l^{-1} as $CaCO_3$	Ca hardness mg l^{-1} as $CaCO_3$	Mg mg l^{-1}	Ca mg l^{-1}
K1	Karati Mission	33.5 (0)	2745	7.00 ± 0.3	85.5 ± 15.1	39.4 ± 5.5	25.5 ± 5.1	10.5 ± 2.1	15.2
K2	Karati centre	15 (18.5)	2134	6.69 ± 0.4	96.8 ± 4.5	44.3 ± 6.8	33.3 ± 8.7	11.0 ± 1.7	17.8
K3	Nakuru road	5.5 (28.0)	1903	7.27 ± 0.3	104.2 ± 13.5	48.9 ± 5.5	35.8 ± 5.8	10.6 ± 2.1	19.6
K4	Delamere	0.5 (33.0)	1890	7.49 ± 0.4	140.2 ± 18.9	48.9 ± 8.2	35.8 ± 8.2	18.5 ± 3.6	19.6

Mean ± standard error (n = 16).

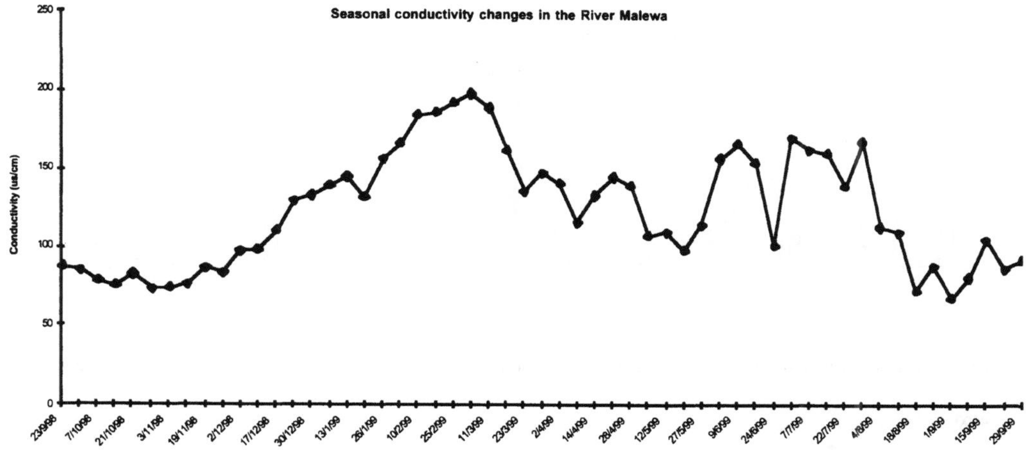

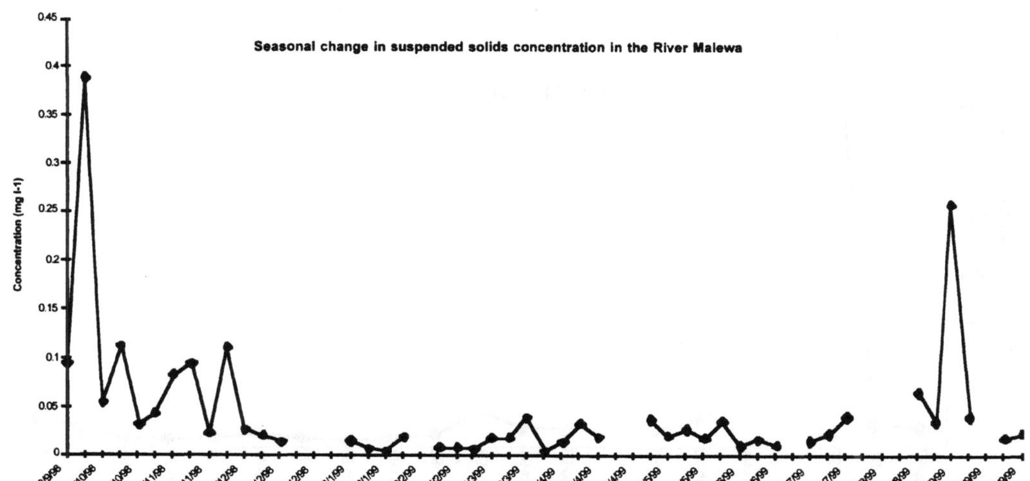

Figure 13. The seasonal pattern of conductivity and TSS in the river Malewa, between 1998 and 1999. The rainy season and high discharges were at the beginning (the heavy rains) and end of this period ('normal' rains), the low discharge in the middle (see Fig. 1).

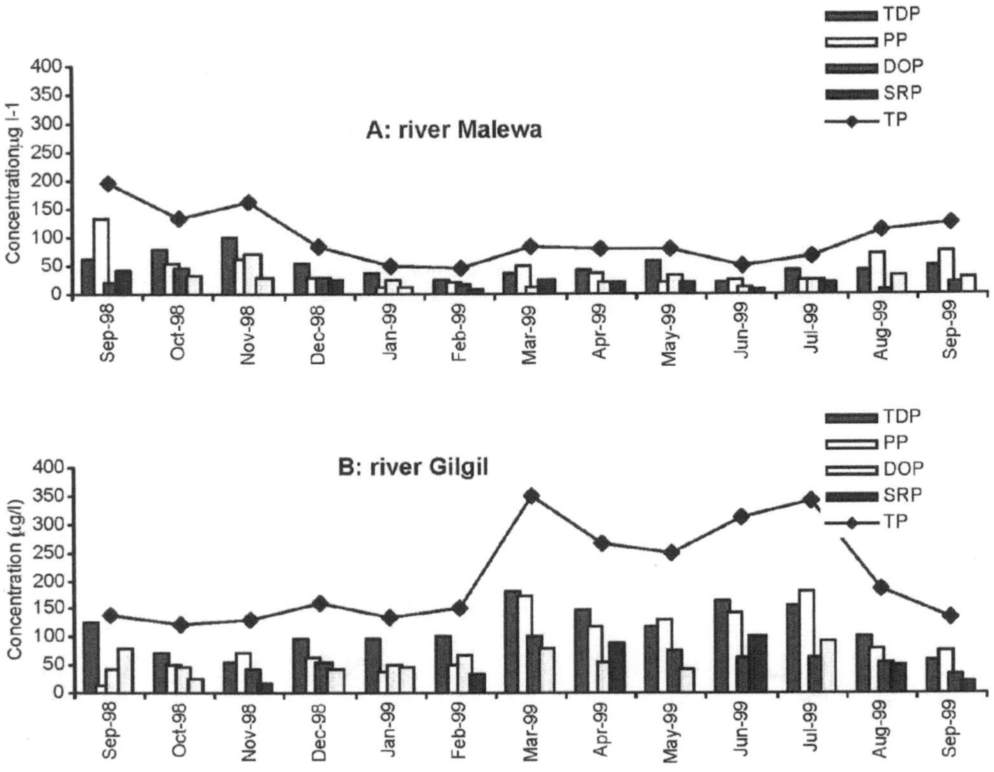

Figure 14. Seasonal pattern of TP and the different forms of phosphorus in the rivers Malewa and Gilgil (lower sites), 1998–99.

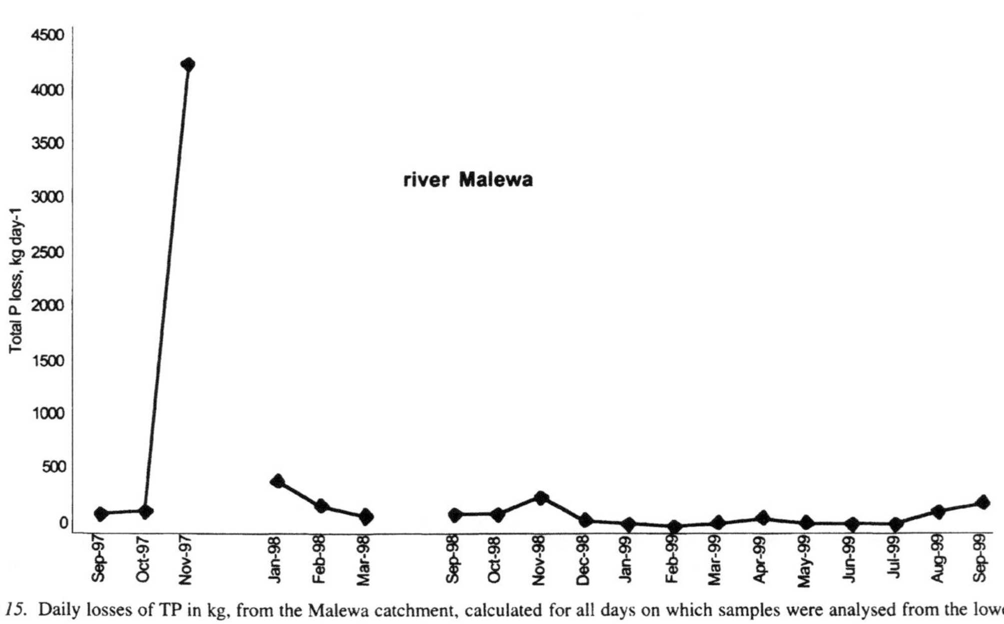

Figure 15. Daily losses of TP in kg, from the Malewa catchment, calculated for all days on which samples were analysed from the lower site, M5.

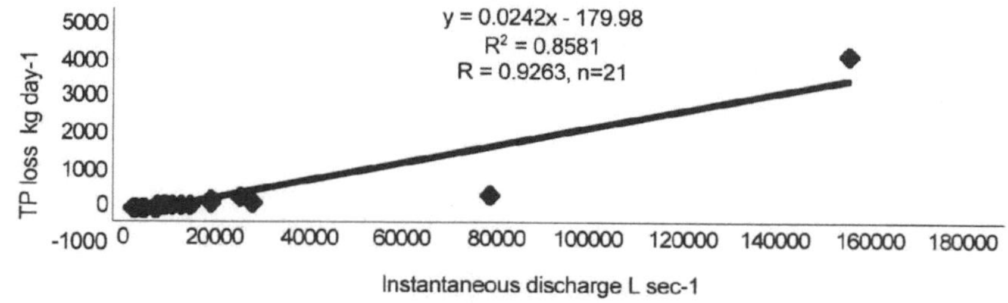

Figure 16. Correlation of daily TP with instantaneous discharge for the river Malewa.

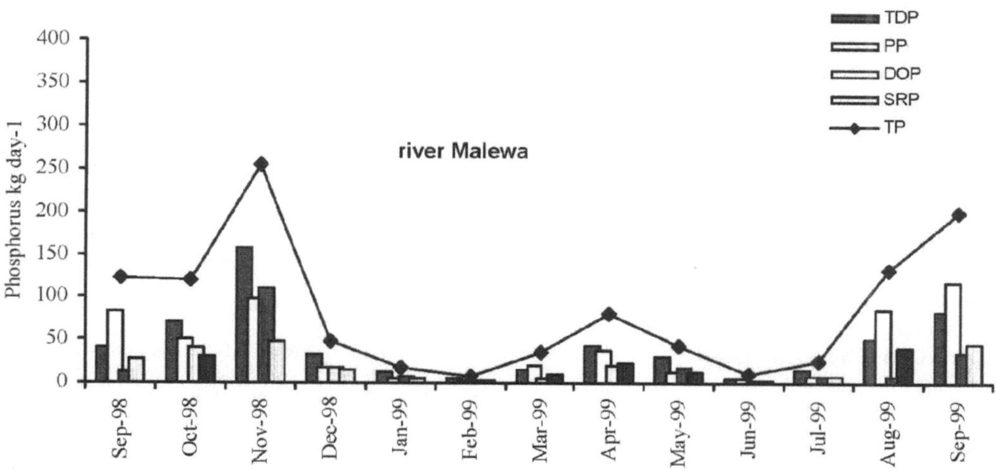

Figure 17. Transport of the different forms of phosphorus in the river Malewa, kg day^{-1} 1998–99.

Naivasha catchment has not previously been studied and so comparisons are barely possible. Gaudet (1979) had measured the concentration of TP and PO_4-P of the lower Malewa, as part of a large study of the chemical budget of the erstwhile *C. papyrus* North swamp, through which the river waters entered lake Naivasha. In 1975, his mean TP values were 40 μg l^{-1}, soluble 3 μg l^{-1}. His study spanned a period of moderate river discharges (maximum monthly mean 24 m^3 s^{-1}) more comparable with 1998–1999, when annual mean TP was 3x Gaudet's and SRP 6 × Gaudet's PO_4-P.

The rapid increase of TP at the start of the 1997 rains yet decline during equally high discharge a month or two later, suggests a phosphorus flush into the rivers and streams as a result of surface erosion. Three quarters of the land in the catchment is cultivated, intensively up to and even down the riverbanks. High surface runoff results in a rate of surface erosion increased tenfold. The significant linear relationship obtained between PP and TSS, as found elsewhere (Burwell et al., 1977; Nelson et al., 1979; Kronvang et al., 1996), supports this. The importance of vegetation cover for reducing soil mobilization is indicated by the lower TP and higher proportion of SRP in higher catchment streams, which compare with 62% SRP from grass watersheds on the southern plains area of Oklahoma and Texas (Sharply & Smith, 1990). In the Kinangop plateau, only 1/4 of the land is pasture, carrying livestock with a density estimated at 0.62 cows per hectare, which is heavy (Simpson, pers. comm.). The lower catchment has a slightly lighter stocking density.

The TP export for the heavy rains (1997–1998) are at the upper end of figures produced for temperate agricultural land (Reckhow & Chapra, 1983) while the normal year (1998–1999) gave more acceptable losses.

The catchment presently lies outside the designated RAMSAR site of the lake, and outside any present land use management policies other than Kenya government advice to small farmers, which has minimal environmental component. The future needs of the lake basin can only be met if sustainable use of the catchment's soil resource is an integral part of its management; the future needs of the rural population can only be met if they receive environmental as well as economic advice for maintaining their agricultural resource base. The two are identical and should be tackled together.

Acknowledgements

This study formed past of the work of the University of Leicester research project at Lake Naivasha, which since 1987 has been authorised by the Office of the President, Government of Kenya under research permit to Dr D. M. Harper no. OP 13/001/12C 46. The project was funded by the Earthwatch Institute, Boston U.S.A. and Oxford, England. The data collection would not have been possible without the assistance of numerous Earthwatch volunteers or the facilities at the erstwhile Elsamere Conservation Centre. Our sincere thanks go to the numerous colleagues in Kenya for logistical help, in particular Jill & Angus Simpson and Velia Carn, and Chris Campbell Clause.

References

Åse, L-E., K. Sernbo & P. Syren, 1986. Studies of Lake Naivasha, Kenya and its Drainage Area. Forkningsrapporter 63, Stockholms Universitet Naturgeografiska Institutionen, Stockholm: 75 pp.

Burwell, R. E., G. E. Schuman, H. G. Heinemann & R. G. Spomer, 1977. Nitrogen and phosphorus movement from the agricultural watersheds. J. Soil Wat. Conserv. 32: 226–230.

Calamari, D., 1987. Water quality problems in Africa: a brief report. Int. J. Environ. Studies 29: 3–8.

Dean, W. E., 1974. Determination of carbonate and organic matter in calcareous sediments and sedimentary rocks by loss on ignition. Comparison with other methods. Journal of Sedimentary Petrology 44: 242–248.

Dejoux, C., 1988. La Pollution des Eaux Continentales Africaines. Éditions de l'ORSTOM, Collection Travaux et documents no 213, Paris: 513 pp.

Downing, T. E., S. Lezberg, C. Williams & L. Berry, 1990. Population change and environment in central and eastern Kenya. Environ. Conserv. 17 (2): 123–133.

Dregne, H. E., 1990. Erosion and soil productivity in Africa. J. Soil Wat. Conserv. 45 (4): 431–436.

Faniran, A. & O. Areola, 1978. Essentials of Soil Study (with special reference to tropical areas). Heinemann, London: 278 pp.

Fardeau, J. C. & E. Frossard, 1991. Processus de transformations du phosphore dans les sols de l'Afrique de l'ouest semi-aride: application du phosphore assimilable. In Tiessen H. & Frossard (eds), Phosphorus Cycling in Terrestrial and Aquatic Ecosystems, Proceedings of a SCOPE/UNEP Workshop (Nairobi 18–22 March 1991). Vol. 4 Africa: 108–128.

Gaudet, J. J., 1976. Nutrient relationships in the detritus of a tropical swamp. Archiv. Hybrobiol. 78: 213–239.

Gaudet, J. J., 1977. Natural draw-down of Lake Naivasha, Kenya and the origin of papyrus swamps. Aquat. Bot. 3: 1–17.

Gaudet, J. J., 1979. Seasonal changes in nutrients in tropical swamp water. J. Ecol. 67: 953–981.

Gibbs, R., 1970. Mechanisms controlling world water chemistry. Science 170: 1088–1090.

Harper, D. M., G. Phillips, A. Clivers, N. Kitaka & K. M. Mavuti, 1993. Eutrophication prognosis for Lake Naivasha, Kenya. Verh. int. Ver. für Theor. Ang. Limnol. 14: 192–195.

Hecky, R. E., 1993. The eutrophication of lake Victoria. Verh. int. Ver. für Theor. Ang. Limnol. 25: 39–48.

Hinga, G., 1973. Phosphate sorption capacity in relation to properties of several types of Kenyan soil. East African Agricultural and Forestry Journal 38: 400–404.

Jones, M. E., 1979. Rainfall intensities of East African storms. East African Agriculture and Forestry Journal 43: 261–264.

Kilham, P., 1972. Biogeochemistry of African Lakes and Rivers. PhD thesis, Duke University, Durham, North Carolina: 199 pp.

Kronvang, B., A. Laubel & R. Grant, 1996. Transport of sediment and phosphorus in the Aarable Gelbaæk catchment, Denmark. In Proceeding of Sediment and Phosphorus Workshop in Denmark. N.E.R.I Technical Report 178: 89–92.

Lennox, S. D., R. H. Foy, R. V. Smith & C. Jordan, 1997. Estimating the contribution from agriculture to the phosphorus load in surface water. In Tunney, H. O. T. Caton, P. C. Brookes & A. E. Johnston (eds), Phosphorus Loss from Soil to Water. CAB International: 55–75.

Mackereth, F. J. H., J. Heron & J. F. Talling, 1989. Water Analysis. Freshwater Biological Association, Windermere, U.K.: 120 pp.

Melack, J. M., 1976. Limnology and Dynamics of Phytoplankton in Equatorial African Lakes. PhD thesis, Duke University, Durham: 453 pp.

Melack, J. M., P. Kilham & T. R Fisher, 1982. Responses of phytoplankton to experimental fertilization with ammonium and phosphate in an African soda lake. Oecologia 52: 321–326.

Melack, J. M. & S. MacIntyre, 1991. Phosphorus concentrations, supply and limitation in tropical african lakes and rivers in Phosphorus cycling in terrestrial and aquatic ecosystems. In Tissen, H. & H. Frossard (eds), Proceedings of a SCOPE/UNEP Workshop (Nairobi 18–22 March 1991). Vol. 4 (Africa): 1–17.

Meybeck, M., 1982. Carbon, nitrogen & phosphorus transport by world rivers. Am. J. Sci. 282: 401–450.

Nelson, D. W., E. J. Monke, A. D. Bottcher & L. E. Sommers, 1979. Sediment and nutrient contributions to the Maumee River from an agricultural watershed. In: Loehr, R. G., D. A. Haith, M. F. Walter & C. S. Martin (eds), Best Management Practises of Agriculture and Salviculture. Ann. Abor. New York: 491–505.

Njuguna, S. G., 1982. Nutrient-Productivity Relationship in Tropical Naivasha Basin Lakes, Kenya. PhD thesis, University of Nairobi: 300 pp.

Nürnberg, G. K., 1988. Prediction of phosphorus release rates from total and reductant-soluble phosphorus in anoxic lake sediments. Can. J. Fish. aquat. Sci. 45: 453–462.

Nyandat, N., 1981. The primary minerals in some Kenya's topsoil and their significance to inherent soil fertility. East Africa Agriculture and Forestry Journal 46: 71–76.

Odingo, R. S., 1971. The Kenya Highlands: Land-use and Agricultural Development. East Africa Publishers House, Nairobi: 226 pp.

Pacini, N., 1994. Coupling of Land and Water; Phosphorus Fluxes in the Upper Tana RiverCatchment, Kenya. Ph.D. Thesis. University of Leicester. U.K.: 176 pp.

Pacini, N. & D. M. Harper, 1999. Phosphorus transport from a tropical catchment and implications for sustainable catchment management. In Harper, D. M. & A. G. Brown (eds), The Sustainable Management of Tropical Catchments. Wiley, Chichester: 61–76

Peters, R. H. & S. MacIntyre 1976. Orthophosphate turnover in East African lakes. Oecologia (Berl) 25: 313–319.

Reckhow, K. H. & S. C. Chapra 1983. Engineering Approaches for Lake Management: Vol.1. Data Analysis and Empirical Modeling, Butterworth Publishers, Boston.

Roseboom & Lotriet, 1993 A new sediment yield map for southern Africa, erosion and transport monitoring programmes in river basins. Proceedings of the Oslo Symposium, August 1992, IAHS Publ. No. 210: 527–538.

Sharpley, A. N. & S. J. Smith, 1990. Phosphorus transport in agricultural runoff. The role of erosion. In Bardman, J., I. D. L. Foster & J. A. Dearing (eds), Soil Erosion on Agricultural Land. J. Wiley and Sons Ltd: 352–366.

Sharpley, A. N. & Syers, 1979. Phosphorus inputs into a stream draining an agricultural watershed. Wat. Air Soil Poll. 11: 417–428.

Smaling, E. M. A., 1993. An Agro-ecological Framework for Integrated Nutrient Management with Special Reference to Kenya. Ph.D. Thesis, Agricultural University of Wageningen, Netherlands.

Smaling, E. M. A. & J. Bouma, 1982. Bypass flow and leaching of nitrogen in a Kenyan vertisol at the onset of the growing season. Soil Use and Management 8: 44–48.

Talling, J. F. & I. B. Talling, 1965. The chemical composition of African Lakes waters. Int. Rev. ges. Hydrobiol. 50: 421–463.

Viner, A. B., 1975. The supply of minerals to tropical rivers and lakes (Uganda). In Hasler, A. D. (ed.), Coupling of Land and water systems. Springer Verlag, New York, Ecological Studies 10: 227–261.

Viner, A. B., 1987. Nutrients transported on silt in rivers. Archiv. Hydrobiol. Beih. (Ergebnisse der Limnologie) 28: 63–71.

Wolf, A. M., D. E. Baker, H. B. Pionke & H. M. Kunishi, 1985. Soil tests for estimating labile, soluble, and algae-available phosphorus in agricultural soils. J. Environ. Qual. 14 (3): 341–348.

Hydrobiologia **488:** 73–80, 2002.
D.M. Harper, R. Boar, M. Everard & P. Hickley (eds), Lake Naivasha, Kenya.

Phosphorus inputs to Lake Naivasha, Kenya, from its catchment and the trophic state of the lake

Nzula Kitaka[1], David M. Harper[2,*] & Kenneth M. Mavuti[3]
[1]*Department of Zoology, Egerton University, Box 536, Njoro, Kenya*
[2]*Department of Biology, University of Leicester, Leicester LE1 7RH, U.K.*
[3]*Department of Zoology, University of Nairobi, Box 30197, Nairobi, Kenya*
**Author for correspondence. E-mail: dmh@le.ac.uk*

Key words: TP, Chlorophyll *a*, Secchi transparency, eutrophication

Abstract

The main river supplying Lake Naivasha, Kenya, the Malewa, drains a catchment given over to largely subsistence cultivation and animal husbandry. The lake itself is the focus for an intensive horticultural industry based upon irrigation from the lake. The Malewa, however, is relatively independent of the impact of industry, and so its contribution to eutrophication of the lake was evaluated. Two periods of study, a very wet-dry and a 'normal' wet-dry season showed that the river contribution of phosphorus led to a total phosphorus loading of 1.4 g m^{-2} lake surface ann^{-1} in the very wet period compared to 0.2 in the 'normal'. Chlorophyll '*a*' in the open water of the lake was significantly related to soluble reactive phosphorus. The lake is now eutrophic by normal limnological criteria.

Introduction

Eutrophication in tropical lakes is not as well understood than it is in the temperate zone. In general tropical/sub-tropical systems appear to tolerate higher phosphorus loads for similar algal biomass and, of course, experience a more uniform light climate but a more dramatic wet/dry hydrological regime (Talling, 2001). Total phosphorus concentration of 50–60 μg l^{-1} has been suggested as the boundary between mesotrophic and eutrophic for Lake Chivero (formerly Lake Mcllwaine) in Zimbabwe (Thornton & Nduku, 1982), similar to the concentration reported for Australian water bodies (McDougall & Ho, 1991) and higher than that of the temperate zone. Ryding & Rast (1989) suggest that tropical systems often develop low N/P ratios, thereby favouring the dominance of nitrogen-fixing cyanobacteria. Toerien (1975) suggests that these are more dominant in tropical lakes, but a change in phytoplankton either from diatom to cyanobacteria dominance and *vice versa*, induced by wind regime, has been observed in lakes Naivasha (Kalff & Brumelis, 1993), Volta (Biswas, 1972; Talling, 1986), Tanganyika, Malawi (Talling, 1966), Victoria (Hecky, 1993; Lehman, 1996) and Kariba (Cronberg, 1997) as a result of seasonal water mixing.

Generally low levels of industrialization in Africa, means that eutrophication does not present the same problems as it does in temperate countries. The leading causes of environmental degradation in Africa are over-cultivation, overgrazing, and deforestation. Deforestation takes place as a result of clearing agricultural land, wood for fuel (about 90% of the population use fuel wood), building material and a source of income (UNEP-IETC, 1999). Elsewhere in the tropics, construction of new reservoirs and increase in uses of natural lakes for water supply combined with settlement in the catchments has resulted in extensive problems following accelerated nutrient inputs (Thornton, 1987a). Nutrient runoff and sedimentation may become serious problems and widespread in future if they are not properly addressed now.

Many lakes,reservoirs and rivers studied within the tropical region have shown an increase in eutrophication, for example in southern Africa (Twinch, 1986; Thornton, 1987b) although the benefits of deliberate eutrophication to increase edible fish yield from ponds fertilized with human wastes has been enjoyed for centuries in Asia (Payne, 1984). According to UNEP-IETC (1999), Lake Malawi has already given an early warning of degradation in water quality using Eccles's (1974) and Bootsma's (1993) data. Subsequent surveys a decade later including Malawi, noted a marked reduction in water quality in the region (ILEC, 1994).

In Lake Kariba (the world's largest man-made lake by storage), the levels of total phosphorus in some parts are approaching thresholds with respect to future eutrophication (Cronberg, 1997). Lake Chivero became hypereutrophic in the 1960s with phytoplankton species *Anabaena* and *Microcystis* dominating. After installation of a Biological Nutrient Removal (BNR) sewage treatment plant in 1974, the lake showed a recovery (Thornton, 1982), but due to further increase in population within the surrounding area, it reverted to hypereutrophic status in the mid 1980s (Moyo, 1997) up to the present day (UNEP-IETC, 1999) which currently has chronic water hyacinth problems and repeated fish kills.

Naivasha in the past was more phosphorus limited than nitrogen, (Talling & Talling, 1965; Kalff, 1983). There is no study undertaken on the trophic state of the rivers flowing into Lake Naivasha and although phosphorus is generally low in Kenyan soils (Hinga, 1973; Nyandat, 1981), erosion which follows ploughing or over-grazing may bring sediment-bound phosphorus into the lake in higher quantities.

The ecological stability of Lake Naivasha is threatened by the impact of both internal and external factors. Internal factors are exotic species' introductions and accidental arrival. External factors are the impact of an intensive horticultural industry, which has grown up on the shores over the last 15 years, based upon lake water for irrigation. The impact is believed to be both runoff of pesticides, effluents and nutrients and the pressure of tens of thousands of people who have migrated into the area for work. The catchment, however, is given over to mainly subsistence cultivation and animal husbandry.

This study examines the sources and quantity of phosphorus entering the lake from the Malewa. It was carried out from September 1997, when heavy rains fell over the country, believed to be a consequence of the 'El Niño' earlier in that year in the southern Atlantic, to March 1998, after Naivasha lake level had risen by 2 vertical metres (Fig. 1). This period was then compared with a more 'normal' period, between 1998 and 1999.

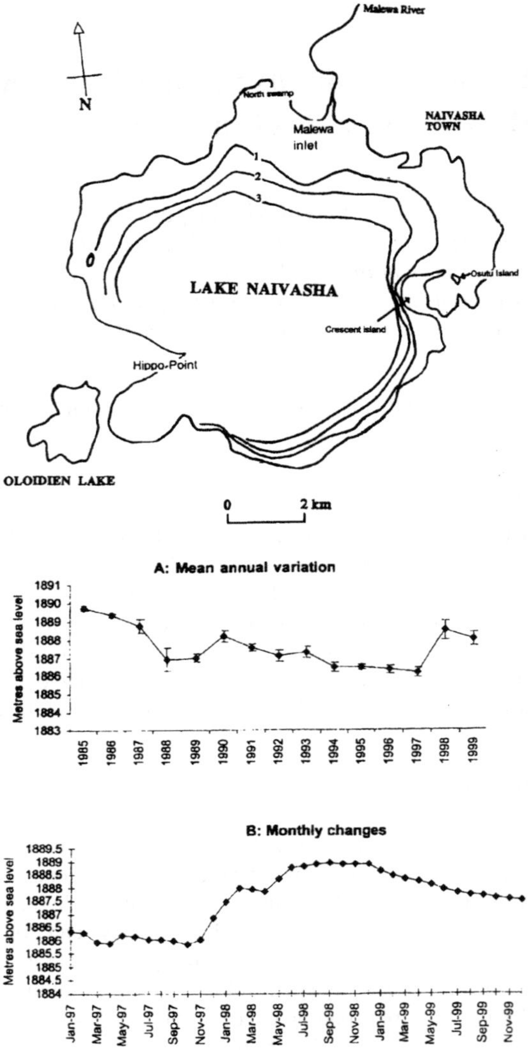

Figure 1. Location of Lake Naivasha, Kenya and its lake level change, 1985–99 & 1997–99. The depth contours on the map refer to 1991 (Hickley et al., 2002).

Methods

In the field, weekly samples were taken along a transect from the river inlet to the open water of the lake, 8.5 km to the south. Five stations were sampled – at the inlet, then 100 m, 500 m, 5 km and 8.5 km south. Secchi disc transparency and water conductivity were recorded directly from a boat (conductivity using YSI conductivity meter). Water samples were collected just below the surface of the lake using a

1.5 l messenger-closed plastic sample bottle (Institute of Oceanographic Sciences, Surrey). Water was transferred into acid-washed polythene 1 l bottles and a portion immediately filtered using an acid-washed 60 ml disposable syringe to push the water through a pre-weighed GF/C glass fibre filter (Whatman, U.K.). Both filtered and unfiltered samples were transported in a cool-box and analysed within 1–3 h on return to a lake-side laboratory; a further unfiltered sample was transported in the same way for chlorophyll '*a*' analysis.

In the laboratory, chlorophyll *a* as a measure of phytoplankton biomass was analysed using 90% ethanol as solvent and biomass calculated from absorbancies of the extract before and after acidification with HCl. Soluble reactive phosphorus was determined in filtered water samples using the molybdenum blue technique after Mackereth et al. (1979). Total phosphorus (TP) and total dissolved phosphorus (TDP) were obtained by measuring P after digestion of 20 ml unfiltered and filtered water samples respectively with 0.5 g potassium persulphate and 1.2 ml of 10N sulphuric acid. Particulate phosphorus (PP) was calculated as the difference between TP and TDP; dissolved organic phosphorus as the difference between TDP and SRP (Lennox et al., 1997). The filer papers were oven dried and re-weighed to give total suspended solids (TSS).

Data were analysed using Minitab. Phosphorus input from the catchment was calculated from the river inlet site and expressed as loading per unit area of lake surface. The loading rates were related to mean depth of the lake to estimate its trophic status (Vollenweider, 1968). Internal loading was estimated from Cooke et al. (1993) following a phosphorus sedimentation calculation, to arrive at a predicted TP concentration.

Results

The chemical influence of the river Malewa on the lake was most pronounced at the beginning of the heavy rainy season between October and November 1997 for up to 0.5 km into the lake, shown by the spatial variation in conductivity (Fig. 2), Secchi transparency (Fig. 3), and chlorophyll *a* (Fig. 4).

The effect of the rain on the lake was not apparent until January 1998, (see Fig. 1) when the river inlet was almost 2.5 km further north of its normal position as the water rose. This dramatically changed the riverine conductivity characteristics of the inlet

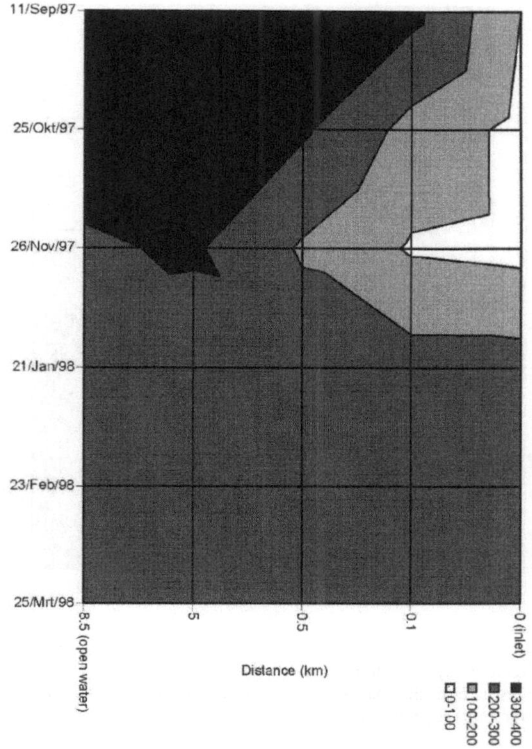

Figure 2. Spatial variation of conductivity, μS cm^{-1}, on a transect from the Malewa inlet SE to Hippo Point (vertical axis) over time (horizontal axis). Note the scale changes on the axes.

September–November 1997 into lake conditions in January 1998 (Fig. 5).

A significant negative correlation between Secchi tranparency and TSS (Fig. 6) occurred both in the inlet and the open water, but the much steeper relationship in the inlet demonstrates the effect here of inorganic silt, the major contributor to TSS. A strong negative correlation between chlorophyll *a* and Secchi depth in the open water (compared to no relationship at the inlet; Fig. 7) shows that transparency in the open water is caused by variations in phytoplankton biomass, Phytoplankton are of course part of the lacustrine TSS (a strong significant positive correlation between TSS and chlorophyll *a* (R = 0.871, R^2 = 0.758, n = 20 p < 0.0001)), which is why TSS influenced transparency more significantly (R^2 = 0.67, p < 0.0001) in the open water than did chlorophyll *a* (R^2 = 0.003, p > 0.05).

Mean TP concentrations of 83.9 +/- 33.2 and 52.2 +/- 18 occurred at the inlet and the open water respect-

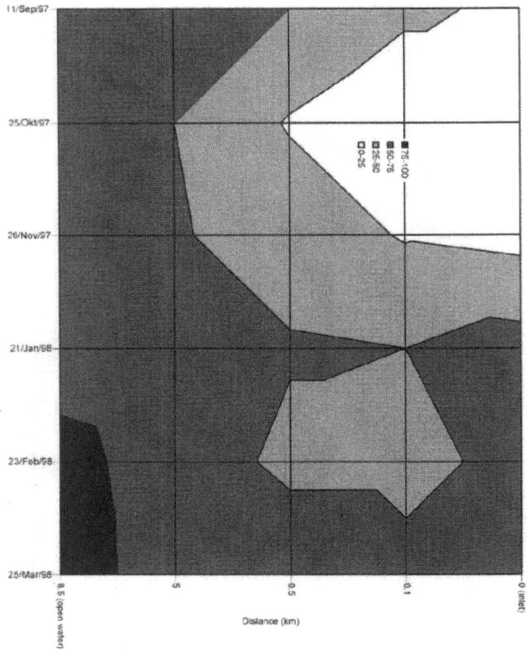

Figure 3. Spatial variation of Secchi disc transparency, cm, on a transect from the Malewa inlet SE to Hippo Point (vertical axis) over time (horizontal axis). Note the scale changes on the axes.

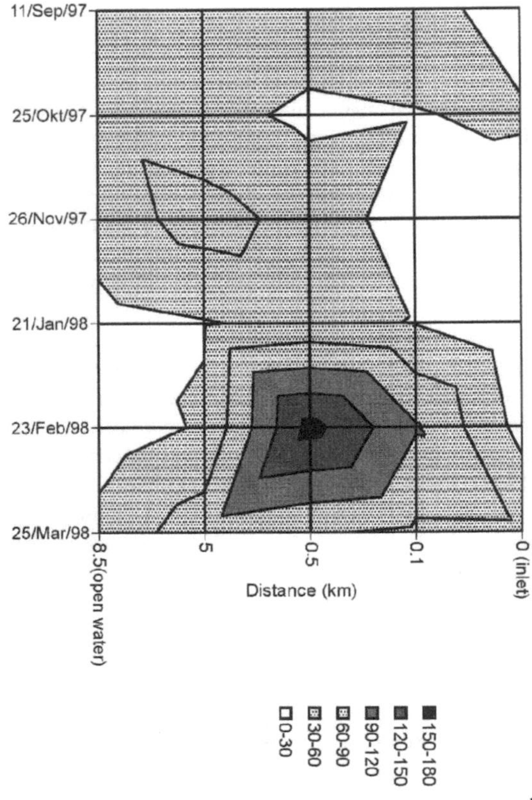

Figure 4. Spatial variation of chlorophyll *a* concentration, $\mu g\ L^{-1}$ on a transect from the Malewa inlet SE to Hippo Point (vertical axis) over time (horizontal axis). Note the scale changes on the axes.

ively. In both locations the means hide an early peak from the river at the time of the rains but before the volume increase (Fig. 8). There was a much stronger positive correlation between TSS and TP at the inlet ($R = 0.92$, $R^2 = 0.84$, $n = 20$, $p < 0.001$) than in the open water ($R = 0.62$, $R^2 = 0.38$, $n = 20$, $p < 0.05$), which indicates the phosphorus is entering the lake bound to the sediment particles. Once in the lake, PP is reduced by sedimentation but also transformed as SRP is turned into phytoplankton P. In both periods chlorophyll *a* increased in proportion to TP – 8 fold during the very wet period, 2-fold during the 'normal' (Table 1). Chlorophyll *a* was correlated most strongly with TP and PP in the open water (Chl $a = 0.226$TP $+ 10.3$ $R = 0.67$, $R^2 = 0.44$, $n = 18$, $p < 0.005$) and with nothing at the inlet.

The TP in the open water predicted from Cooke et al. (1993) matches its measured mean concentration, but the estimated lake loading derived from these figures do not agree, with the predicted figures 2–3 fold lower (Table 2). When chlorophyll *a* was predicted from the many north-temperate lake models, reasonable concentrations were obtained in the open water but higher predictions in the inlet. This is not surprising considering the light restriction that the suspended sediment caused despite the TP value. Only two models seriously under- (OECD, 1982) or over- (Schindler et al., 1978) estimated chlorophyll (Table 3), which considering the major difference of an endorrheic lake in an 'enless summer' climate (Kilham & Kilham, 1990) is encouraging.

Discussion

Temperate analogues are often necessary for understanding tropical waters when an inadequate data base exists (Kalff & Watson, 1986). Temperate trophic status classification of Lake Naivasha, estimated for the heavy rain period and the normal period, is eu-polytrophic and meso-eutrophic, respectively, using Vollenweider's five-class tentative classification (Vol-

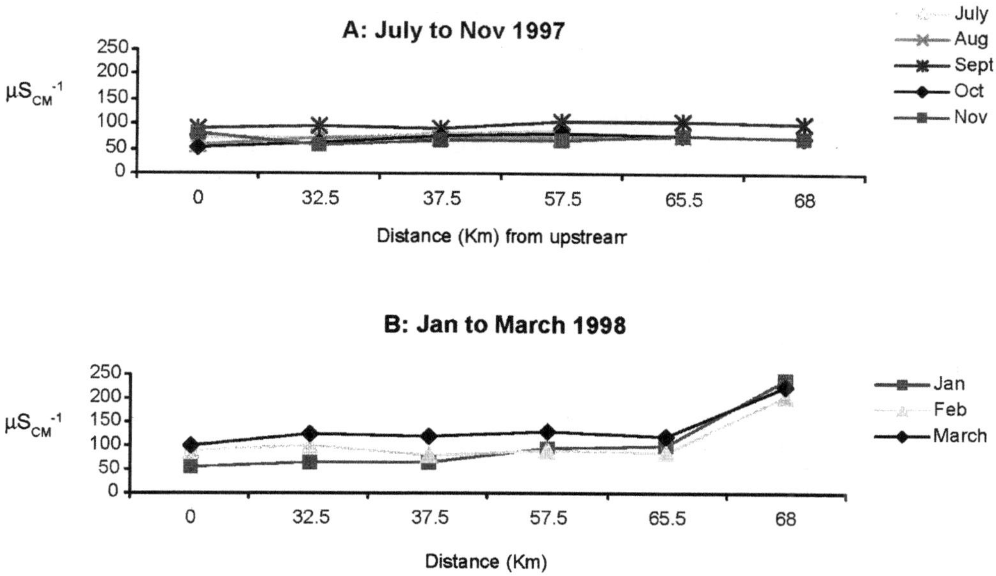

Figure 5. Longitudinal pattern of conductivity in the river Malewa during the very heavy wet season (upper) and following dry season when river discharge increased (lower) when the 'inlet site' (distance 68 km from source) became influenced by lake conditions.

Table 1. Comparison of different phosphorus fractions and ratios between the inlet and the open water

Station	PP%	SRP/TP	TP/Chla
September 1997–March 1998			
Malewa inlet	62.1 ± 7.8	0.38 ± 0.1	16.2 ± 7.4
Open water (Hippo point)	54.0 ± 10.5	0.30 ± 0.1	2.2 ± 0.5
November 1998–September 1999			
Malewa inlet	66.1 ± 7.0	0.11 ±0.02	5.3 ± 0.8
Open Water (Hippo point)	59.7 ± 7.1	0.12 ±0.02	2.7 ± 0.4

Table 2. The observed and predicted TP concentrations and lake loadings for Naivasha

	1997–8	1998–9	Overall mean
Measured TP mg l^{-1}	45.9	37.2	52.2
Measured TP mg l^{-1}	45.0	37.7	51.2
Estimated lake loading g m^{-2} ann^{-1}	1.41	0.21	0.6
Predicted lake loading g m^{-2} ann^{-1}	0.39	0.14	0.23

lenweider, 1968). Plotting the quotient between in-lake phosphorus and mean inflow phosphorus concentration P/P_I against P_I on a logarithmic scale and extrapolated using Ahlgren et al. (1988), Lake Naivasha is also eutrophic (Fig. 9).

TP concentrations in a water body should not be considered as the only indicator of water quality without considering the consequences for algae (Jones & Lee, 1982). Naivasha is eutrophic with chlorophyll *a* concentration >10 $\mu g/l^{-1}$, Secchi depth <1.7 m and TP concentration of >40 $\mu g/l^{-1}$ using the Lee et al. (1980) and Jones & Lee (1982) categories. The lake is further classified into various categories according to the OECD 'open boundary' system (OECD, 1982) summarised in Table 4. An alternative classification of Carlson (1977) used the calculation of a 'Trophic State Index' (TSI) was also considered and using Kratzer & Brezonik's (1981) index scale, Naivasha was approaching hyper-eutrophic status during the heavy rainy season with TSI scale values range of 58–66.

Lake Naivasha is now a eutrophic lake; a condition which was probably reached in the late 1970s although the recorded evidence is sparse (Hubble & Harper, 2001). It appears to have had a high transparency (several metres) up until the mid 1970s (Melack, 1976) but

Table 3. The trophic state of Lake Naivasha calculated on OECD and TSI indices at two different periods of heavy rainfall and normal rainfall based on three parameters

	TP ($\mu g\ l^{-1}$)	Chlorophyll *a* ($\mu g\ l^{-1}$)	Secchi depth (m)
Heavy rainfall (Oct 97–Feb 98)			
Measured value in the lake	54.75	31.32	0.57
OECD	eutrophic	hyper-trophic	hyper-trophic
TSI index	59.3	58.04	65.6
Normal rainfall (Sept 98–Feb 99)			
Measured value in the lake	46.93	11.5	1.26
OECD	eutrophic	eutrophic	eutrophic
TSI index	59.2	56.1	57.7

KEY
TSI scale
< 20 ultra-oligotrophic,
30–40 oligotrophic,
45–50 meso-trophic,
50–60 eutrophic
and above 70 hyper-trophic (Kratzer & Brezonik, 1981).

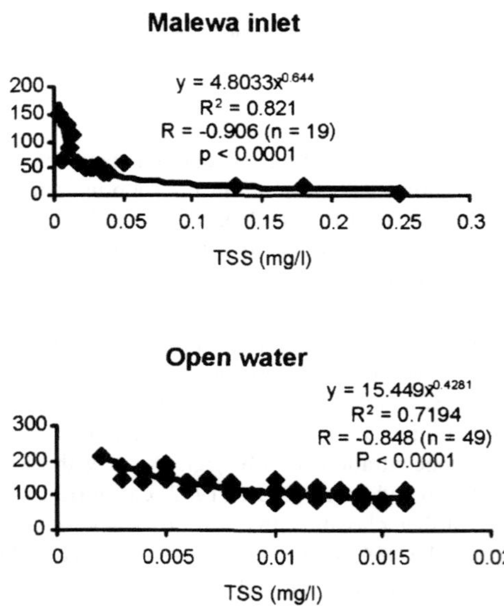

Figure 6. Correlation relationships between Secchi transparency and TSS at the inlet site and the open water site.

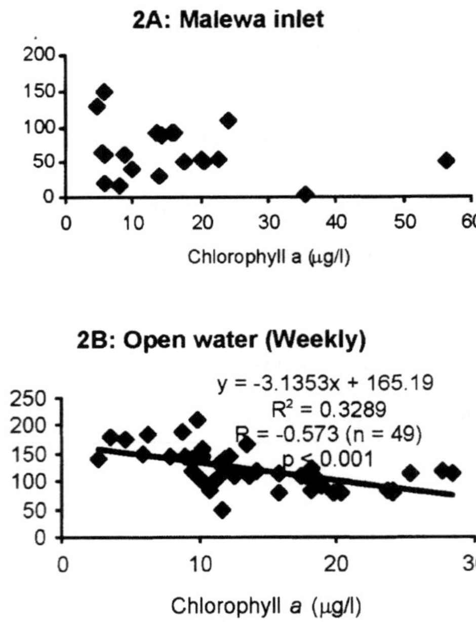

Figure 7. Correlation relationship between Secchi transparency and chlorophyll *a* at the inlet site and the open water site.

not always (Jenkin, 1936). It is tempting to suggest that its current eutrophic state has been accelerated by the loss of any natural buffering of the *Cyperus papyrus* fringe (Gaudet, 1979) following the decline of the area covered by the north swamp (Boar et al., 1999), but there is no direct evidence for this. Equally, there is no evidence of the earlier condition of the suspended sediment or the phosphorus in the rivers; only the certainty that the river did not run directly into the lake when the 'North Swamp' was intact. Evidence from the Malewa input is adequate to explain the increased trophic state of the lake without citing any

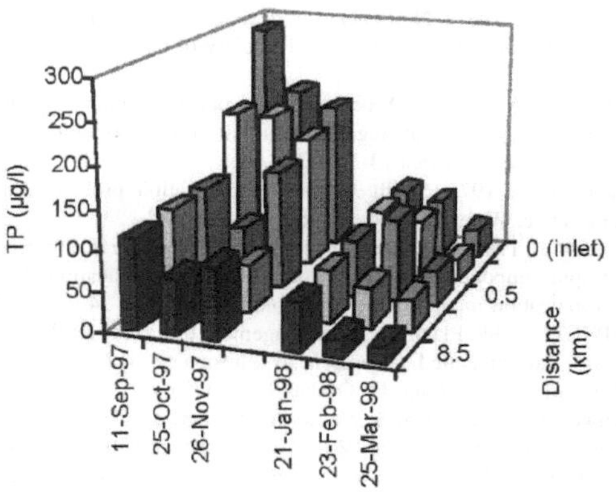

Figure 8. Spatial and temporal change in TP concentrations along the transect from the Malewa inlet (0 km) to open water at Hippo Point (8.5 km).

additional nutrient inputs from urban or horticultural sources.

Acknowledgements

This study formed past of the work of the University of Leicester research project at Lake Naivasha, which since 1987 has been authorised by the Office of the President, Government of Kenya under research permit to Dr D. M. Harper no. OP 13/001/12C 46. The project was funded by the Earthwatch Institute, Boston U.S.A. and Oxford, England. The data collection would not have been possible without the assistance of numerous Earthwatch volunteers or the facilities at the erstwhile Elsamere Conservation Centre. Our sincere thanks go to the numerous colleagues in Kenya for logistical help, in particular Jill & Angus Simpson and Velia Carn, and Chris Campbell Clause.

References

Ahlgren, I., T. Frisk & L. Kamp-Nielsen, 1988. Empirical and theoretical models of phosphorus loading, retention and concentration vs. lake trophic state. Hydrobiologia 170: 285–303.

Biswas, S., 1972. Ecology of phytoplankton of the Volta lake. Hydrobiologia 39: 277–288.

Boar, R. R., D. M. Harper & C. Adams, 1999. Biomass allocation in *Cyperus papyrus* in a tropical wetland, Lake Naivasha, Kenya. Biotropica 31: 422–421.

Bootsma, H. A., 1993: Spatio-temporal variation of phytoplankton biomass in Lake Malawi, Central Africa. Verh. int. Ver. Theor. Angew. Limnol. 25: 882–886.

Carlson, R. E., 1977. A trophic state index for lakes. Limnol. Oceanogr. 22: 361–368.

Cronberg, G., 1997. Phytoplankton in Lake Kariba 1989–1990. In Moreau, J. (ed.), Advances in the Ecology of Lake Kariba. Univ. of Zimbabwe Publication, Harare: 66–101.

Cooke, G. D., E. B. Welch, S. A. Peterson & P. R. Newroth, 1993. Restoration and Management of Lakes and Reservoirs, 2nd edn. Lewis Publishers Boca Raton: 548 pp.

Eccles, D. H., 1974. An outline of the physical limnology of Lake Malawi (Lake Nyasa). Limnol. Oceanogr. 19: 730–742.

Gaudet, J. J., 1979. Seasonal changes in nutrients in tropical swamp water. J. Ecol. 67: 953–981.

Hecky, R. E., 1993. The eutrophication of lake Victoria. Verh. int. Ver. Theoret. Angew. Limnol. 25: 39–48.

Hinga, G., 1973. Phosphate sorption capacity in relation to properties of several types of Kenyan soil. East African Agricultural and Forestry Journal 38: 400–404.

ILEC, 1994. Data book of the World Lakes Environments 2 Africa and Europe. International Lake Environment Committee (ILEC), Kosatu, Japan 852.

Jenkin, P. M., 1936. Reports on the Percy Sladen Expedition to some Rift Valley lakes in Kenya 1929. VII. Summary of the ecological results. Ann. Magazine Nat Hist 9: 133–181.

Jones, R. A. & G. F. Lee, 1982. Recent advances in assessing impact of phosphorus loads on the eutrophication related water quality. Wat. Res. 16: 503–515.

Kalff, J., 1983. Phosphorus limitation in some tropical African Lakes. Hydrobiologia 138: 161–176.

Kalff, J. & S. Watson, 1986. Phytoplankton and its dynamics in two tropical lakes: a tropical and temperate zone comparison. Hydrobiologia 138: 161–176.

Kalff, J. & D. Brumelis, 1993. Nutrient loading, wind speed and phytoplankton in a tropical African lake. Verh. int. Ver. Theor. Angew. Limnol. 25: 860.

Kilham, P. & S. S. Kilham, 1990. Endless summer: internal loading processes dominate nutrient cycling in tropical lakes. Freshwat. Biol. 23: 379–389.

Kratzer, C. R. & P. L. Brezonik, 1981. A Carlson-type trophic state index for nitrogen in Florida lakes. Wat. Res. Bull. 17: 713–717.

Lee, G. F., R. A. Jones & W. Rast, 1980. Availability of phosphorus to phytoplankton and its implications for phosphorus management strategies. In Loehr R. C., C. S. Martin & W. Rast (eds), Phosphorus Management strategies of Lakes. Ann Arbor Science, Ann Arbor: 259–308.

Lehman, J. T., 1996. Pelagic food webs of the East African Great Lakes. In Johnson T. C. & E. O. Odada (eds), The Limnology, Climatology, and Paleoclimatology of the East African Lakes. Gordon & Breach Amsterdam: 281–301.

Lennox, S. D., R. H. Foy, R. V. Smith & C. Jordan, 1997. Estimating the contribution from agriculture to the phosphorus load in surface water. In Tunney H., O. T. Caton, P. C. Brookes & A. E. Johnston (eds), Phosphorus Loss from Soil to Water. CAB International: 55–75.

Mackereth, F. J. H., J. Heron & J. F. Talling, 1989. Water analysis. Freshwater Biological Association, Windermere, U.K.: 120 pp.

McDougall, B. K. & G. E. Ho, 1991. A study of eutrophication of North lake Western Australia. Wat. Sci. Technol. 23: 163–173.

Melack, J. M., 1976. Limnology and Dynamics of Phytoplankton in Equatorial African Lakes. PhD thesis, Duke University, Durham: 453 pp.

Moyo, N. A. G., 1997. Lake Chivero: a Polluted Lake. Harare: IUCN & University of Zimbabwe Publications.

OECD, 1980. Eutrophication of Waters: Monitoring, Assessment and Control. Organisation for Economic Co-operation and development. Paris: 154 pp.

Nyandat, N., 1981. The primary minerals in some Kenya's topsoil and their significance to inherent soil fertility. East Africa Agriculture and Forestry Journal 46: 71–76.

Payne, I., 1984. Tropical Lakes and Rivers. Butterworths, London.

Ryding, S. O. & W. Rast, 1989. The Control of Eutrophication of Lakes and Reservoirs. Vol. 1. Man and the Biosphere series. UNESCO, Paris.

Talling, J. F., 1966. The annual cycle of stratification and phytoplankton growth in Lake Victoria (East Africa). Int. Rev. ges. Hydrobiol. 51: 545–621.

Talling, J. F., 1986. The seasonality of phytoplankton in African lakes. Hydrobiologia 138: 139–160.

Talling, J. F. & I. B. Talling, 1965. The chemical composition of African Lakes waters. Int. Rev. ges. Hydrobiol. 50: 421–463.

Talling, J. F., 2001. Environmental controls on the functioning of shallow tropical lakes. Hydrobiologia 458: 1–8.

Thornton, J. A. & W. K. Nduku, 1982. Lake McIlwaine. Dr W. Junk Publishers, The Hague: 251 pp.

Thornton, J. A., 1987a. Nutrients in African lake ecosystems: do we know all? Journal of Limnological Society of South Africa 12: 25–43.

Thornton, J. A., 1987b. Aspects of eutrophication management in tropical/sub-tropical regions. Journal of the Limnological Society of South Africa 13: 25–43.

Toerien, D. F., 1975. South African eutrophication problems: a perspective. Water Pollution Control 74: 134–142.

Twinch, A. J., 1986. The phosphorus status of sediments in a hyper-eutrophic impoundment. (Hartbeespoort Dam): implications for eutrophication management. Hydrobiologia 135: 23–34.

UNEP-IETC, 1999. Planning and Management of Lakes and Reservoirs: An Integrated Approach to Eutrophication. Technical Publication Series Issue 11, Osaka/Shiga: 375 pp.

Vollenweider, R. A., 1968. Scientific Fundamentals of Eutrophication of Lakes and Flowing Waters, with Particular Reference to Nitrogen and Phosphorus as Factors in Eutrophication. OECD Paris, DAS/C31/68.27.

Hydrobiologia **488:** 81–88, 2002.
D.M. Harper, R. Boar, M. Everard & P. Hickley (eds), Lake Naivasha, Kenya.

Magnetic susceptibilities of lake sediment and soils on the shoreline of Lake Naivasha, Kenya

R.R. Boar[1,*] & David M. Harper[2]
[1]*School of Environmental Sciences, University of East Anglia, Norwich NR4 7TJ, U.K.*
[2]*Department of Biology, University of Leicester, Leicester LE1 7RH, U.K.*
**Author for correspondence. E-mail: r.boar@uea.ac.uk*

Key words: environmental magnetism, *Cyperus papyrus*, soil erosion, buffer zones, RAMSAR

Abstract

On steep, unvegetated slopes, sands (particle sizes 0.063 to 2 mm) and gravels (2–64 mm) erode from the shoreline of Lake Naivasha (Kenya) and enter the lake basin. This occurred freely where fringing papyrus *(Cyperus papyrus)* swamp had been cleared in favour of landing jetties or agriculture. Gravel-sized particles have been recovered up to 60 m offshore and sands to 80 m. In an area where papyrus was undisturbed and the swamp margin was 52 m wide, gravel did not enter the lake and sands penetrated to around 35 m. Large particles are much easier to trace to source and to manage than the finer silts and clays (<0.63 μm) that form the bulk of Lake Naivasha's sediment. The pattern of mass specific magnetic susceptibilities for the <63 μm fraction of lake sediment around the southern shoreline of the lake suggests that fine particles enter open water directly from the shoreline and are not transported from the lake's only perennial inflow, the River Malewa in the north. Such particles originate from a hinterland that supports high intensity horticulture and are therefore a potential source of contamination. Mean $\pm$ SD susceptibility (χ_{lf}) immediately offshore papyrus fringe was $0.49 \pm 0.08 \times 10^{-6}$ m^3 kg^{-1} compared with higher values of $1.33 \pm 0.14 \times 10^{-6}$ m^3 kg^{-1} where there was no papyrus barrier ($P < 0.0001$). The value for five sites in the middle parts of the lake was $0.45 \pm 0.02 \times 10^{-6}$ m^3 kg^{-1} with $1.38 \pm 0.10 \times 10^{-6}$ m^3 kg^{-1} near the mouth of the River Malewa. The results of this study are evidence, therefore, that conservation of a continuous papyrus margin of about 50 m width is a priority for intercepting particulate material.

Introduction

Lake Naivasha is a shallow freshwater lake in southern Kenya (0° 45′ S and 36° 20′ E) with a surface area that has fluctuated around 150 km^2. The only perennial inflow to the north of the lake, the River Malewa, drains a largely volcanic catchment of 1730 km^2. The other main river of the drainage basin, the Gilgil, distributes into irrigation ditches about 5 km north of the lake shore and does not flow directly into the lake. The Karati River to the northeast is an ephemeral stream that flows into Lake Naivasha on about 100 days each year. Several other small ephemeral streams, mostly along the southern shore, carry storm runoff toward open water during heavy rains. With an average mean daily flow of around 20 m^3 s^{-1} recorded between April 1994 and March 1995 (P. Obade, pers. com.), the River Malewa is likely to contribute by far the major part of Lake Naivasha's external supply of suspended and dissolved materials. Lake Naivasha has no surface outflow. The lake itself lies in a shallow basin dominated by Tertiary and Quaternary pyroclastic and lacustrine deposits and slope varies between about 1° and 48° over a 2000-m belt around the shore.

Annual rainfall in the area averages 680 mm (1951–1982) and annual evaporation is about 1865 mm (1966–1982; Åse et al., 1986). Lake Naivasha is a RAMSAR site of international importance for wildlife conservation, supports a local fishery and supplies water for arable agriculture, and geothermal power generation and horticulture.

Although the lake has international conservation status, its ecology is compromised by cultivation of its immediate shoreline, particularly in the south and in the northern part of the lake where large tracts of papyrus (*Cyperus papyrus* L.) swamp have been cleared for intensive agriculture. Undesirable influences of shoreline development on the ecology of the lake include: changes in water chemistry in the direction of increases in nutrient status (Harper, 1992); losses in lake volume due to water abstraction (LNRA, 1999; Becht & Harper, 2002); loss of papyrus swamps (Boar et al., 1999) and, introduction of alien fish (Muchiri & Hickley, 1991), crayfish and water plants (Harper et al., 1995). Potential influences of shoreline development include soil erosion and transport to the lake basin of soil particles on which pesticides; nutrients, particularly phosphorus and other agrochemicals may be adsorbed.

The lithosols and solentz soils around the southern shore of the lake are derived from volcanic ashes and other pyroclastic rocks (e.g. tuff and pumice) formed during recent volcanic activity (Kenya Soil Survey, 1982). A survey of the geochemistry of sediments from inflowing rivers (Tarras-Wahlberg, 1998) using X-Ray fluorescence analyses, has shown that river sands from the lower River Malewa contain around 9% haematite (Fe_2O_3). The ferrimagnetic content of the material is thus high and the natural magnetic content of catchment soils likely, therefore, to be detectable using standard methods in environmental magnetism (Smith, 1999). Organic sediments formed from primary production within lakes have very low, or sometimes negative magnetism and are thus likely to be quite distinct from sediment containing iron-rich minerals that may have eroded from lake catchments (Dearing, 1999). Environmental magnetism has been used since the mid 1970s to separate the origins of lake sediments and to trace their actual origin from known source domains (Thompson et al., 1975; Dearing, 1979; Oldfield et al., 1989; David et al., 1998). However there are other influences that complicate, and can invalidate, interpretation. For example, magnetic properties of materials vary with particle size (Thompson & Morton, 1979) and with variations in the size and shape of crystals (Maher, 1988). A particular feature of lake sediments is that in conditions of low redox potential, iron dissolution may occur and so reduce magnetic signatures (Hilton & Lishman, 1985). Conversely, authigenic iron sulphides (e.g. greigite) might be formed (Hilton, 1990) or iron nodules (magnetosomes) may be formed through intense bacterial activity (Snowball, 1994) and may enhance magnetic signatures significantly. Casual observations of sediment in Lake Naivasha over the period 1993–1998 suggest, however, that lake sediment is always oxic and that reduction of iron is not, therefore, a dominant process. This means that particles that have eroded from the immediate shoreline and that have entered the lake are likely to have retained largely unaltered magnetic properties.

This work set out to discover: (i) if sediment in the lake's littoral zone receives soil that has eroded from the lake shore; (ii) if magnetic properties of such sediment can be used to distinguish it from material that has come from the main inflowing river and, (iii) the extent to which papyrus swamp intercepts finer size classes of soil that, potentially, carry contaminants into the lake. The wider aim of the work is to help establish quantitatively the role of papyrus in trapping particulate material that may be moving from the immediate shoreline into the lake's basin.

Methods

Samples of topsoil were taken at 50-m or 10-m intervals along three transects of 420 m, 575 m and 920 m from upland (1960 m.a.s.l.) to the lake edge on the southern shore of Lake Naivasha (water level 1890 m.a.s.l.) during August 1998, (Fig. 1). One year later, samples of surface lake sediment were taken with an Ekman grab at intervals of up to 30 m to extend the three transects into the open lake. Transects were chosen to represent areas with a high risk of soil erosion where continuous papyrus swamp was present on the lake shore; areas of similarly high erosion risk, but with a discontinuous papyrus fringe and an area of low erosion risk. Erosion risk was taken as high on slopes with a gradient of greater than 30° with shrubs (*Tarchonanthus camphoratus* L. and *Acacia drepanolobium* Sjøstedt) and very sparse (c.1°) cover by ground vegetation and low where gradients were low (c. 1°) with continuous ground cover (*Cynodon* sp. and other grasses, *Senna didymobotrya* (Fresen.) Irwin & Barneby and *A. xanthophloea* Benth.)

Additional samples of lake sediment were taken at the waterward edge of papyrus swamp in 16 other sampling areas distributed around the lake, from five sites in the middle of the lake and, to represent the drainage catchment of the inflowing river, from two sites at the mouth of the River Malewa and four areas further south of the inflow. Samples were taken around the mouth of the River Malewa since it is possible that minerals from the inflow could have been transported and then redistributed along the southern shoreline.

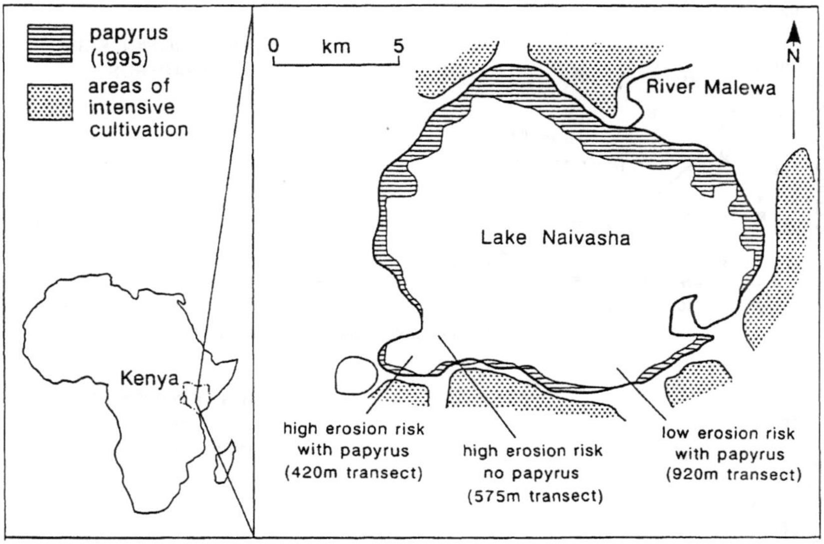

Figure 1. Schematic map of Lake Naivasha, Kenya, showing the locations of three sampling transects. Transect lengths are not shown to scale.

Particle size analysis was carried out on lake sediment and soil samples from the transects. Magnetic analyses have been carried out on the < 63 μm fraction (silts and clays) of all lake samples and on soil samples from the high erosion risk transect where there was no shoreline papyrus. For particle size analysis, 100 ml of each lake sediment sample was wet sieved into four size classes: >2000 μm (gravel); 2000–500μm (very coarse and coarse sand); 500–250 μm (medium sand) and 250–63 μm (fine and very fine sand). Each fraction was dried and weighed to the nearest 0.001 g. Replicate 100-ml bulk samples were dried and weighed and the mass of the <63 μm fraction calculated by difference.

Magnetic analyses were carried out on the <63 μm size fraction of topsoils and lake sediment of five extra replicate samples taken in each sampling area and dried for subsequent sieving. Samples of 10 cm^3 of the <63μm fraction were weighed in plastic pots to the nearest 0.0001 g. Their magnetic susceptibility was measured in a weak magnetic field (0.1 mT) at 0.46 kHz (low frequency, lf) and 4.6 kHz (high frequency, hf) on a Bartington MS2B single sample dual frequency sensor. Low frequency mass specific magnetic susceptibility (χ_{lf}), in units corrected for variations in bulk density (m^3 kg^{-1}) and percentage mass specific frequency dependent susceptibility (χ_{fd}%) were calculated according to Walden (1999a).

Further magnetic analyses were carried out to correct for variations in magnetic behaviour due only to variations in the content of ultrafine material within samples. Samples were first demagnetised by tumbling them at 80 mT in a Molspin AF (alternating field) demagnetiser. This was carried out inside a magnetic shield that reduces the Earth's magnetic field to near zero. Samples were then fixed in one position relative to the coil and remagnetised in an applied alternating field (0.08 mT) so that they acquired an anhysteretic remanent magnetism (ARM). The remanence of the acquired magnetism was then measured (Am^{-2} kg^{-1}) immediately in a Molspin Spinner and values corrected to mass specific ARM (χ_{ARM}) in m^{-3} kg^{-1}. The method for this followed Walden (1999b).

Results

The particle size analysis of sediments along the southern shoreline shows the permeability of the shoreline to different size classes of particles (Fig. 2) and a marked effect of removal of papyrus swamp on the movement of topsoil into open water. Where the risk of soil erosion was high, gravel-sized particles entered the lake from its shoreline and were found at up to 60 m offshore, but only where papyrus had been removed (Fig. 2a). Gravel was not found in the lake where a 52-m fringe of papyrus was present (Fig.

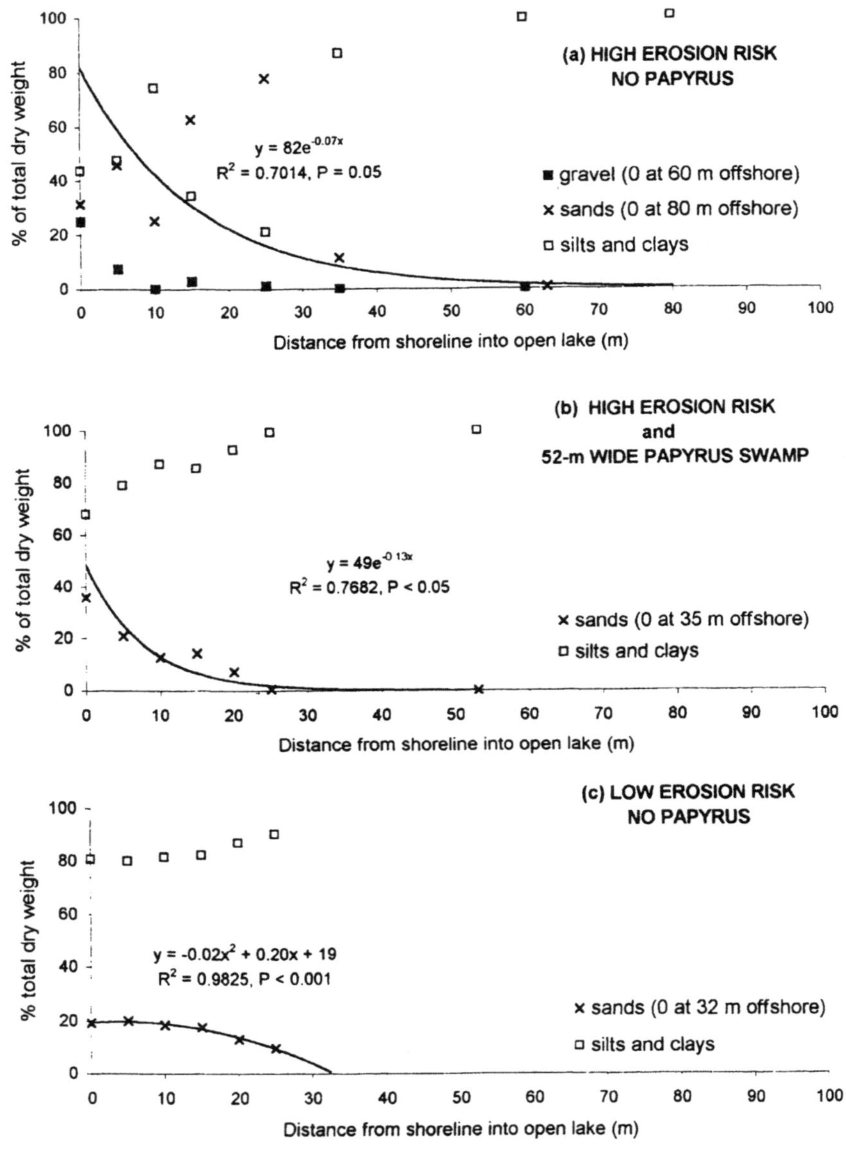

Figure 2. Particle size distributions in lake sediment from three transects along the southern shore of Lake Naivasha, Kenya, during 1998. Transects begin at the lake's shoreline and papyrus had been removed from the shorelines of (a) and (c).

2b). Sands penetrated much further (up to 80 m) into open water where fringing swamp had been removed. Where the risk of soil erosion was low (Fig. 2c) sands penetrated to about 30 m offshore. The difference made by papyrus in high risk areas was thus the interception of gravel and a reduction in the lakeward spreading of sands. The control area, with no papyrus and at low risk of soil erosion, did not contribute gravel to open water but did contribute sands to the predicted c. 30 m. In this area, sands contributed about 20% of the dry weight of sediment on the immediate shoreline. The most disturbed sampling area (Fig. 2a) appeared to contribute much larger loads of sand (and

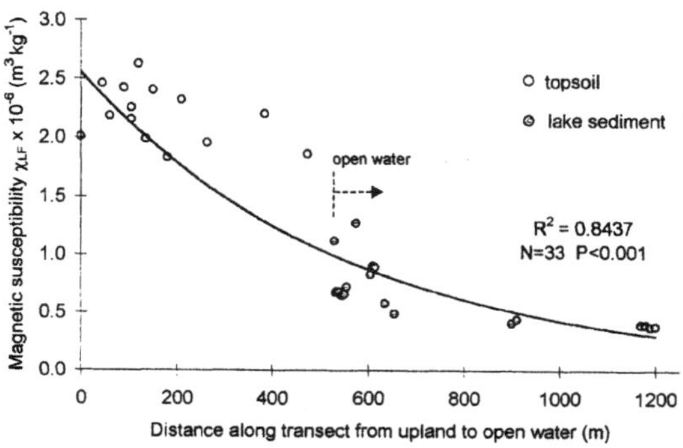

Figure 3. Mass specific magnetic susceptibility of silts and clays (<63 μm) in topsoil and lake sediment along a transect from upland into Lake Naivasha, Kenya where marginal *Cyperus papyrus* swamp had been removed from the lake's shoreline. The lakeward end of the same transect is shown in Figure 2a.

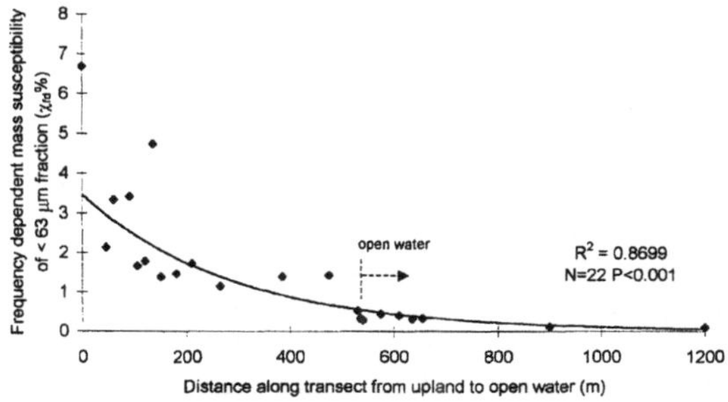

Figure 4. Frequency dependent mass specific magnetic susceptibility of silts and clays (<63 μm) in topsoil and lake sediment along the transect shown in Figure 3.

gravel) than either of the other two sites although the material penetrated little further into open water.

The dry weight of sediment that was not accounted for by sand or by gravel was much finer (<63 μm) silts and clays. Figure 3 shows that the mass specific magnetic content of the <63 μm size fraction of topsoils and sediments fell more than fourfold ($P < 0.001$) from upland to lake bed along a 1200-m transect from an eroding upland to lake sediment. Particle sizes of the lake sediments on this transect are shown in Figure 2a. The water's edge was 575 m from the start of the transect, and the average $\pm$ SD mass specific magnetic susceptibility of fine sediment sieved from seven lake sediment samples taken at 5-m intervals within 35 m of the edge was $0.83 \pm 0.26 \times 10^{-6}$ m^3 kg^{-1}. At 575 m, magnetic susceptibility reached its peak of $1.28 \pm 0.04 \times 10^{-6}$ m^3 kg^{-1}. This sampling site was thus taken to represent extreme erosion of the shoreline with values higher ($P < 0.01$) than the mean value in the five sites sampled in the middle of the open lake ($0.45 \pm 0.02 \times 10^{-6}$ m^3 kg^{-1}).

The data in Figure 4 help show whether the origin of the sediment sampled along the southern shoreline was from material eroded from the immediate catchment or if it was transported from the northern part of the basin. Large values for frequency dependent susceptibility are typical of material that has been weathered little, and low values suggest that soil minerals have undergone physical and chemical weathering. Figure 4 thus suggests that products of soil weathering had been transported down slope and accumulated on the lower slopes and shoreline of the sampling site. Frequency dependent susceptibility reached almost 7% on the upper slopes and fell to between 1 and 2% on lower slopes. Values varied quite closely with distance along the transect ($R^2 = 0.87$, $P < 0.001$), suggesting that material on lower slopes had originated from higher ground.

Since magnetic susceptibility depends upon grain size as well as the concentration of magnetic minerals within samples, magnetic analyses that correct for grain size had to be carried out. Figure 5 shows the results of ARM analysis on samples from the transect. The importance of Figure 5 is that, after correction for variations in ultra-grain size, the pattern is very similar to the pattern shown in Figure 3. On Figure 5, ARM values are shown also for samples taken from the edge of open water at five other sites on the southern shore where, in contrast to the sampling transect, the papyrus fringe was undisturbed. The mean $\pm$ SD ARM (χ_{ARM}) of sediment from these five sites ($2.53 \pm 0.56 \times 10^{-3}$ m^3 kg^{-1}) was lower ($P < 0.01$) than along the sampling transect and was the same as sediment sampled from the middle of the lake basin. Together,

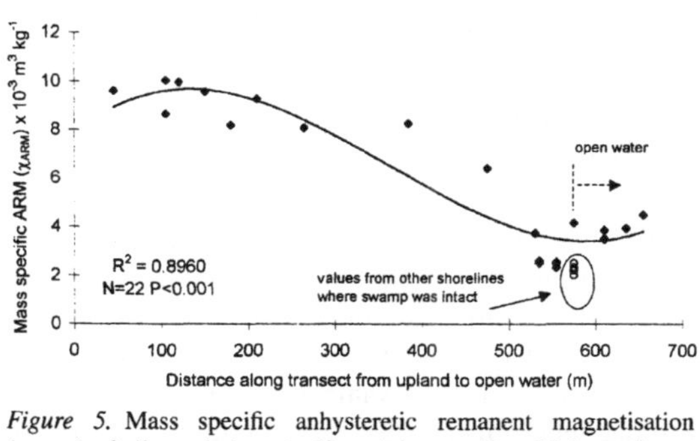

Figure 5. Mass specific anhysteretic remanent magnetisation (χ_{ARM}) of silts and clays (<63 μm) in topsoil and lake sediment along the transect shown in Figure 3 extending the transect to 655 m into the lake.

these results indicate that papyrus forms a barrier to the entry of fine particles from the shoreline.

Figure 6 is based upon values for the magnetic susceptibility of all the sediments that were sampled and shows that the strongest present-day magnetic susceptibility of sediment in the lake basin is at the mouth of the River Malewa and the weakest toward the middle of the lake. If the middle of the lake basin is taken to represent a background concentration for the lake (of $<0.50 \times 10^{-6}$ m^3 kg^{-1}), then magnetic susceptibilities well above background occurred along the southern shore where papyrus was absent and where the risk of soil erosion was high. Sediment with high magnetic concentration in the north of the lake appeared to reflect inputs from the River Malewa, rather than from the much smaller Karati. Patches of relatively high magnestism along the western shores of the lake may be associated with very steep shoreline gradients in these areas.

Discussion

A papyrus margin of 52-m width formed an effective barrier to the movement of gravel from the southern shore of Lake Naivasha into open water and appeared also to reduce to around half the rate of entry of sands. The pattern of results for these larger particles thus shows that clearance of swamp in areas at high risk of soil erosion results in transport of topsoil into the lake. However, it is the entry of smaller particles that is the more likely to influence the ecology of lake sediments and, potentially, the water column.

At around 10%, the magnetite content of sediment in Lake Naivasha is high (Tarras-Wahlberg, 1999) and this was reflected in this work by values for magnetic susceptibility that were high for lake sediment. The magnetic susceptibility of topsoil from the Lake Naivasha catchment was around $2\chi_{lf}$ 10^{-6} m^3 kg^{-1} and so at the upper end of published ranges of between 0.01 χ_{lf} 10^{-6} m^3 kg^{-1} to 10 χ_{lf} 10^{-6} m^3 kg^{-1} (Dearing, 1999). Magnetic susceptibility appears, therefore, to form a good basis for tracing the transport of fine material into the lake. Since the magnetic strength of eroded materials may be altered by both diagenetic and authigenic processes, the reliability of the method would be reduced if there were evidence of these from the lake. Since Tarras-Wahlberg (1998) did not find ferrimagnetic minerals (such as greigite) and negligible rates of phosphorus release from lake sediment have been measured (N. Kitaka pers. com.), it is unlikely that ferrimagnetic minerals of catchment origin were transformed significantly during their residence in the lake's sediment.

Variations in frequency dependent susceptibility between 1% and 10% are accepted as interpretable in terms of extent of soil weathering with lower values typical of more weathered material. This means that the range of values from 0.1% to 7% in topsoil taken from the main sampling transect probably does indicate that the magnetic content of the material was washed down slope rather than formed *in situ*.

At the five sites where papyrus swamp was intact, values for ARM were significantly lower that where papyrus formed a continuous barrier. Thus, both magnetic susceptibilities and remanence measures have indicated that silts and clays have entered open water directly from the southern shoreline and have not been transported from the inflow of the River Malewa.

The ecological implications of the results involve potential increases in the rate of sediment accumulation in the lake. Rates of sedimentation may be reconstructed using environmental magnetism as well as a variety of methods in paleolimnology. Paleolimnological approaches may include analysis of the diatom record and of zooplankton and other invertebrate remains. Such methods have been used in separate basins within Lake Naivasha (Verschuren et al., 1999). Using stratigraphic and geochronological analyses of sediment cores from the middle of Lake Naivasha, Verschuren (1999) estimated that the linear sedimentation rate in the middle of the lake is around 1 cm per year (0.1 g dry matter cm^{-2} yr^{-1}) and that the depositional environment has been relatively stable over the last 40 years. On the basis of the present study, such stability is unlikely to persist, at least at eroding lake margins, given the penetration of sands into open water where shoreline papyrus has been removed. The actual areal extent of papyrus swamp, although fluctuating naturally with lake level, has been in a period of net decline over the last 40 years with the area changing from 48 km^2 during 1960 to 14 km^2 in 1995 (Boar et al., 1999) with a recent estimate for 1999 based upon maps made from aerial photographs of only 8 km^2 (R.R. Boar, unpublished data). Moreover, in his recent study of the geochemistry of sediment cores taken in the south-eastern part of Lake Naivasha and thus in the same part of the lake as the main transect in this study, Tarras-Wahlberg (1999) has estimated a sedimentation rate of 3 cm per year which is much higher than elsewhere in the lake. Nutrients may accumulate also as a result of increases in external

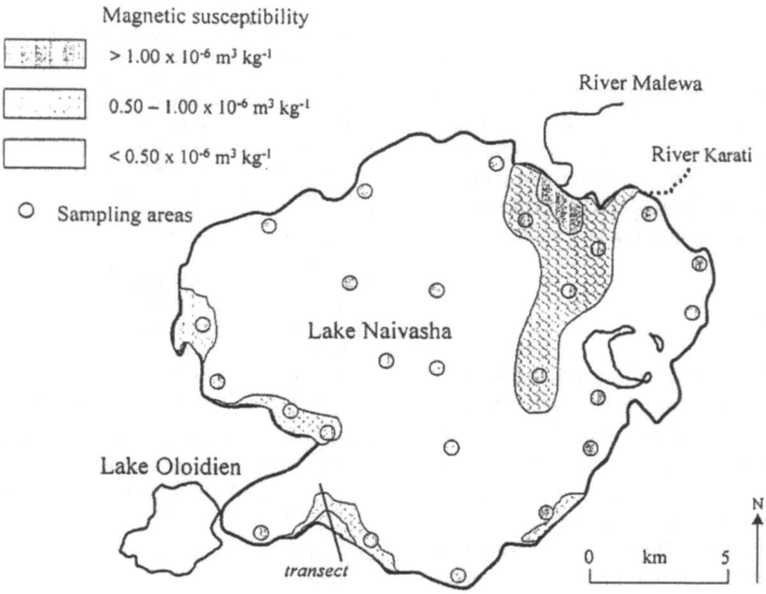

Figure 6. Mass specific magnetic susceptibility of silts and clays (<63 μm) in surface sediment sampled around the shoreline of Lake Naivasha, Kenya, during 1999. The position of the transect referred to in Figures 3–5 is shown.

loading of sediment to the lake. A recent phosphorus budget shows that at least 70% of the external load of phosphorus to Lake Naivasha is associated with particulate material (N. Kitaka, pers. com.), some of which may enter from the shoreline rather than from inflowing rivers.

The importance of conservation of papyrus swamp is thus twofold; to capture particulate material, and to intercept dissolved materials that enter the lake from its margins. Although a swamp margin of about 50 m does, to an extent, buffer the lake against entry of particulates, such a margin may be insufficient to remove or transform nutrients or other ions that are dissolved in swamp throughflow. There is some evidence for this from a transect through a 50-m wide swamp in Lake Naivasha in that neither electrical conductivity of standing water nor the size or density of papyrus stems changed with distance from the shore to open water (Boar et al., 1999). This is in contrast to much larger papyrus swamps where important changes in nutrient chemistry occur in through flow under floating papyrus mats (Gaudet, 1979).

There is potential for transport of other contaminants (such as pesticides) to open water where silts and clays eroded from intensively cultivated shorelines enter the lake. However, the extent to which contaminants adsorbed onto silts and clays would be transferred to higher trophic levels is limited in Lake Naivasha since the benthic community is patchy (J.E.P.C. Darlington, pers. com.) and in some years sparse (Clarke et al., 1989), and none of the members of the lake's species-poor fish community are benthivorous (Muchiri et al., 1994). The biological community of Lake Naivasha has simplified over the past decades. There have been large-scale losses in cover by submerged water plants that have corresponded to nutrient enrichment as well as disturbance by introduced Louisiana crayfish (*Pracambarus clarkii* Girard), water hyacinth (*Eichhornia crassipes* (Mart) Solms) and salvinia (*Salvinia molesta* Mitchell). The future of the lake's ecology would appear, at least in part, to depend also upon rates of soil erosion in the catchment and the conservation of papyrus swamp.

Conclusion

The main outcome of this work has been to show that, in the absence of papyrus, fine silts and clays are transported into the main basin of Lake Naivasha directly from its shoreline. Where the shoreline is cultivated intensively, any contaminants adsorbed onto such particles would be transported also. Papyrus swamps of at least 50-m width appear to form an effective barrier to gravel and to reduce to at least 50% the distance that sands penetrate into the lake. Such a buffer zone would not necessarily absorb nutrients or other dissolved salts. Until the width of papyrus needed for adequate nutrient buffering is quantified for Lake Naivasha, conservation of a minimum papyrus margin of 50 m that is continuous and not broken must be a priority for continuing lake management.

Acknowledgements

The fieldwork was funded by the Earthwatch Institute, Boston, U.S.A. and Oxford, England. Earthwatch volunteers on Team IV, 1998 and Team V, 1999, collected and helped prepare samples and thanks are due also to Mr Ruben Kamau Chege. Sampling was carried out with the permission of the Osieran Development Company, Mrs A. Root and The Elsa Trust. Magnetic analyses were carried out in the School of Environmental Sciences, UEA with the help of Mr Tom Mutch and Prof. Barbara Maher. We are grateful to our colleague Prof. Ken Mavuti at the University of Nairobi. The research was authorized by the Office of the President of the Government of Kenya under permit no. OP 13/001/12C46.

References

Åse, L-E., K. Sernbo & P. Syrén, 1986. Studies of Lake Naivasha, Kenya, and its drainage area. Forsk. Naturgeogr. Inst. Stockholm University 33: 1–77.

Boar, R. R., D. M. Harper & C. S. Adams, 1999. Biomass and its allocation in *Cyperus papyrus* in a tropical wetland, Lake Naivasha, Kenya. Biotropica 31: 411–421.

Clark, F., A. Beeby & P. Kirby, 1998. A study of the macroinvertebrates of Lakes Naivasha, Oloidien and Sonachi, Kenya. Rev. Hydrobiol. Trop. 22: 21–33.

David, C., J. A. Dearing & N. Roberts, 1998. Land use history and sediment flux in a lowland catchment, Groby Pool, Leicestershire, U.K. Holocene 8: 383–394.

Dearing, J. A., 1979. The Use of Magnetic Measurements to Study Particulate Flux in Lake-Watershed Ecosystems. PhD thesis, University of Liverpool, U.K.

Dearing, J. A., 1999. Holocene environmental change from magnetic proxies in lake sediments. In Maher, B. A. & R. Thompson (eds), Quaternary Climates, Environments and Magnetism. Cambridge University Press: 354 pp.

Gaudet, J. J., 1979. Seasonal changes in nutrients in a tropical swamp: North Swamp, Lake Naivasha, Kenya. J. Ecol. 67: 953–981.

Harper, D. M., 1992. The ecological relationships of aquatic plants at Lake Naivasha, Kenya. Hydrobiologia 232: 65–71.

Harper, D. M., C. Adams & K. M. Mavuti, 1995. The aquatic plant communities of the Lake Naivasha wetland, Kenya: pattern, dynamics and conservation. Wetland Ecol. Manage. 3: 111–123.

Hilton, J., 1990. Greigite and the magnetic properties of sediments. Limnol. Oceanogr, 35: 509–520.

Hilton, J. & J. P. Lishman, 1985. The effect of redox changes on the magnetic susceptibility of sediments from a seasonally anoxic lake. Limnol. Oceanogr. 30: 907–909.

Kenya Soil Survey, 1982. Exploratory soil map and agro-climatic zone map of Kenya, 1980. Exploratory Soil Survey Report No. E1. Kenya Soil Survey, Nairobi: 56 pp.

LNRA, 1999. Lake Naivasha Management Plan. The Lake Naivasha Riparian Association (LNRA), P.O. Box 1011, Naivasha, Kenya: 77 pp.

Maher, B. A., 1988. Magnetic properties of some synthetic submicron magnetites. Geophy. J. Royal Astron. Soc. 94: 83–96.

Muchiri, S. M. & P. Hickley, 1991. The fishery of Lake Naivasha, Kenya. In Cowx, I. G. (ed.), Catch Effort Sampling Strategies: their Application in Freshwater Fisheries Management. Fishing News Books, Blackwell Scientific Publications, Oxford: 382–392.

Muchiri, S. M., P. Hickley, D. M. Harper & E. North, 1994. The potential for enhancing the fishery of Lake Naivasha, Kenya. In Cowx, I. G. (ed.), Rehabilitation of Freshwater Fisheries. Fishing News Books, Blackwell Scientific Publications, Oxford: 348–357.

Oldfield, F., B. A. Maher & P. G. Appleby, 1989. Sediment source variations and lead-210 inventories in recent Potomac Estuary sediment cores. J. Quat. Sci. 4: 189–200.

Smith, J. P., 1999. An introduction to the magnetic properties of natural materials. In Walden, J., F. Oldfield & J. P. Smith (eds), Environmental Magnetism: a Practical Guide. Technical Guide, No. 6, Quaternary Research Association, London: 5–34.

Snowball, I. F., 1994. Bacterial magnetite and the magnetic properties of sediments in a Swedish lake. Earth Planet. Sci. Lett. 126: 129–142.

Tarras-Wahlberg, H., 1998. Investigations of Sediment Geochemistry in Rivers in the Lake Naivasha Catchment. University of Leicester/Earthwatch Institute Report: 11 pp.

Tarras-Wahlberg, H., 1999. Investigations of Sediment Stratigraphy and Sediment Geochemistry of Lake Naivasha, Kenya. University of Leicester/Earthwatch Institute Report: 14 pp.

Thompson, R. & D. J. Morton, 1979. Magnetic susceptibility and particle-size distributions in recent sediments of the Loch Lomond drainage basin, Scotland. J. Sed. Petrol. 49: 801–811.

Thompson, R., R. W. Battarbee, P. E. O'Sullivan & F. Oldfield, 1975. Magnetic susceptibility of lake sediments. Limnol. Oceanogr. 20: 687–698.

Verschuren, D., 1999. Sedimentation controls on the preservation and time resolution of sedimentary climate-proxy records from shallow fluctuating lakes. Quaternary Science Reviews 18: 821–837.

Verschuren, D., C. Cocquyt, J. Tibby, C. N. Roberts & P. R. Leavitt, 1999. Long-term dynamics of algal and invertebrate communities in a small, fluctuating tropical soda lake. Limnol. Oceanogr. 44: 1216–1231.

Walden, J., 1999a. Magnetic susceptibility. In Walden, J. F. Oldfield & J. P. Smith (eds), Environmental Magnetism: a Practical Guide. Technical Guide, No. 6, Quaternary Research Association, London: 35–62.

Walden, J., 1999b. Remanence measurements. In Walden, J., Oldfield F. & J. P. Smith (eds), Environmental Magnetism: a Practical Guide. Technical Guide, No. 6, Quaternary Research Association, London: 63–88.

Hydrobiologia **488**: 89–98, 2002.
D.M. Harper, R. Boar, M. Everard & P. Hickley (eds), Lake Naivasha, Kenya.

Phytoplankton community structure and succession in the water column of Lake Naivasha, Kenya: a shallow tropical lake

David S. Hubble[1,*] & David M. Harper
Department of Biology, University of Leicester, Leicester LE1 7RH, U.K.
[1]*Ecological Monitoring & Research, 7 Ainsley Gardens, Eastleigh, Hampshire SO50 4NX, U.K.*
**Author for correspondence. E-mail: dshubble@yahoo.com*

Key words: succession, cyanobacteria, diatom, *Aulacoseira*

Abstract

Lake Naivasha is a shallow freshwater lake in the Rift Valley of Kenya. Since the 1980s, when the lake showed a seasonal shift between diatom and cyanobacterial dominance it has become moderately eutrophic. Its algal assemblage is now dominated by a persistent *Aulacoseira italica* population both numerically and in terms of contribution to overall primary production. Algal and cyanobacterial counts were used to derive Simpson's diversity, succession rate and total community succession, focusing on the 10 most numerically abundant taxa. 170 species were identified, 43 of which were in common with the 143 found in 1979–80, before the increase in trophic state. Most diatoms are indicators of moderate to high nutrient conditions. There is little horizontal or vertical variation in successional processes throughout the lake and although the absolute abundance of cells varies widely, proportional composition is relatively stable. In Crescent Island lagoon, the only regularly stratified site, hypolimnetic succession rates are lower than those in the epilimnion. Overall, community composition is controlled by mixing (and hence light regime) and nutrient availability. With 'endless summer' conditions and full mixing, there is a successional pattern of 'muted seasonality' adapted to physical instability and environmental stability.

Introduction

Lake Naivasha, in common with many tropical African lakes, experiences fluctuating water-levels which influence its area and productivity (Harper, 1991). It also shows continually changing limnology through rainfall fluctuations in the catchment (Vincent et al., 1979). Its limnology and productivity may also be affected by human intervention in the form of lakeside agriculture through papyrus clearance, drainage and application of fertilisers (Johnson et al., 1998) coupled with the effects of species introductions (Harper, 1991). These changes affect the aquatic plant ecology in terms of macrophyte species and distribution and phytoplankton biomass (Harper, 1992). Due to nutrient inputs, Lake Naivasha can now also be considered moderately eutrophic (Harper et al., 1993). Before eutrophication in the 1980s it exhibited lower but more seasonally variable biomass compared with temperate lakes, with a seasonal shift between diatoms and cyanobacteria (Kalff & Watson, 1986). This paper examines the planktonic community and the degree of succession which now occurs.

In general, tropical lakes show cyanobacterial dominance during droughts and falling water levels, and diatom dominance during periods of inflow and mixing (Harris & Baxter, 1996). Rapid monsoon dilution may affect community structure as physicochemical conditions are rapidly changed. The work of Lewis (1978) on Lake Lanao in the Philippines gives a clear description of how a small number of chemical and physical factors may control succession in a deeper, stratified seasonal lake. The succession from dominance by diatoms and cryptomonads to chlorophytes, followed by cyanobacteria and finally dinoflagellates was seen to be promoted by increased sunlight availability, reduced nutrient availability and greater water-column stability.

Site description

Lake Naivasha is a shallow freshwater lake situated in warm and semi-arid conditions in the eastern Rift Valley of Kenya, lying within an enclosed basin at an altitude of approximately 1890 m. Rainfall is bimodal with peaks during April and October but shows inter-annual irregularity and an annual evapotranspiration rate which exceeds local rainfall (Harper et al., 1995). Regular sampling sites on Lake Naivasha (Fig. 1) were chosen to provide a range of potential influences on light regime and resulting conditions.

The sample sites were as follows:

1. Crescent Island Lagoon – a deep (ca. 18 m) almost completely enclosed sub-basin.
2. Malewa River inflow – a shallow (ca. 2 m) turbid area where the River Malewa enters Lake Naivasha.
3. Rema Island – an inshore area about 4 m deep away from any inflow.
4. Open Water – the central area of the lake which is well mixed with less shore effects than the rest of the lake. It is about 5 m deep.
5. Hippo Point – A well-mixed inshore site, about 6 m deep.
6. Oloidien Bay – a well-mixed bay average depth 3 m with little agriculture or horticulture nearby, and furthest the main sediment input from the River Malewa.

The final four sites – RI, OW, HP and Bay – are used as replicates of the 'Main lake'.

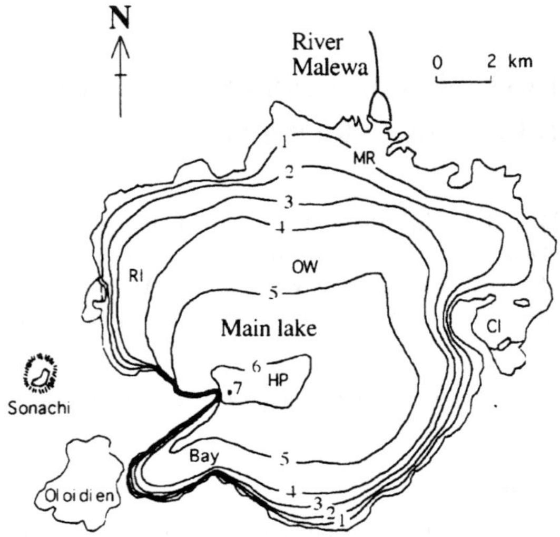

Key to sites	
CI - Crescent Island Lake	OW - Open Water
MR - Malewa inflow	HP - Hippo Point
RI - Rema Island	Bay - Oloidien Bay

Figure 1. Map of Lake Naivasha with sampling sites and 1991 depth contours (m). From Hickley et al. (2002).

Methods

Samples were taken from 4/12/97 to 21/10/98, at depths of 0, 1, 3 and 7 m using a 1.5-l messenger-activated Institute of Oceanographic Sciences (IOS) sampler. This was used to over-fill sample bottles to prevent trapping of air. Water samples were passed through a 200-μm screen in the laboratory, and then all algal and cyanobacterial taxa were identified to species where possible with a compound microscope at $\times$400 magnification. Counts were made to genus level in most cases as identification of all individuals to species was often impossible and/or excessively time-consuming. Two taxa were grouped more broadly (flagellates, cyanobacteria), as the aim was a broad overview of planktonic succession rather than an in-depth study of community structure. Previous work, albeit on British rivers (Hawkes, 1979), has shown that generic groupings can clarify otherwise complex community changes by reducing the number of groups requiring monitoring. This reduces complexity due to species-specific environmental requirements, although the species forming a group must show broadly similar characteristics to avoid oversimplification. Counting was undertaken using an eyepiece graticule and sedimentation chamber with a Zeiss Axiovert 100 inverted microscope (Zeiss, Germany). Cells were identified and counted along random transects covering 200 whole graticule areas. Cells were counted which were wholly within the area or touching the right or bottom edge. A multiplication factor then gave counts ml^{-1} Flagellates and cyanobacteria form two broad functional groups. The most important groups (those which form 10% or more of total cell numbers at any time) were considered separately to identify any succession for a given sample site. Cyanobacteria were counted as numbers of colonies. Some species were

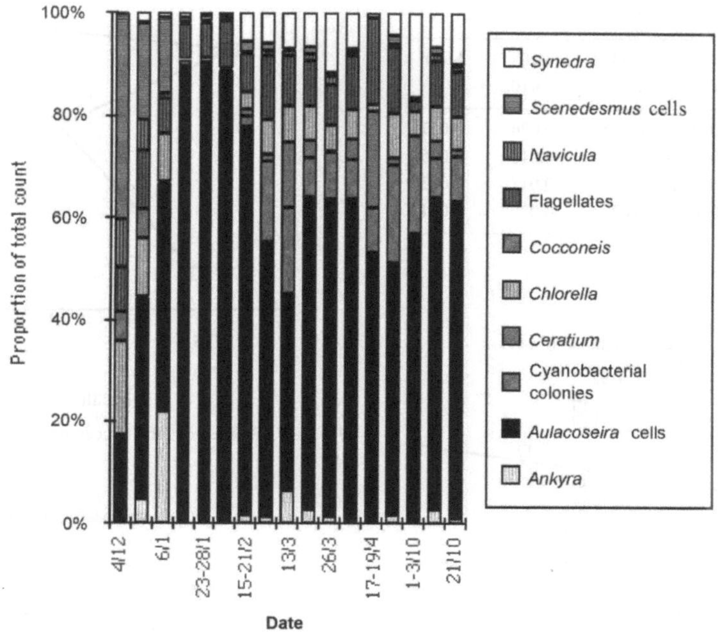

Figure 2. Main lake algal succession4/12/97–21/10/98.

present in such small numbers that they did not feature in analysis of succession.

Simpson's Diversity scores were calculated for all samples counted and analyses performed between sites and depths to investigate both horizontal and vertical spatial variation in the algal community. The index (Simpson, 1949) was

$$D = \Sigma P_i^2,$$

where P_i is the proportional numerical abundance of each species i.

To measure community change, estimates of succession rate were calculated (Jassby & Goldman, 1974; Lewis, 1976). These integrate changes in biomass or abundance over time through summation of such changes for each species as a proportion of the whole community. The Jassby & Goldman (1974) succession rate (R_s) was

$$R_s = \surd\{\Sigma(dc_i/dt)^2\},$$

where

$$c_i(t) = b_i(t)/\surd\{\Sigma b_i(t)^2\}$$

and

$$b_i(t)$$

is the numerical abundance of species i at time t with n possible species.

Jassby-Goldman measurements do not include directional aspects of succession, therefore using the above equations but weighting all species equally gives the formula derived by Lewis (1976):

$$R_s = \Sigma i \frac{|\{b_i(t_1)/B(t_1)\} - \{b_i(t_2)/B(t_2)\}|}{(t_2 - t_1)},$$

where $b_i(t)$ is the abundance of species i at time t and $B(t)$ is the size of the community at time t.

Results

The 170 identified species (and eight identified to genus only) are listed in the Appendix. The taxa most abundant were; *Ankyra, Aulacoseira*, cyanobacteria, *Ceratium, Chlorella, Cocconeis*, flagellates, *Navicula, Scenedesmus* and *Synedra* (Fig. 2). This shows *Aulacoseira* dominance in main lake surface water for

Table 1. Succession rate profiles for site CI

Depth (m)	12/12–6/1	6/1–16/1	17/4–12/5	3/10–13/10	Mean
0	0.055	0.062	0.035	0.076	0.057
1		0.066	0.022	0.018	0.035
3	0.052	0.067		0.011	0.043
7	0.027	0.040	0.027	0.029	0.031

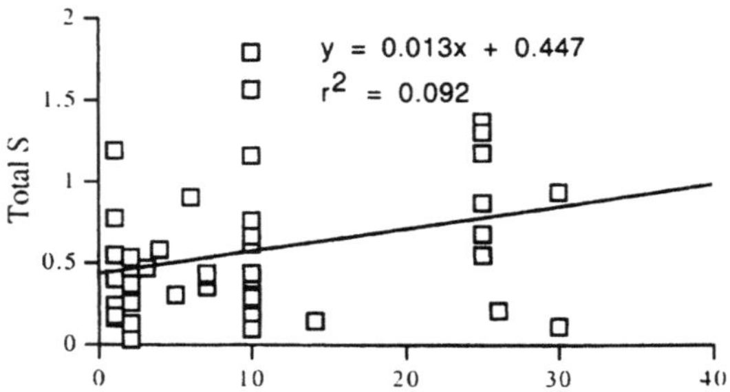

Figure 3. Regression of succession against time.

most of the sampling period, although early in this period *Scenedesmus* was more important whilst other taxa contributed variably to the algal assemblage.

Succession rate

Figure 3 illustrates the weak relationship ($r^2 = 0.092$) and non-significant correlation ($C_p = 0.304$, df = 40, $p > 0.05$) between total succession values (*S*-total) and period. For site CI, a profile of *S*-total values was developed (Fig. 4), showing that community fluctuations were greater at the surface than deeper in the water column. This is also clearly illustrated by successional rate (Table 1) indicating that surface fluctuations were not only greater in magnitude, but also occured more rapidly, than those lower down the water column.

Table 2. Mean planktonic diversity values

Site	Mean diversity	Site	Mean diversity
Main, surface	0.55	CI, 1 m	0.48
MR, surface	0.64	CI, 3 m	0.44
CI, surface	0.62	CI, 7 m	0.42

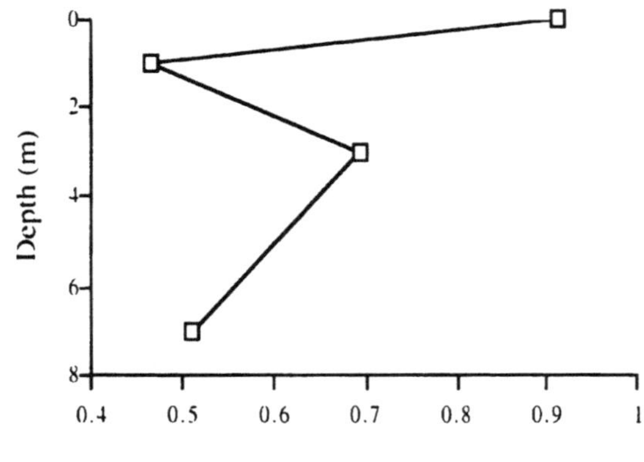

Figure 4. Profiles of total succession for Crescent island lagoon (CI).

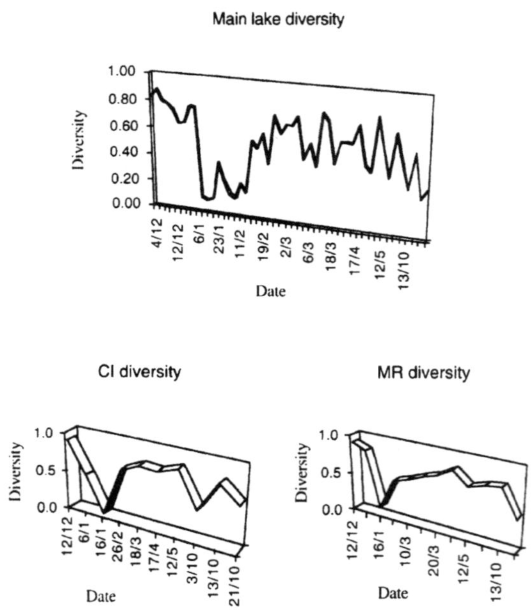

Figure 5. Simpson's Diversity, 4/12/97–21/10/98.

Diversity

Figure 5 shows Simpson's diversity over time for all sites. Table 2 shows mean values for a number of sites and depths. Site CI is investigated in profile due to its temporarily stratified nature (Hubble, 2000), whilst MR and Main sites are fully mixed (Hubble, 2000), showing no significant vertical variation.

All sites show a similar pattern of temporal diversity change, with troughs in January 1998 when

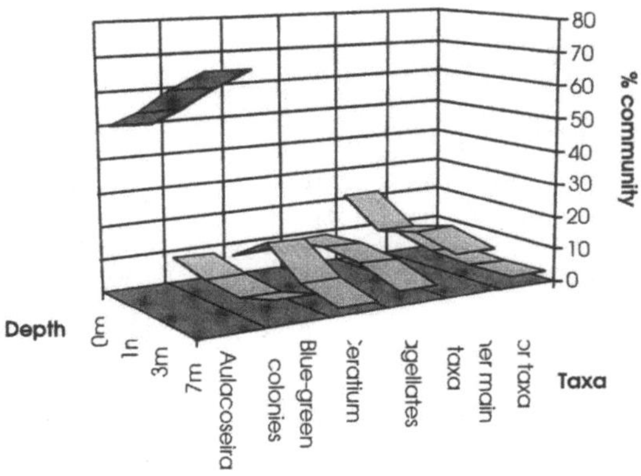

Figure 6. Phytoplankton community composition at Crescent Island lagoon (site CI).

there was almost total domination by an abundant population of *Aulacoseira*. Diversity was however similar between sites, with an overall mean of 0.60, and maxima generally occurring when *Aulacoseira* densities were relatively low.

There are no significant differences in diversity between sites, suggesting that the processes affecting diversity act equally throughout the lake. Sites CI and Main lake are not significantly correlated, and regression is closest between sites MR and Main lake, possibly due to the different hydrological regime found at site CI. Figure 6 shows depth profiles of community composition at site CI. The percentage composition of main taxa in the main lake correlates with that at site CI at all depths ($p < 0.001$). Although absolute abundance of taxa may vary, proportional community structure is therefore closely linked both vertically and between sites.

Aulacoseira formed approximately 50% of the surface community and 70% at a depth of 7 m. 'Other main taxa' (those which are sometimes dominant or sub-dominant) formed 10–20% of the community, with the remaining groups in small, fairly equal proportions. Comparing Fig. 5 with Fig. 7, community succession, shows that low diversity periods equated to high *Aulacoseira* periods as this genus comprised over 90% of the algal community at times. Such periods occurred during January and October 1998, whilst high diversity – low *Aulacoseira* periods are seen during December 1997 and April–May 1998. Vertical differences at site CI (Tables 2 and 3) show that hypolimnetic diversity is lower than that in the epilimnion, probably due to the more physically stressed conditions caused primarily by light attenuation. The most significant differences are between 0m and the other depths although diversity reduced with depth as *Aulacoseira* dominance increased.

Discussion

Comparison with the 170 species and eight genera identified during this study and the 143 taxa identified by Kalff & Watson (1986) during 1979–80 shows 43 species common to both studies. No relationship was found between species abundance and algal biomass, the latter used as a measure of lake trophy in the earlier study. More species have been found in this study than by Kalff & Watson (loc cit), but it is important to note that species richness may be underestimated even for large samples of a speciose community (Lande, 1996). This is because in natural assemblages, increased competition amongst individuals in a community can lead to competitive exclusion. Thus as the density of individuals increases, richness does not increase with sample size as much as expected (Goldberg & Esterbrook, 1998) as seen in dense *Ceratium* populations in Lake Naivasha (Hubble, 2000). Cox (1996) gives details of diatom species' environmental requirements. Of the 47 diatom species identified with known environmental requirements, 35 are either widespread or found in waters of moderate to enriched electrolyte or nutrient conditions (eutrophic). Of the remainder, 5 indicate moderate electrolyte or nutrient conditions (mesotrophic) and 7 indicate low to moderate electrolyte or nutrient conditions (oligotrophic). The oligotrophic species such as *Fragilaria capucina* were rarely found, and only in small numbers. Oligotrophic conditions probably no longer exist following the water level decline between 1980 and 1987, which allowed land to be cultivated with subsequent inundation of that land (Harper et al., 1993), hence such populations may be merely relicts, although *F. capucina* is known to exist as a number of ecological races.

Overall succession from diatoms and cryptomonads to chlorophytes, cyanobacteria and finally dinoflagellates may be largely controlled by a small number of key factors (Lewis, 1979) and promoted by high insolation, reduced nutrient availability and reduced mixing. Work at Lake Kinneret, Israel (Pollingher 1981) also indicates that low sinking rates and high nutrient availability are favourable as diversity increased during mixed and high nutrient periods, whilst it was minimal during periods of strong physical, chemical and biological stress. This indicates that in

Table 3. Site CI diversity comparison between sampling depths

Cp (prob)	r^2 (one-tailed probability)			
	CI 0 m	CI 1 m	CI 3 m	CI 7 m
CI 0 m		0.736 (0.029)	0.576 (0.035)	0.564 (0.021)
CI 1 m	0.858 (0.01)		0.946 (0.045)	0.876 (0.078)
CI 3 m	0.759 (0.05)	0.972 (<0.001)		0.888 (0.387)
CI 7 m	0.751 (0.05)	0.936 (<0.001)	0.942 (0.01)	

df = 5 for comparisons with CI 3 m; df = 6 for other comparisons.

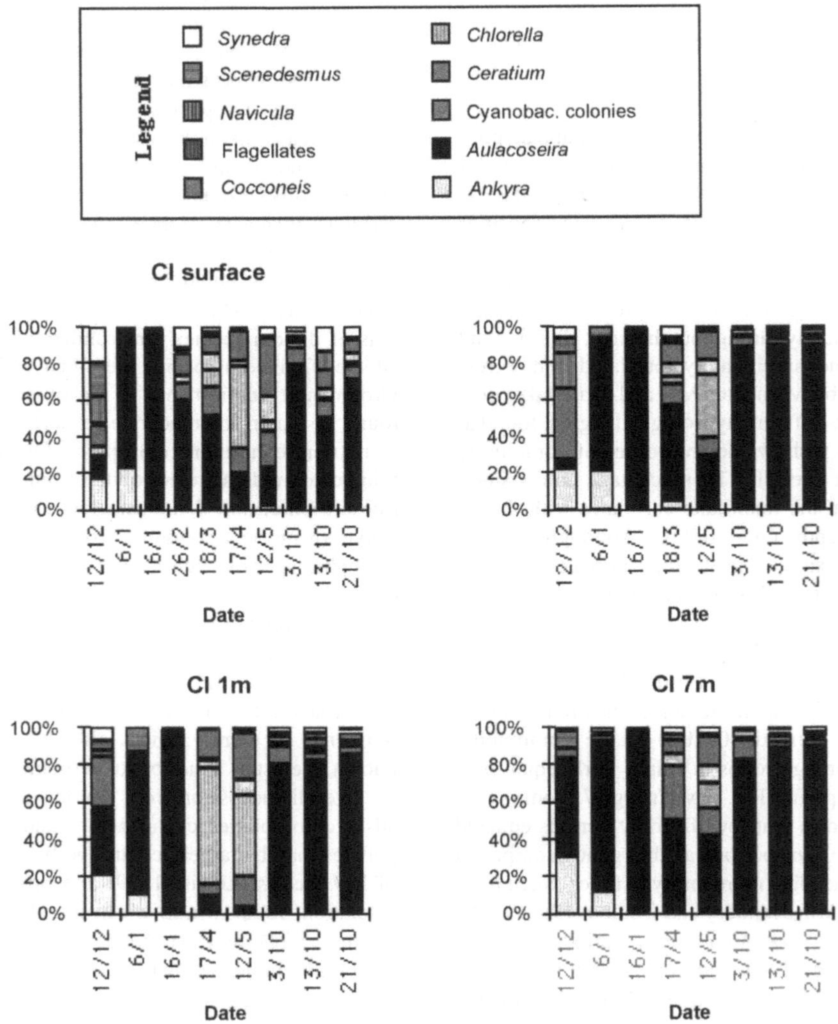

Figure 7. Community succession at Crescent Island lagoon (site CI).

early succession, low sinking rates and high nutrient availability favour low SA:V species (e.g. *Aulacoseira*) whereas nutrient depletion and low turbulence favour high SA:V species (e.g., *Microcystis*) which can scavenge scarce nutrients. The favouring of diatoms is seen both from palaeolimnology (Richardson & Richardson, 1972) and more recently (Melack, 1979, Kalff & Watson, 1986) in Lake Naivasha where there was an almost permanent early successional state. Lind (1968) stated that Lake Naivasha was dominated by *Aulacoseira ambigua* but no explanation of phytoplankton periodicity was made, although for the Sasumua and Ruiru reservoirs in Kenya it was considered to be controlled by rainfall. *Aulacoseira* is now the most important algal genus, both numerically and in terms of its contribution to overall primary productivity (Hubble, 2000).

Melack (1979) separated phytoplankton seasonality in tropical lakes into three broad categories. The most common pattern is that of pronounced seasonal fluctuations corresponding with differences due to rains, rivers or mixing. This was seen in Lake Oloidien and sometimes Crescent Island lagoon where mixing was predominant with minimal river inflow. Developing this idea, a second pattern consists of 'muted seasonality' due to the presence of buffer fringes, perennial rivers and sufficient internal recycling of nutrients. Diel variations are thus greater than longer-term changes, as conditions of 'endless summer' are found (Kilham & Kilham, 1990); this pattern is seen in Lake Naivasha. This means that as the same phytoplankton assemblages can persist for many days, species must be adapted to the full range of environmental variation. Phytoplankton communities may therefore experience both physical *instability* due to the mixed state of the water column and environmental *stability* as this mixing provides a certain light regime and prevents sedimentation loss. This form of 'continuous disturbance' is paralleled in ecosystems such as continually grazed grasslands where succession may be considered a function of 'vital attributes' (Noble & Slatyer, 1980). Around its persistent assemblages, Lake Naivasha did show wide short-term community variations such as temporary *Chlorella* dominance and wind-driven concentrations of *Ceratium* (Hubble, 2000). These occurred despite rapid physical and chemical turnover, and the idea that brevity of the mixing period may restrict species fluctuations (Ganf & Horne, 1975). The dominant or 'central' form of the assemblage however remained stable. This form of seasonality is contrary to the first pattern where there is seasonal succession because species composition changes as conditions change, with species better adapted to the new conditions becoming dominant. A final pattern, not seen in Lake Naivasha, is that of 'multiple stable points' in the form of persistent assemblages with occasional abrupt shifts between them.

Hypolimnetic succession rates are also notably lower and less variable than those of surface waters due to two main factors:

1. There is greater physical stability found below the surface as the largely wind-driven mixing processes have a smaller effect deeper in the water column. In the temporarily stratified Crescent Island lagoon, diversity is however significantly correlated between samples from 0 and 7 m, suggesting that there is only partial separation.
2. There is reduced productivity below the optimal photosynthetic depth of approximately 0.5–0.75 m, and therefore community fluctuations are slowed by the reduced rates of cellular division, and overall lower numbers of algal cells.

Such rates of succession (as well as indices of ecosystem stability) change as succession progresses as the resulting increased order leads to system predictability and pre-adaptation. There are changes in phytoplankton community structure; however, no stages are seen which correlate with nutrient levels. Lake Naivasha is moderately eutrophic and shows diatom dominance with dinoflagellate and cyanobacterial subdominance. It therefore fits with the mesotrophic stage expected in temperate waters as many phytoplanktonic species are found across a range of latitudes. Considerable fluctuations in nutrient levels were measured (Hubble, 2000), but as the pattern of temperate oligotrophic–eutrophic succession is not seen, they are not large enough to constitute a change in trophic state.

In summary, the lake is now dominated by *Aulacoseira* following eutrophication, probably due to land use changes in the early 1980s. As well as the seasonal pattern of dominance, the species present have also largely changed since eutrophication began to occur, and most diatoms found are indicative of moderate to high nutrient levels. Large community fluctuations are seen but these are temporary with the community rapidly reverting to its *Aulacoseira*-dominated state. As the lake experiences the 'endless summer' conditions of the tropics and is well-mixed, there is strong spatial homogeneity in the phytoplankton community as its composition is controlled by light regime and nutrient availability. Changes in successional processes and community structure are therefore determined by the combination of tropical conditions and human intervention, with the phytoplankton community acting as an indicator of trophic state.

References

Cox, E. J., 1996. Identification of Freshwater Diatoms from Live Material. Chapman & Hall, London.

Ganf, G. G. & A.J. Horne, 1975. Diurnal stratification, photosynthesis and nitrogen-fixation in a shallow, equatorial lake (Lake George, Uganda). Freshwat. Biol. 5: 13–39.

Goldberg, D. E. & G. F. Estabrook, 1998. Separating the effects of number of individuals sampled and competition on species diversity: an experimental and analytic approach. J. Ecol. 86: 983–988.

Harper, D. M., 1991. Primary production in Lake Naivasha, Kenya. Verh. Int. Ver. Theor. angew. Limnol. 24: 1112–1116.

Harper, D. M., 1992. The ecological relationships of aquatic plants at Lake Naivasha, Kenya. Hydrobiologia 232: 65–71.

Harper, D. M., G. Phillips, A. Chilvers, N. Kitaka & K. Mavuti, 1993. Eutrophication prognosis for Lake Naivasha, Kenya. Verh. Int. Ver. theor. angew. Limnol. 25: 861–865.

Harper, D. M., C. Adams & K. Mavuti, 1995. The aquatic plant communities of the Lake Naivasha wetland, Kenya: pattern, dynamics and conservation. Wetlands Ecol. Manage. 3: 111–123.

Harris, G. P. & G. Baxter, 1996. Interannual variability in phytoplankton biomass and species composition in a subtropical reservoir. Freshwat. Biol. 35: 545–560.

Hawkes, H. A., 1979. Invertebrates as indicators of river quality. In James, A. & L. Evison (eds), Biological Indicators of Water Quality. Wiley, Chichester.

Hubble, D. S., 2000. Controls on primary production in Lake Naivasha, a shallow tropical freshwater. PhD thesis, Leicester University, UK.

Jassby, A. D. & C. R Goldman, 1974. A quantitative measure of succession rate and its application to the phytoplankton of lakes. Am. Nat. 108: 688–693.

Johnson, G., D. M. Harper & K. Mavuti, 1998. Information for the sustainable management of shallow lakes: Lake Naivasha, Kenya. In Harper, D. M. & T. Brown (eds), The Sustainable Management of Tropical Catchments. Wiley, Chichester.

Kalff, J. & S. Watson, 1986. Phytoplankton and its dynamics in two tropical lakes: a tropical and temperate zone comparison. Hydrobiologia 138: 161–176.

Kilham, P. & S. S. Kilham, 1990. Endless summer: internal loading processes dominate nutrient cycling in tropical lakes. Freshwat. Biol. 23: 379–389.

Lande, R., 1996. Statistics and partitioning of species diversity, and similarity among multiple communities. Oikos 76: 5–13.

Lewis, W. M., 1976. Analysis of succession in a tropical phytoplankton community and a new measure of succession rate. Am. Nat. 112: 401–14.

Lewis, W. M., 1978. Dynamics and succession of the phytoplankton in a tropical lake. Journal of Ecology 66: 849–80.

Lewis, W. M., 1979. Zooplankton Community Analysis: Studies on a Tropical System. Springer, New York.

Lind, E. M., 1968. Notes on the distribution of phytoplankton in some Kenya waters. Br. Phycol. Bull. 3(3): 481–493.

Melack, J. M., 1979. Temporal variability of phytoplankton in tropical lakes. Oecologia 44: 1–7.

Noble, I. R. & R. O. Slatyer, 1980. The use of vital attributes to predict successional changes in plant communities subject to recurrent disturbances. Vegetation 43: 5–21.

Pollingher, U., 1981. The structure and dynamics of the phytoplankton assemblages in Lake Kinneret, Israel. J. Plankton Res. 3(1): 93–105.

Richardson, J. L. & A. E. Richardson, 1972. History of an African Rift lake and its climatic implications. Ecol. Monogr. 42: 499–534.

Simpson, E. H., 1949. Measurement of diversity. Nature 163: 688.

Vincent, C., T. Davies & A. Beresford, 1979. Recent changes in the level of Lake Naivasha, Kenya, as an indicator of equatorial westerlies over East Africa. Climate Change 2: 175–189.

Appendix 1. Algal species identified 1996–8

Phylum Cyanophyta

Order Chroococcales (14 spp.)

Aphanothece caldariorum Richter
Aphanothece ellipsoidea Schröder
Aphanothece nidulans Richter
Chlorogloea microcystoides Geitler
Chroococcus limneticus Lemmermann
Chroococcus turgidus (Kützing) Nägeli
Gloeothece rupestris (Lyngbye) Bornet
Merismopedia convoluta Brébisson
Merismopedia geminata Lagerstedt
Merismopedia tenuissima Lemmermann
Microcystis aeruginosa (Kützing) Kützing
Microcystis biformis (A. Braun) Rabenhorst
Microcystis elachista (W. & G.S.West) Starmach
Microcystis viridis (A. Braun) Lemmermann

Order Nostocales (= Oscillatoriales) (11 spp.)

Anabaena sphaerica Bornet & Flahault
Anabaena spiroides (Lemmermann) Elenkin
Anabaenopsis circularis (G.S. West) Müller
Borzia trilocularis Cohn
Lyngbya epiphytica Gardner
Lyngbya muralis Kützing
Lyngbya sp.
Nostoc parmelioides Kützing
Oscillatoria jenneri (Hassall) Kützing
Oscillatoria sp.
Spirulina subsalsa Oersted

Phylum Chrysophyta

Order Ochromonadales (2 spp.)

Didymochrysis paradoxa Pascher
Mallomonas leboimei Bourrelly

Order Chromulinales (7 spp.)

Chromulina pyriformis Playfair
Chrysococcus rufescens Skuja
Chrysomonas ellipsoidea Skvortzov
Chrysospora fenestrata Pascher
Epicystis peridinearum Pascher
Gloeochrysis pyrenigerum Pascher
Ochromonas viridis Bourrelly

Phylum Bacillariophyta

Order Achnanthales (5 spp.)

Achnanthes brevipes Agardh
Achnanthes exigua Grunow
Achnanthes inflata Kützing
Achnanthes minutissima Kützing
Diatomella hustedtii Manguin

Appendix 1. Continued

Order Coscinodiscales (6 spp.)

Aulacoseira ambigua (Grunow) Simonsen
Aulacoseira granulata (Ehrenberg) Simonsen
Aulacoseira italica (Ehrenberg) Simonsen
Cyclotella meneghiniana Kützing
Cyclotella stelligera Cleve & Grunow
Stephanodiscus neoastraea Håkannson & Hickel

Order Diatomales (11 spp.)

Cocconeis placentula Ehrenberg
Diatoma ehrenbergii Kützing
Diatoma tenuis Agardh
Fragilaria capucina Desmazières
Rhoicosphenia curvata (Kützing) Grunow
Synedra acus Kützing
Synedra berolinensis Lemmermann
Synedra capitata Ehrenberg
Synedra pulchella Kützing
Synedra ulna (Nitzsch) Ehrenberg
Synedra sp.

Order Eunotiales (2 spp.)

Eunotia denticula (Brébisson) Rabenhorst
Eunotia monodon (Gregory) W.Smith

Order Naviculales (45 spp.)

Amphora communata Grunow
Amphora ovalis Kützing
Caloneis amphisbaena (Bory) Cleve
Cymatopleura solea (Brébisson) W.Smith
Cymbella lacustris (Agardh) Cleve
Cymbella prostrata (Berkeley) Cleve
Denticula pelagica Hustedt
Denticula thermalis Kützing
Diploneis didyma (Ehrenberg) Cleve
Encyonema minutum (Hilse in Rabenhorst) D.G.Mann
Encyonema prostratum (Berkeley) Kützing
Epithemia turgida (Ehrenberg) Kützing
Gomphonema acuminatum Ehrenberg
Gomphonema augur Ehrenberg
Gomphonema truncatum (= constrictum) Ehrenberg
Gomphonema olivaceum (Hornemann) Brébisson
Gyrosigma attenuatum (Kützing) Rabenhorst
Navicula capitata Ehrenberg
Navicula cryptocephala Kützing
Navicula cuspidata Kützing
Navicula gibbula Cleve
Navicula lanceolata (C. Agardh) Kützing
Navicula pygmaea Kützing
Navicula protractoides Hustedt
Navicula radiosa Kützing
Navicula schoenfeldii Hustedt
Navicula scutelloides W. Smith
Navicula trivialis Lange-Bertalot

Appendix 1. Continued

Navicula tuscula (Ehrenberg) Grunow
Navicula sp.
Neidium ampliatum (Ehrenberg) Krammer
Nitzschia acicularis (Kützing) W. Smith
Nitzschia amphibia Grunow
Nitzschia fonticola Grunow
Nitzschia linearis (Agardh) W. Smith
Nitzschia palea (Kützing) W. Smith
Nitzschia pusilla Grunow
Nitzschia sp.
Pinnularia maior Ehrenberg
Pinnularia viridis (Nitzsch) Ehrenberg
Rhopalodia gibba (Ehrenberg) O. Müller
Stauroneis anceps Ehrenberg
Surirella angusta Kützing
Surirella linearis W. Smith
Tryblionella apiculata Gregory

Phylum Chlorophyta
Order Volvocales (4 spp.)
Chlamydomonas debaryana Gorosch
Chlamydomonas ovalis Pascher
Chlorotriangulum minutum Kufferath
Dunaliella lateralis Pascher & Jahoda
Order Chlorococcales (37 spp.)
Acanthosphaera zachariasi (Geitler) Lemmermann
Ankistrodesmus falcatus (Chodat) Lemmermann
Ankistrodesmus gelifactum (Chodat) Bourrelly
Ankyra ancora (G.M. Smith) Fott
Ankyra judai (G.M. Smith) Fott
Botryococcus braunii Kützing
Botryosphaera sudetica (Lemmermann) Chodat
Characium ornithocephalum A. Braun
Chlorella miniata (Nägeli) Oltmanns
Chlorella vulgaris Beijerinck
Chlorococcum wimmeri (Rabenhorst) Starr
Chodatella ciliata (Lagerheim) Lemmermann
Chodatella quadriseta Lemmermann
Coelastrum microsporum Nägeli
Coelastrum reticulatum (Dangeard) Senn
Crucigenia quadrata Morren
Crucigenia tetrapedia (Korchikoff) Bourrelly
Dictyosphaerum pulchellum Wood
Euastropsis richteri (Schmidle) Lagerheim
Micractinium pusillum (Lund) Fresenius
Oocystidium ovale Korchikoff
Oocystis lacustris Chodat
Pediastrum boryanum Meyer
Pediastrum clathratum (Schröder) Lemmermann
Pediastrum duplex Meyer
Pediastrum tetras (Corda) Rabenhorst
Selenastrum bibraianum Reinsch
Scenedesmus crassus Chodat
Scenedesmus falcatus Chodat

Appendix 1. Continued

Scenedesmus flexuosus (Lemmermann) Ahlstrom
Scenedesmus protuberans Fritsch & Rich
Scenedesmus tenuispina Chodat
Scenedesmus sp.
Sphaerocystis schroeteri Chodat
Tetraedron caudatum Lemmermann
Tetraedron limneticum (Skuja) Borge
Tetrastrum heteracanthum (Nordstedt) Chodat

Order Oedogoniales (3 spp.)
Oedogonium mammiferum Wittrock
Oedogonium minus Wittrock
Oedogonium sudanense Gautlièvre
Order Desmidiales (11 spp.)
Closterium acerosum (Schrank) Ehrenberg
Closterium acutum Brébisson
Closterium setaceum Grönblad
Cosmarium depressum Riverdin
Cosmarium meneghinii Brébisson
Gonatozygon monotaenium De Bary
Micrasterias tropica Nordstedt
Pleurodiscus africanus Bourrelly
Staurastrum lunatum Ralfs
Staurastrum sebaldi Krieger & Bourrelly
Staurodesmus dickei (Ralfs) Lillieroth

Phylum Euglenophyta
Order Euglenales (7 spp.)
Euglena polymorpha Dangeard
Euglena variabilis Klebs
Euglena viridis Ehrenberg
Phacus tortus (Lemmermann) Swirenko
Scytomonas pusillus Stein
Sphenomonas quadrangularis Stein
Strombomonas gibberosa (Playfair) Deflandre

Phylum Pyrrhophyta
Order Prorocentrales (2 spp.)
Gymnodinium inversum Nygaard
Gymnodinium sp.
Order Peridiniales (4 spp.)
Ceratium cornutum (Ehrenberg) Chapman & Bachmann
Ceratium hirundinella (O.F. Müller) Schrank
Peridiniopsis borgei Lemmermann
Peridinium volzii Lemmermann
Order Cryptomonadales (7 spp.)
Chroomonas acuta Ütermohl
Chroomonas minuta Skuja
Chroomonas rubra Geitler
Cryptomonas marssonii Skuja
Cryptomonas ovata Ehrenberg
Cryptomonas tetrapyrenoidosa Skuja
Tetragonidium verrucatum Pascher

Hydrobiologia **488:** 99–105, 2002.
D.M. Harper, R. Boar, M. Everard & P. Hickley (eds), Lake Naivasha, Kenya.

Nutrient control of phytoplankton production in Lake Naivasha, Kenya

David S. Hubble[1,*] & David M. Harper
Department of Biology, University of Leicester, Leicester LE1 7RH, UK
[1]*Ecological Monitoring & Research, 7 Ainsley Gardens, Eastleigh, Hampshire SO50 4NX, U.K.*
**Author for correspondence. E-mail: dshubble@yahoo.com*

Key words: nutrients, tropical, shallow, lake, eutrophication

Abstract

Lake Naivasha, a shallow tropical lake in Kenya's Rift Valley, has an unstable water column and is moderately eutrophic. Nutrient (bottom-up) control of primary production is more important than grazing (top-down) control. Experimental nutrient enrichment was used to investigate bottom-up control in more detail. Minor nutrients were not found to be limiting, whilst nitrogen was more limiting than phosphorus with an algal preference for ammonium over nitrate. Sediments form a phosphorus sink but there is hypolimnetic release from the one area showing regular temporary stratification. This indicates that the rate of primary production in the water column could double if conditions change to allow lake-wide nutrient release from sediments. Both external and recycled nutrient regeneration are important.

Introduction

Lake Naivasha is a shallow freshwater lake situated in the eastern Rift Valley of Kenya, lying within an enclosed basin at an altitude of approximately 1890 m. Rainfall is bimodal with peaks in April and October, but shows inter-annual irregularity and annual evapotranspiration exceeds rainfall. There is one main inflowing river (Malewa) and a number of smaller, seasonal ones, but no visible outflow, although it is generally believed to have an underground outflow keeping it fresh. In common with many shallow tropical lakes, Naivasha is generally well-mixed (wind and convection) without persistent stratification in the main basin (Hubble, 2000). This creates a situation where algal cells undergo continuous vertical circulation which determines the prevailing light regime under conditions of intense solar irradiance. Lake Naivasha also experiences fluctuating water levels (Vincent et al., 1979) which influence its area, limnology and hence productivity (Becht & Harper, 2002). Recent human-induced changes have increased nutrient concentrations, and this has affected phytoplankton biomass (Hubble & Harper, 2002), so that the lake can now be considered moderately eutrophic with a state of 'endless summer' (Kilham & Kilham, 1990).

Nutrient (bottom-up) controls have been considered more important than top-down controls in tropical lakes, with nitrogen more limiting than phosphorus in Lake Naivasha (Talling, 1965). Concern about the effects of eutrophication and increased exploitation of water resources leads to the need to understand the control of phytoplankton primary production in order to accurately predict the effects of any changes. The objectives of the present study were to understand the present controls on primary production in Lake Naivasha.

Methods

Three study stations were used on the lake; a main lake station Hippo Point (HP); Malewa rivermouth (MR) close to the main inflow; and Crescent Island lagoon (CI), a partially separated crater within the main lake.

Light-dark bottles were incubated *in situ* at the three stations, at depths of 0, 0.5, 1, 2, 3, 5 and 7 metres usually for 1 h and the oxygen before and after measured by a B.O.D. probe on a YSI Oxygen meter (Yellow Springs, Ohio). The accuracy and repeatability of the method was found to be comparable to the more normal titration methods (Hubble, 2000) but gave a distinct advantage in field measuring.

Experimental nutrient enrichment was used to investigate whether nutrient (bottom-up) control was more important than grazing (top-down) control. Experimental enrichments were made to incubation bottles, measuring productivity in mg C m^{-3} h^{-1}. Serial nutrient enrichment experiments were carried out in four different sets, each with a different nutrient. Four minor nutrients: silicon (as $Si(OH)_4$), manganese and chloride (as $MnCl_2$), iron and sulphate (as $FeSO_4$), and sodium (as Na_2EDTA) were tested. For each of these, a range of concentrations were added with the most enriched condition equating to the saturation level in Jaworski's medium. Four major nutrients: soluble reactive phosphorus (SRP), nitrate, ammonium and nitrogen (equal amounts of nitrate and ammonium added) were used. For each of these, samples were enriched with up to 10 μM of the relevant nutrient fraction prior to productivity incubation.

To investigate different nutrient pools, grazing dilutions were prepared and nutrients were added to produce three different conditions: +N+P (all nutrients added), –N (nitrogen as NH_4Cl omitted) and –P (phosphorus as KH_2PO_4 omitted). These experiments were run in pairs with one having the whole plankton community diluted, and the other only having the grazers diluted. Using productivity measurements to measure apparent growth rate, the contributions of 'Internal', 'External' and 'Recycled' nitrogen and phosphorus nutrient pools were calculated for each nutrient condition from the equation used by Andersen et al., (1991):

$$m(x) = {}^{1}/_{t}\cdot\ln[P_{(xt)}/P_{(xo)}],$$

where $m(x)$ = the apparent specific growth rate at dilution level x, t = time (h), $P_{(xo)}$ = initial productivity at dilution level x, $P_{(xt)}$ = productivity at dilution level x at time t.

Results

The minor nutrient series showed no response of productivity to enrichment, indicating that none of these nutrients were limiting (Fig. 1). This is especially important for silicon, as silicaceous diatoms are an important, often dominant, part of the algal assemblage in Lake Naivasha (Hubble & Harper, 2002). The major nutrients, nitrogen (separately as nitrate and ammonium) and phosphorus (as soluble reactive phosphate), produced a clear increase in primary productivity, with nitrate and ammonium increasing productivity more than SRP ($p = 0.005$ and $p = 0.003$ respectively) (Fig. 2).

There was an overall algal preference for ammonium over nitrate shown as higher productivity with ammonium enrichment ($p = 0.001$) as the latter requires more energy and the use of nitrate reductase. Productivity reached a plateau with enrichment of approximately 1 μM as nutrient levels became saturated given the algal population present.

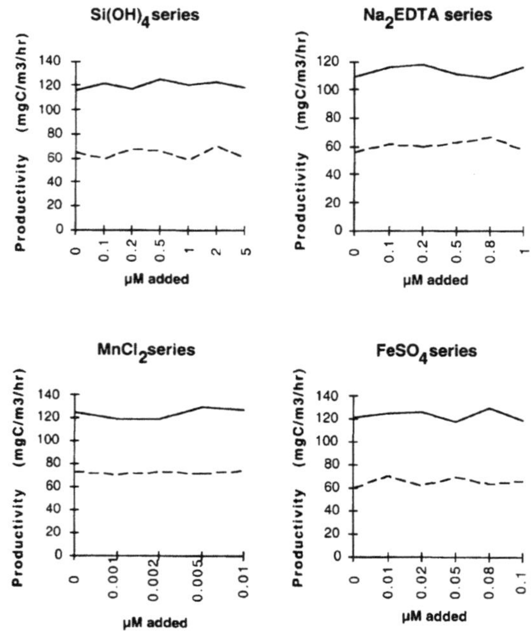

Figure 1. Response of primary productivity to enrichment of minor nutrients.

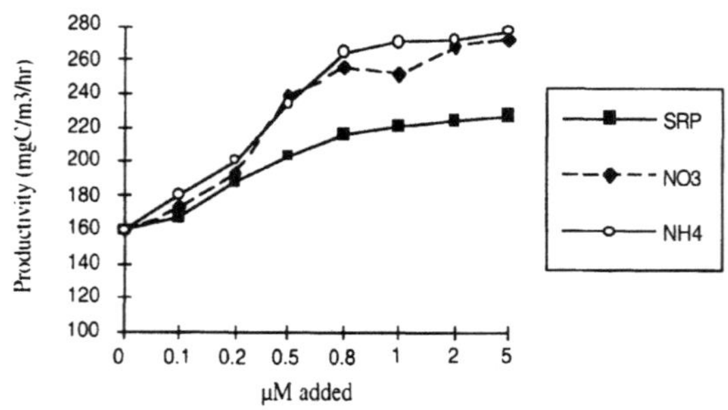

Figure 2. Response of primary productivity of surface lakewater to enrichment of major nutrients.

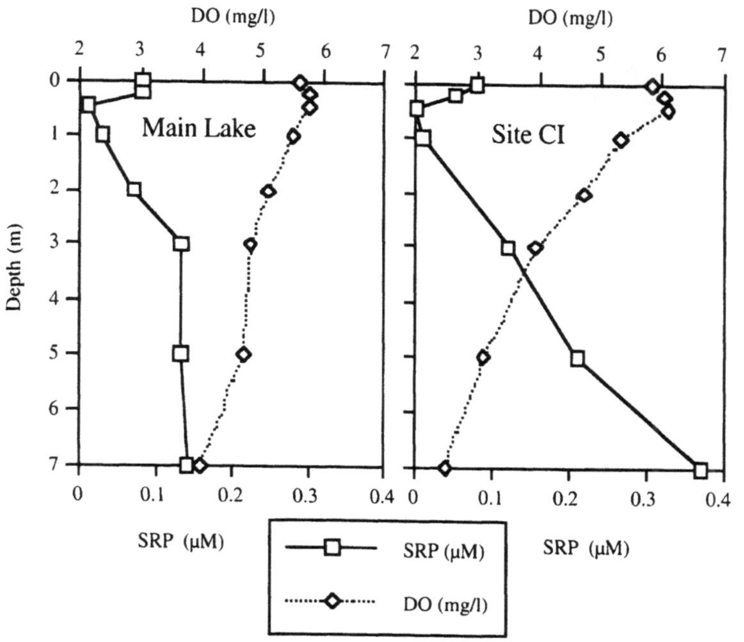

Figure 3. SRP & DO profiles at the main lake and Crescent Island lagoon, 21/10/98.

Experimental enrichments gave productivity responses under artificial nutrient conditions. The sediments form a sink for phosphorus (Kitaka et al., 2002), but the lake is well mixed (Hubble, 2000) and does not deoxygenate enough to release this store of nutrients. However, in the small Crescent Island lagoon, where the lake lies in a smaller volcanic crater, it is approximately 20 m deep and does stratify (Hubble, 2000). Here, hypolimnetic deoxygenation did occur (Fig. 3) and phosphorus was then released from the sediments, a process not seen in the main lake.

The main lake and the inflow showed little difference in the contribution of external (E) nutrients but the remaining productivity was mostly fuelled by recycled (R) nutrients at the inflow. In both cases there was little difference between nitrogen- and phosphorus-limited conditions with the fully enriched condition showing a greater contribution of internal (I) nutrients (Fig. 4).

In Crescent Island lagoon there was little difference between the contributions of nutrient pools under the three experimental conditions, whereas both the main lake and the inflow showed reduced 'Internal' contribution under the nutrient-limited -N and -P conditions, but increased 'Internal' contribution under the +N+P condition. Changes in the contributions of the three nutrient pools occurred through variations in 'Internal' nutrient supply as the 'Recycled' and 'External' contributions remained equal (though variable) at all sites and under all conditions. For the +N+P condition, there was no significant variation between sites or pools, although inter-pool variation is greater ($p = 0.112$) than inter-site variation ($p = 0.839$). This is similar in both the –N condition ($p_{\text{pool}} = 0.077$ and $p_{\text{site}} = 0.816$) and –P condition ($p_{\text{pool}} = 0.132$ and $p_{\text{site}} = 0.764$).

Variation between conditions (for both absolute and percentage contributions) is indicated by two-tailed paired-sample *t*-tests at $p \leq 0.05$ for significance. There is a significant difference between +N+P and –N conditions for all cases except the 'Recycled' nutrient pool, which does not show a difference between any two conditions, as the rate of nutrient recycling by planktonic primary producers is extremely rapid, requiring only small quantities of nutrient unless a balancing nutrient is in excess, causing increased limitation. Hence the percentage contribution of the 'Recycled' nutrient pool does show a significant difference between conditions +N+P and both –N and –P, but not between –N and –P separately as both con-

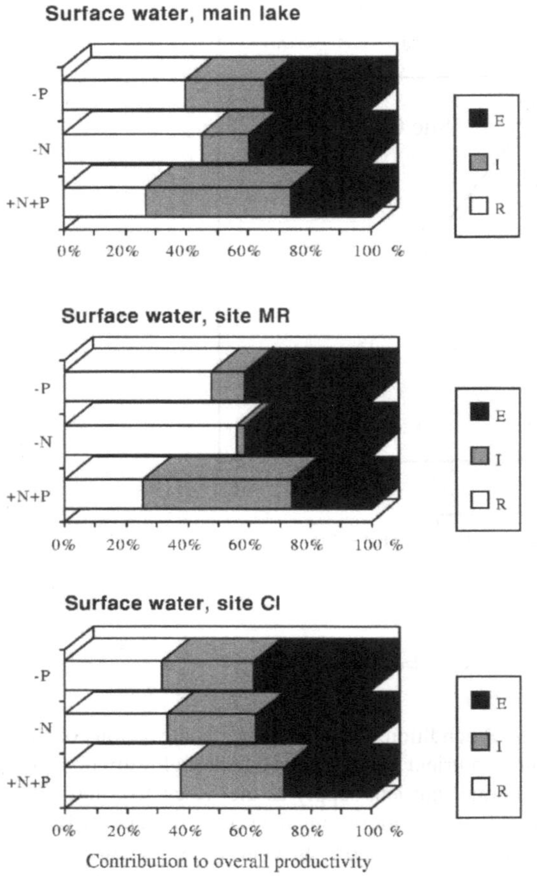

Figure 4. Separation of nutrient pool contributions to productivity in the main lake, Malewa River inflow (MR).

ditions experience opposite nutrient limiting effects. All other cases show significant differences except the percentage contribution of the 'Internal' nutrient pool between –N and –P, and absolute contribution of the 'External' pool between +N+P and –P. Considering each combination of two pools, significant differences are seen between pool contributions to productivity for all pairs, sites and conditions except between the 'Recycled' and 'External' nutrient pools in the main lake for the fully enriched +N+P condition. No significant inter-site differences are revealed. Crescent Island lagoon depth profiles (Fig. 5) show a peak of 'Internal' productivity at 0.5 m depth for the fully enriched and phosphorus-limited conditions. The nitrogen-limited condition shows a peak of 'Recycled' productivity at

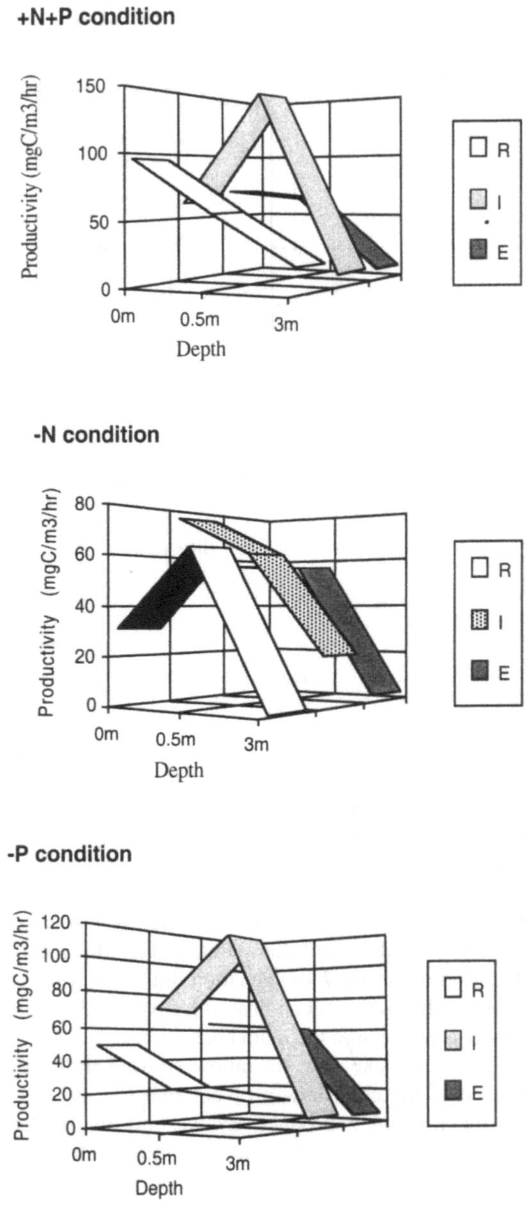

Figure 5. Crescent Island lagoon profiles of nutrient pool contributions to productivity.

this depth although it is no higher than that of the other pools.

Separation ofnutrient pools indicates that both 'External' and 'Recycled' regeneration are important, although further analysis will be required to link this to energy flows through classic and microbial food webs. Analysis of inter-pool and inter-site differences between nutrient pools indicate that the variations in contribution of 'Recycled', 'Internal' and 'External' nutrient pools are not largely due to spatial heterogeneity within the lake despite differences in nutrient regime, hydrological conditions and sometimes overall productivity levels. Main lake and Malewa inflow sites showed little difference in the contribution of external nutrients but the remaining productivity was mostly fuelled by recycled nutrients at the Malewa inflow. This suggests that the effects of inflow is spread across the lake, and not merely confined to the river mouth. In both cases there was little difference between nitrogen- and phosphorus-limited conditions. Serial nutrient enrichment showed nitrogen to be more limiting than phosphorus and that ammonium was generally preferred over nitrate as a form of usable nitrogen. Productivity increase through nutrient enrichment varied between experiments but tended to reach a plateau at higher nutrient levels once saturation occurred, at which point a near doubling of productivity was often shown. This indicates the potential importance of nutrient release from sediments. This nutrient release does not appear, however, to have affected productivity either locally in Crescent Island lagoon or in whole lake terms, especially as any nutrient-rich waters from below 3 m in the lagoon would have been greatly diluted once transported elsewhere by currents in the lake. Even within the lagoon itself (diameter 1.3 km, depth 18 m) the water below 3 m only constitutes approximately 57% of the total lagoon volume (0.008 km^3) and thus this relatively nutrient-rich water would have been diluted by half if the lagoon mixed. Phosphorus resupply from sediments therefore occurs, but due to the lack of permanent stratification, is less important than water-column recycling.

Lake Naivasha underwent a 3 m depth increase during late 1998 and early 1999, and if this deepening continued, or wind-driven mixing became reduced, stratification could form more widely and with greater stability (Hubble, 2000). If this occurs, the resulting phosphorus release could accelerate eutrophication and increase productivity throughout the lake. There are a number of pressures upon the lake affecting the nutrient status (Johnson et al., 1998) including the development of intensive agriculture in the catchment right down to the lake edge (which removes the buffering papyrus fringe) and small-scale farming along the River Malewa increasing nutrient runoff.

Major nutrient input is due largely to riverine transport, with flow and nutrient load regulated by climatic and anthropogenic factors. The rivers Gilgil and Karati are seasonal depending on rains and show much smaller volumes of flow, but they may transport more concentrated materials than the Malewa (Hubble, 2000). Sediment is transported into the lake with a noticeable underflow (Hubble, loc cit) beneath the main lake surface waters as particles sink whilst still flowing away from the river mouth, especially during high rainfall – erosivity periods. Increased SRP levels lead to greater reduction in nitrate concentrations as phosphate enrichment occurs and nitrogen becomes limiting. The results of the experiments showing that the nitrate series correlated less closely with changes in SRP than changes in NH_4 indicate that the increased availability of nitrate due to enrichment outweighs some of the algal preference for ammonium. The concentrations of SRP in the lake indicate that, as well as differences in processes above and below the oxycline, there are processes causing variations in conditions within mixed layers. For instance at approximately 0.5–1 m depth there is maximal photosynthesis (Hubble, 2000) causing depletion of nutrients, often to zero or undetectable levels. Above and below this region, photosynthesis (and hence nutrient depletion) is reduced by photoinhibition and light attenuation respectively.

Discussion

Nutrient pool dynamics are more important in tropical lakes, especially in shallow waters, than temperate. Minor nutrients are not limiting, but may show variation with depth due to differential use by algae (Hubble, 2000). For example, silicates are partially depleted in areas of high diatom productivity (Talling, 1966), although this depletion is not limiting (Talling, 1993). Neither nitrate nor ammonium levels showed much difference between sites, remaining low (<20 μM) from the surface to 2 m depth, with a small steady increase below this as there was reduced algal activity and nutrient uptake due to light attenuation (Hubble, 2000). This effect outweighs that of mixing as the rates of nutrient uptake and cycling are greater than those of hydrological turnover (Hubble, 2000). As nitrogen fractions are transported in solution, not bound to sediment particles, any actual increase in nitrogen ions with depth (i.e., not merely a lack of algal utilisation) is likely to be due to settling of organic material, or release of dissolved organic material from the lake sediments.

Much of the former papyrus fringe has been cleared from the northern shore (Harper et al., 1990), and so there is reduced buffering and consequently a gradient of reduced nutrient concentration moving away from the northern inflow of the River Malewa. In shallow water, wind, waves and currents lead to repeated resuspension of fine material (Knighton, 1998) and may contribute to concentrations at a given point, such as a *Ceratium* mass (122.5 μg L^{-1} chl *a*) at the Malewa inflow in May 1998 (personal observation). Once particles have settled, variations in fetch and depth affect wave generation and hence the resuspending shear stress at the lakebed (Giller et al., 1992). Reduced oxygen levels are found as sampling profiles meet the sediment–water interface (personal observation), however temperature reduction is sometimes detected without recording deoxygenation. Transects (Hubble, 2000) have shown that this occurs when an underflow of oxygenated water from the Malewa separates some relatively anoxic water from the lakebed which also has the effect of reducing temperature. These effects combine to partially determine vertical and horizontal variation in limnological parameters. The differences between enrichment series are seen at both low and saturated concentrations, at all sites and depths, and over time. This indicates that nutrient limitation is a major productivity control throughout the different conditions found in shallow tropical freshwater, assuming that Lake Naivasha is typical of such systems. Increased nutrient inputs both from changes in catchment land use via river inflow and sediment release therefore have the potential to greatly increase algal productivity and accelerate the processes of eutrophication. Although sediment phosphorus release currently occurs only in Crescent Island lagoon, if the lakewater becomes less oxygenated, there is potential for phosphorus to be released throughout the main lake basin. This situation may be more likely if nutrient inputs from the catchment continue to be enhanced by human activity and the lake becomes more eutrophic. Such concerns are applicable to many tropical lakes which are also vital water resources, therefore consideration of 'bottom-up' productivity controls may be an essential management tool.

Acknowledgements

This study formed past of the work of the University of Leicester research project at Lake Naivasha, which since 1987 has been authorised by the Office of the President, Government of Kenya under research permit to Dr D. M. Harper no. OP 13/001/12C 46. The project was funded by the Earthwatch Institute, Boston USA and Oxford, England and the studentship to the first author by the BBSRC as a CASE award with the Elsa Trust. The data collection would not have been possible without the assistance of numerous Earthwatch volunteers or the facilities at the erstwhile Elsamere Conservation Centre. Our sincere thanks go to the numerous colleagues in Kenya for logistical help, in particular Jill & Angus Simpson and Velia Carn, and Chris Campbell Clause.

References

Andersen, T., A. K. Schartau & E. Paasche, 1991. Quantifying external and internal nitrogen and phosphorus pools, as well as nitrogen and phosphorus supplied through remineralisation, in coastal marine plankton by means of a dilution technique. Mar. Ecol. Prog. Ser. 69: 67–80.

Elser, J. J. & D. C. Frees, 1995. Microconsumer grazing and sources of limiting nutrients for phytoplankton growth. Application and complications of a nutrient deletion/dilution gradient technique. Limnol. Oceanogr. 40(1): 1–16.

Giller, P. S., A. G. Hildrew & D. G. Raffaelli (eds), 1992. Aquatic Ecology: Scale, Pattern and Process. Blackwell Science, Oxford.

Gouder de Beauregard, A.-C., D. M. Harper, F. Malaisse & J.-J. Symoens, 1998. Dynamique récente et cartographie de la végétation aquatique (1960-1996) du lac Naivasha (Rift Valley, Kenya). Bull. Séanc. l'Acad. r. Sci. Outre-Mer 44(3): 373–389.

Harper, D. M., C. Adams & K. Mavuti, 1995. The aquatic plant communities of the Lake Naivasha wetland, Kenya: pattern, dynamics and conservation. Wetlands Ecol. Manage. 3(2): 111–123.

Harper, D. M. & K. Mavuti, 1995. Freshwater wetlands and marshes. In McClanahan, T. R. & T. Young (eds), Ecosystems of East Africa and their Conservation. Oxford University Press, London.

Harper, D. M., K. M. Mavuti & S. M. Muchiri, 1990. Ecology and management of Lake Naivasha, Kenya, in relation to climatic change, alien species' introductions, and agricultural development. Environ. Conserv. 17 (4): 328–36.

Hubble, D. S., 2000. Controls on primary production in Lake Naivasha, a shallow tropical freshwater. PhD thesis, Leicester University, UK.

Johnson, G., D. M. Harper & K. Mavuti, 1998. Information for the sustainable management of shallow lakes: Lake Naivasha, Kenya. In Harper D. M. & Brown T. (ed), The Sustainable Management of Tropical Catchments, Wiley, Chichester.

Kalff, J., 1983. Phosphorus limitation in some tropical African lakes. Hydrobiologia 100: 101–12.

Kilham, P. & S. S. Kilham, 1990. Endless summer: internal loading processes dominate nutrient cycling in tropical lakes. Freshwat. Biol. 23: 379–389.

Kitaka, N., D. M. Harper, & K. M. Mavuti, 2002. Phosphorus inputs to Lake Naivasha, Kenya, from its catchment and the trophic state of the lake. Hydrobiologia 488 (Dev. Hydrobiol. 168): 73–80.

Knighton, D., 1998. Fluvial Forms and Processes. Arnold, London.

Landry, M. R., J. Kirshtein & J. Constantinou, 1995. A refined dilution technique for measuring the community grazing impact of microzooplankton, with experimental tests in the central equatorial Pacific. Mar. Ecol. Progr. Ser. 120: 53–63.

Olila, O. G., K. R. Reddy & W. G. Harris, 1995. Forms and distribution of inorganic phosphorus in sediments of two shallow eutrophic lakes in Florida. Hydrobiologia 302: 147–161.

Talling, J. F., 1965. The photosynthetic activity of phytoplankton in East African lakes. Int. Rev. ges. Hydrobiol. 50: 1–32.

Talling, J. F., 1966. The annual cycle of stratification and phytoplankton growth in Lake Victoria (East Africa). Int. Rev. ges. Hydrobiol. 51: 545–621.

Talling, J. F., 1993. Environmental regulation in African shallow lakes and wetlands. Rev. d'Hydrobiol.e Trop. 25 (2): 87–144.

Vincent, C., T. Davies & A. Beresford, 1979. Recent changes in the level of Lake Naivasha, Kenya, as an indicator of equatorial westerlies over East Africa. Climate Change 2: 175–189.

Hydrobiologia **488:** 107–113, 2002.
D.M. Harper, R. Boar, M. Everard & P. Hickley (eds), Lake Naivasha, Kenya.

Impacts of the C_4 sedge *Cyperus papyrus* L. on carbon and water fluxes in an African wetland

M. B. Jones[1,*] & S. W. Humphries[2]
[1]*Botany Department, Trinity College, University of Dublin, Dublin 2, Ireland*
[2]*Department of Biological Sciences, University of Essex, Colchester, U.K.*
**Author for correspondence. E-mail: mike.jones@tcd.ie*

Key words: Africa, *Cyperus papyrus* L., carbon balance, C_4 photosynthesis, evapotranspiration, wetlands

Abstract

Fluxes of CO_2 and H_2O vapour were measured by eddy covariance from a stand of the C_4 emergent sedge *Cyperus papyrus* (papyrus), which formed a fringing swamp on the north-west shore of Lake Naivasha, Kenya. The fluxes of CO_2 and H_2O vapour between the papyrus swamp and the atmosphere were large but variable, depending on the hydrology of the wetland system and the condition of the vegetation. These measurements, combined with simulation modelling of annual fluxes of CO_2, show that papyrus swamps have the potential to sequester large amounts of the carbon ($\sim$1.6 kg C m^{-2} y^{-1}) when detritus accumulates under water in anaerobic conditions, but they are a net source of carbon release to the atmosphere ($\sim$1.0 kg C m^{-2} y^{-1}) when water levels fall to expose detritus and rhizomes to aerobic conditions. Evapotranspiration from papyrus swamps (E) was frequently lower than evaporation from open water surfaces (E_o) and plant factors have a strong influence on the flux of water to the atmosphere. For the period of measurement E/E_o was 0.36.

Introduction

Papyrus (*Cyperus papyrus* L.) forms the dominant emergent vegetation in most permanently flooded wetlands of tropical Africa (Hughes & Hughes, 1992). There are no accurate records of the area covered by papyrus-dominated swamps but one estimate puts it at about 4000 km^2, most of which is in central and east Africa (Thompson & Hamilton, 1983). Papyrus is remarkable because it is one of the largest herbaceous species, with culms growing to a height of 5 m and above-ground standing phytomass which often exceeds of 12.0 t C ha^{-1} (Jones & Muthuri, 1985). The culms are topped by characteristically large, spherically shaped, reproductive umbels that are also the main photosynthetic surface. The umbel is typically about 50 cm in diameter and consists of several hundred cylindrical rays, each of which extends into three to five flattened (leaf-like) bracteoles.

Papyrus, like other productive tropical emergent wetland species such as *Echinochloa polystachya* and *Paspalum repens*, uses the C_4 photosynthetic pathway (Jones, 1987b). The wetland environment appears an unlikely habitat for C_4 species, which are generally assumed to be adapted best to hot and dry locations (Doliner & Joliffe, 1979). However, the high efficiencies of use of radiation, water and nitrogen by C_4 species is hypothesised to have an important impact on the functioning of wetlands dominated by C_4 species such as papyrus (Knapp & Medina, 1999). Two functions of ecosystems which are influenced strongly by the presence of C_4 characteristics are the CO_2 balance between carbon gains in photosynthesis and losses in respiration, and H_2O vapour losses in evapotranspiration.

Living papyrus vegetation often overlies several metres depth of detritus or peat (Jones & Muthuri, 1997). This peat forms in the oxygen depleted environment below established floating rhizome mats as detritus accumulates from the dead and decomposing papyrus vegetation. Few estimates have been made of the amount of peat below papyrus but it is likely that African swamps could form a significant sink for carbon when continuously inundated and carbon gains in photosynthesis exceed carbon losses in respiration (Jones & Muthuri, 1997). Conversely, when the detritus is exposed to the atmosphere, and aerobic conditions, by a hydrological draw-down or drainage then these systems may become a net carbon source for the atmosphere as carbon losses exceed carbon gains.

It is clear that evapotranspiration by emergent vegetation can have a significant impact on the hydrology of wetlands and associated water bodies. However, the question as to whether the presence of aquatic vegetation leads to an increase or decrease in evaporative losses from water bodies has been a topic of debate for many decades, and there still appears to be no clear answer (Idso & Anderson, 1988). Anderson & Idso (1987) conclude that a major factor influencing the rate of evapotranspiration is the height of the vegetation, so that extensive canopies of tall emergent macrophytes such as the grass *Typha latifolia* do tend to increase evaporative loss of water from large water bodies, whereas extensive canopies of short emergent vegetation such as the floating macrophyte *Eichhornia crassipes* tend to reduce evaporative losses from such surfaces. However, these conclusions do not take account of any physiological differences between vegetation types, in that C_4 species have higher water use efficiencies than the C_3 species considered above (Long, 1999). It might be anticipated therefore that evaporative water loss from canopies of C_4 vegetation, like papyrus, would be lower than that from C_3 vegetation of a similar stature. The only previously reported measurements of evapotranspiration from papyrus were made by Rijks (1969), who found that an old stand of papyrus in Uganda lost, on average, 40% less water to the atmosphere than did the open lake.

The aim of the present study was to quantify the effect of papyrus vegetation on two important ecosystem functions viz. carbon balance and evaporation. These processes were measured using the Eddy Covariance (EC) technique and the data were used in conjunction with a 'bottom-up' model of canopy gas exchange which could predict annual cycles of carbon dioxide and water vapour exchange. The measurements enabled us to calculate the strength of papyrus swamps as net sinks for carbon when inundated with water and net sources for carbon when experiencing a draw-down of water levels. The evaporation measurements aimed to provide an answer to the question "does the presence of emergent aquatic vegetation tend to increase or decrease evaporative water losses from water bodies?" (Idso & Anderson, 1988).

Materials and methods

Site description

The experimental work was carried out on a lake-fringing papyrus swamp on the N-W shore of Lake Naivasha, Kenya (0° 45′ S, 36° 20′ E) at an altitude of 1890 m in the Eastern Rift Valley of Africa. The flora and ecology of the lake edge have been described by Gaudet (1977). The papyrus mat at the study site extended 500 m into the lake where it was floating, but it was grounded at the lake margin. During the period of measurements (1995–96), the lake was going through one of its natural draw-downs (Gaudet, 1977; Becht & Harper, 2002). The effect of this was to expose an increasing amount of the rhizome mat and associated detritus to the atmosphere, although the root systems remained continuously waterlogged.

Climate

The equatorial climate at Lake Naivasha is marked by two rainy season peaks in April//May and November, although irregularities from this pattern are common, since rainfall is quite erratic. The average annual rainfall is 620 mm. Ten year, monthly mean records of maximum, mean and minimum temperatures and solar radiation were obtained from a weather station located 5 km from the study site. These have been reported by Muthuri (1985) and Muthuri et al. (1989). Shorter term meteorological measurements were made using an automated weather station (AWS), and data logger (Campbell Scientific Ltd., Leicestershire, U.K.). The weather station was situated on an open site about 50 m from the landward edge of the papyrus and approximately 600 m from the location of the Eddy Convariance (EC) tower. The data average recording interval was set to 15 minutes to coincide with the EC measurements.

Evaporation rates from the lake open water and evapotranspiration from the papyrus canopy were calculated using the Penman formula (Penman, 1948). Meteorological data, collected by the AWS, were used to calculate 15 min averages throughout the day.

CO_2 and H_2O ecosystem measurements

Ecosystem fluxes of CO_2 and H_2O were measured using the EC technique (Moncrieff et al., 1997a, b). The observations were taken over a 4 week period in August 1995 and a 1 week period in March 1996. These periods corresponded with, respectively, the coolest (mean monthly temperature of 15.0 °C) and warmest (mean monthly temperature of 17.9 °C) times of the year. The main components of the EC system were a three-axis symmetrical sonic anemometer (Solent A1012r, Gill Instruments, Lymington, U.K.), an infrared gas analyzer (IRGA) (Li-6262, Li-COR, Lincoln, Nebraska, U.S.A.) in closed path mode and custom written software (Eddy Sol, University of Edinburgh, U.K.) which calculated the fluxes in real time. A full description of the equipment and Eddy Sol software is given by Grace et al. (1995) and Moncrieff et al. (1997a, b).

EC measurements were made by sampling from a mast at a height of 7 m above ground level with air samples drawn down at 6 l min^{-1} down to the IRGA. The adjustable mast was located on a concrete base at a distance of 150 m from the landward edge of the papyrus swamp. The sampling point was 2–3 m above the tallest papyrus culms and gave a flux footprint in which 95% of the footprint area was contained within the region of homogenous papyrus canopy (Schuepp et al., 1990). Averaged fluxes were recorded at 15 min intervals. Span calibration of the CO_2 and H_2O channels of the IRGA was performed at a maximum of 3 day intervals using chemical absorption columns for zero values and a compressed gas source of CO_2 at 600 ± 10 μmol mol^{-1} (Linde Gas, U.K.) and a portable dewpoint generator (Li-610, Li-COR) for span values.

Respiration measurements

Tissue respiration of papyrus was estimated from measurements of respiration rates of material from each of the culm age classes which included juvenile, mature, senescent and standing dead (Muthuri & Jones, 1997). Respiration rates were determined by modifying a 100 mm diameter soil respiration chamber (SRC-1, PP Systems, Hertfordshire, U.K.), used in conjunction with an IRGA (CIRAS, PP Systems), by fixing an air-tight plate to the base of the chamber. This allowed the insertion and sealing of cut culm sections into the chamber and determination of respiration rate by monitoring the rate of increase of CO_2 within the chamber (Parkinson, 1981). Respiration on a dry matter basis was calculated after oven-drying the samples to constant weight.

The respiration rate of papyrus detritus was determined on a ground area basis using the PP Systems soil respiration chamber. Dry and dead papyrus umbels and culms were first cleared from the sampling area and the chamber was pushed firmly down approximately 10 mm into the detritus so that the stainless steel perimeter ring was partially embedded to ensure an airtight seal. Measurements were made at 20 locations around the study site and they were repeated three times at each location. Detritus temperature, measured with a mercury in glass thermometer, was 18 °C at the time of the respiration measurements which were made in the middle to late afternoon. It should be noted that because the detritus contained living root and rhizome material, the measurements made using the chamber included respiration from both detritus and root/rhizome fractions.

Simulation modelling

WIMOVAC (Windows Intuitive Model of Vegetation response to Atmosphere and Climate) is a process-based model of leaf physiology and canopy structure (Humphries & Long, 1995), details of which can be found at the internet site http://www.life.uiuc.edu/plantbio/wimovac/model.htm. This 'bottom-up' model has been modified and parameterised to predict the CO_2 and H_2O exchange of the papyrus canopy (Humphries et al., submitted). This has been done by incorporating the C_4 characteristics of papyrus bracteole photosynthesis and modelling radiation distribution in the papyrus canopy. The model was run to simulate annual cycles of CO_2 exchange under (i) the prevailing climatic conditions during the EC measurements, when the swamp was experiencing a slow draw-down in lake level of about 2 m over about 7 years, which would expose rhizomes and detritus to the atmosphere, and (ii) conditions when the rhizomes and detritus were inundated with water.

Results

Carbon dioxide flux

Eddy Covariance data for 21 days during the periods 4–21 August 1995 and 11–17 March 1996, when continuous EC and meteorological data were collected at 15 min intervals throughout the whole diurnal cycle, were used in the following analysis. The averaged diurnal flux for the whole 21 days is shown in Figure 1(i). Shown also are flux measurements predicted using the WIMOVAC 'bottom-up' model parameterised for papyrus (Humphries et al., submitted). Model CO_2 flux predictions show a close conformity

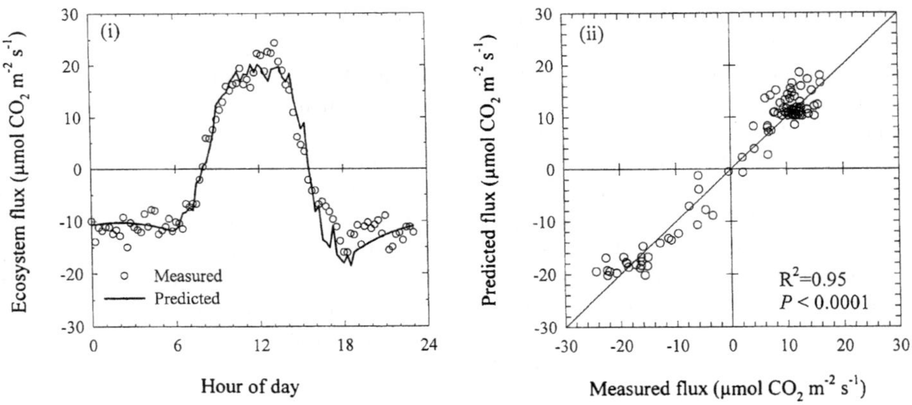

Figure 1. (i) Averaged diurnal cycle of ecosystem CO_2 fluxes from 21 days of Eddy Covariance measurements and predicted (modelled) ecosystem CO_2 fluxes for a stand of papyrus vegetation at Lake Naivasha and (ii) relationship between measured and predicted flux values ($r^2 = 0.95$ $p < 0.0001$). The continuous line shows a 1:1 relationship.

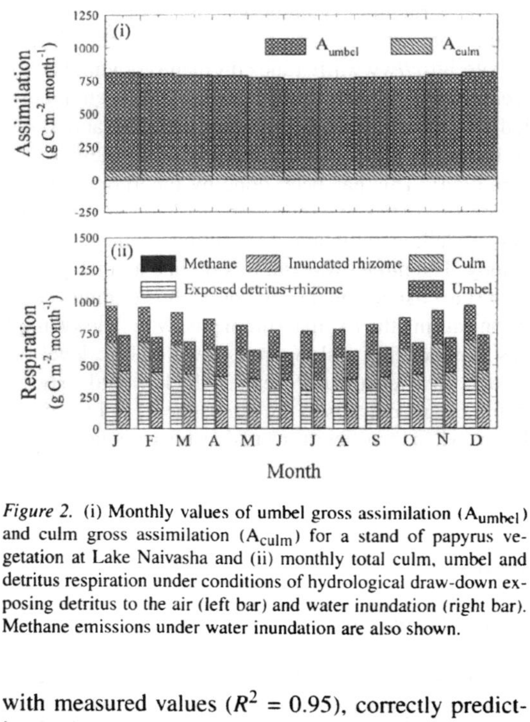

Figure 2. (i) Monthly values of umbel gross assimilation (A_{umbel}) and culm gross assimilation (A_{culm}) for a stand of papyrus vegetation at Lake Naivasha and (ii) monthly total culm, umbel and detritus respiration under conditions of hydrological draw-down exposing detritus to the air (left bar) and water inundation (right bar). Methane emissions under water inundation are also shown.

with measured values ($R^2 = 0.95$), correctly predicting both the magnitude and dynamics of the CO_2 flux (Fig. 1(ii)). Maximum CO_2 gain (24.3 μmol m^{-2} s^{-1}) occurred at approximately 13.00 h and was associated with the highest light intensity and slightly below maximum air temperature. Maximum CO_2 loss (-16 μmol m^{-2} s^{-1}) occurred at 18.00 h and was associated with the low light intensity and high temperatures occurring shortly before sunset (Figure 1(i)).

Table 1. Papyrus ecosystem dry matter distribution and respiration rates

Papyrus ecosystem component	Dry matter (g C m^{-2})	Specific respiration (mg C g^{-1} h^{-1})	Ground area based respiration (g C m^{-2} h^{-1})
Culm	720	0.59	0.424 ± 0.04
Umbel	323	1.28	$0.412 \pm$ n.a.
Rhizome & root	1480	0.10	0.146 ± 0.01
Detritus	64000	n.a.	0.233 ± 0.02

Using monthly mean meteorological data, the WIMOVAC model of canopy photosynthesis was used to predict monthly gross carbon gain in photosynthetic assimilation (A) (Fig. 2(i)). Carbon loss in respiration was derived from measurements of specific respiration of culms, umbels and rhizomes combined with detritus respiration measurements calculated on a ground area basis (Table 1). Specific respiration rates were scaled to canopy values and expressed on a ground area basis from a knowledge of dry matter distribution in the canopy (Muthuri et al., 1989). Monthly values of carbon loss in respiration from culms, umbels and detritus and methanogenesis from detritus (see Jones & Muthuri, 1997) are shown in Figure 2(ii). This was done assuming contrasting conditions of inundation of the rhizomes and detritus by water. The first was when the rhizomes and detritus were partly exposed to the atmosphere, as in the case of the field measurements reported here, when there was a net ecosystem loss of carbon. The second was when the rhizomes and detritus are assumed to be inundated with water. Under these conditions, which rapidly become anaerobic, detritus respiration is strongly inhibited and there is a net ecosystem gain of carbon.

Water vapour fluxes

Eddy Covariance measurements of water vapour flux from the papyrus swamp indicate little loss of water through evaporation at night. During the day, evapotranspiration rose to a peak of about 5.8 mmol H_2O m^{-2} s^{-1} during the early afternoon (Fig. 3). Modelled estimates of canopy evapotranspiration using the Penman equation including stomatal and boundary layer resistance terms gave a close agreement with the measured values, although they tended to underestimate measured values in the morning and overestimate measured values around midday (Fig. 3). The pattern of evaporation from the open lake, calculated using the Penman Equation, showed a similar pattern to the papyrus canopy evapotranspiration (Fig. 3) although there was evaporation of about 2 mmol H_2O m^{-2} s^{-1} throughout the night and evaporation in the middle of the day and into the afternoon exceeded canopy evapotranspiration by as much as 4 mmol H_2O m^{-2} s^{-1}. The averaged total daily flux of papyrus canopy evapotranspiration (E) was 2.4 kg H_2O m^{-2} dy^{-1} while open water evaporation (E_o) was 6.6 kg H_2O m^{-2} d^{-1} so that the ratio of transpiration to evaporation (E/E_o) was 0.36.

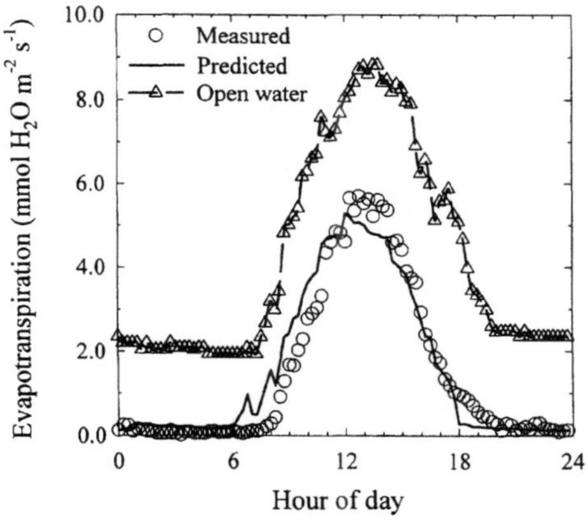

Figure 3. Averaged diurnal cycle from 21 days of Eddy Covariance measurements (circles) and predicted (modelled) values (continuous line) of evapotranspiration for a stand of papyrus vegetation at Lake Naivasha, and open water Penman evaporation (triangles).

Discussion

Emergent macrophytes in swamps, marshes and floodplains form some of the most productive plant communities (Jones, 1987a). We have previously shown, from measurements of plant biomass dynamics, that papyrus on Lake Naivasha has an above-ground net primary production (NPP) of 2.51 kg C m^{-2} yr^{-1} (Muthuri et al., 1989) and Piedade et al. (1991) found that the grass *Echinochloa polystacha*, in the central Amazon region, has a productivity of 3.97 kg C m^{-2} yr^{-1}. It is probably no coincidence that both these species have C_4 photosynthesis as their NPPs are amongst the highest ever recorded for natural communities. These observations suggest that under conditions where water and nutrients are non-limiting, communities of C_4 species have the potential to fix some of the largest amounts of carbon on a ground area basis. Unlike *E. polystacha*, papyrus does not show a marked seasonal pattern of growth, and in fact the standing biomass is remarkably stable at approximately 1.5 kg C m^{-2}, but it has a high annual aerial turnover rate of ~1.7 (NPP divided by standing biomass). The fluxes of carbon in the papyrus swamp at Naivasha, calculated on the basis of the EC measurements and model simulations, are shown in Figure 4. The fluxes are presented for two conditions. In the first case, they show carbon fluxes when the detritus in the swamp is exposed at a time of draw-down (Fig. 4(i)). This was the state of the swamp during the measurements reported here. In the second case, the rhizome and root system is inundated with water (Fig. 4(ii)). Under these circumstances, an oxygen-depleted environment develops rapidly in the water and rates of CO_2 release from detritus decomposition are very low. Measurements of gaseous release by ebullition show that total C release (CO_2 and CH_4) is 0.062 kg C m^{-2} y^{-1} (Jones & Muthuri, 1997) whereas the exposed detritus flux under aerobic conditions is 4.12 kg C m^{-2} y^{-1}. Papyrus wetlands therefore have the potential to be carbon sources or carbon sinks, depending on the hydrological status of the wetlands. The potential strength of the papyrus wetlands carbon sink in Africa is particularly important in relation to the identification and quantification of carbon sinks in the tropics (Schimel, 1995).

The question of whether the presence of emergent aquatic vegetation tends to increase or decrease evap-

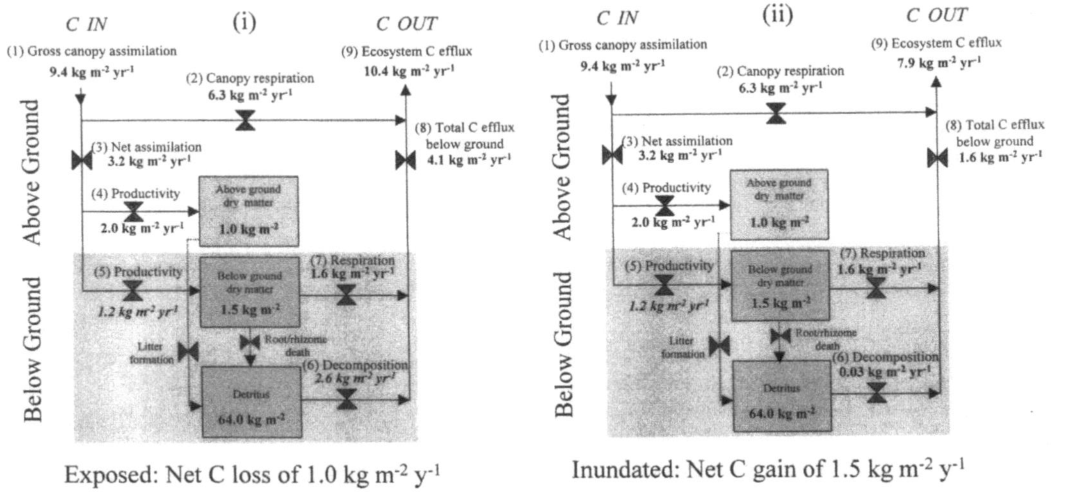

Figure 4. Pools and fluxes of carbon for a papyrus ecosystem (i) when detritus and rhizomes are exposed to the atmosphere under conditions of a hydrological draw-down and (ii) when detritus and rhizomes are inundated by water. Rectangles represent quantities, valve symbols represent flows. Values are expressed in terms of carbon mass on a ground area basis. *Notes*: (1) Gross canopy assimilation predicted by the WIMOVAC model. (2) Canopy respiration predicted by scaling measured values. (3) Net assimilation is predicted from gross assimilation minus canopy respiration. (4) Measured above ground productivity is from Muthuri et al. (1989). (5) Below ground productivity is net assimilation minus above ground productivity. (6) Detritus decomposition in (i) is total detritus carbon flux minus measured and scaled rhizome respiration and in (ii) is measured efflux of CO_2 and CH_4 from Jones & Muthuri (1997). (7) Respiration is scaled measurement of rhizome respiration. (8) Total efflux is below ground respiration plus decomposition. (9) Ecosystem efflux is canopy respiration plus below ground efflux.

orative water loss from wetland areas has been a topic of intense debate for many decades (Idso & Anderson, 1988). However, Idso (1981) pointed out that most data suggestive of a vegetation-induced evaporation increase had been derived from experiments with identical pairs of small-water tanks (with or without vegetation) exposed to the open air. Decreases in evapotranspiration, on the other hand, were suggested by similar experiments which placed the tanks of plants within an extensive natural stand of the vegetation. Relatively few measurements appear to have been made using micrometeorological methods, but those that have generally confirm that swamp evaporation is less than lake evaporation (Rijks, 1969; Linacre et al., 1970). The fact that papyrus is a C_4 species is also of relevance in relation to evapotranspiration. Hasegawa (1977) demonstrated that, in general, the transpiration rate of C_3 plants was higher than C_4 plants and O'Toole & Tomar (1982) supported this view when they demonstrated that, when grown under identical conditions and at a similar leaf area index, C_4 *Echinochloa crus-galli* L. (barnyard grass) transpired about half as much water as C_3 *Oryza sativa* L. (rice).

In conclusion, we have demonstrated that the presence of papyrus with its C_4 characteristics has a major impact on the functioning of tropical wetland ecosystems. The high rates of CO_2 fixation and primary production result in large fluxes of CO_2 which can lead, under suitable hydrological conditions, to correspondingly high rates of carbon sequestration. The C_4 characteristics of papyrus mean that high rates of CO_2 assimilation can occur while, relative to C_3 vegetation, evapotranspiration is comparatively low. Under these circumstances, water loss by evaporation from papyrus vegetation is considerably less than from corresponding area of open water.

Acknowledgements

We thank Dr John Moncrieff for assistance in construction and deployment of the Eddy Covariance system and the Eddy Sol software used in this study. Mr Frances Kariuki is thanked for his assistance with the field measurements. We are grateful to Mr R. Hopcraft for kindly allowing us access to the field site at Loldia Farm, Naivasha. Financial support for this work was received from the EU STD3 R&D Programme under contract no. TS3*-CT92-0149.

References

Anderson, M. G. & S. B. Idso, 1987. Surface geometry and stomatal conductance effects on evaporation from aquatic macrophytes. Wat. Resour. Res. 23: 1037–1042.

Becht, R. & D. M. Harper, 2002. Towards an understanding of human impact upon the hydrology of Lake Naivasha, Kenya. Hydrobiologia 488 (Dev. Hycrobiol. 168): 1–11.

Doliner, L. H. & P. A. Joliffe, 1979. Ecological evidence concerning the adaptive significance of the C_4 dicarboxylic acid pathway of photosynthesis. Oecologia 38: 24–34.

Gaudet, J. J., 1977. Natural draw-down on Lake Naivasha, Kenya, and the formation of papyrus swamps. Aquat. Bot. 3: 1–47.

Grace, J. J., J. Llyod, J. Mcintyre, A. Miranda, P. Meir, H. Miranda, J. Moncrieff, J. Massheder, I. Wright & J. Gash, 1995. Fluxes of carbon dioxide and water vapor over an undisturbed tropical forest in south-west Amazon. Global Change Biol. 1: 1–12.

Hasegawa, S., 1977. Agroclimatological studies on C_3-plants and C_4-plants. Transpiration and leaf temperatures. J. Agric. Meteorol. 33: 129–136.

Hughes, R. H. & J. S. Hughes, 1992. A Directory of African Wetlands. IUCN, Gland, Switzerland and Cambridge, U.K.: 820 pp.

Humphries, S. W. & S. P. Long, 1995. WIMOVAC – a software package for modelling the dynamics of plant leaf and canopy photosynthesis. Computer Applications in the Biosciences 11: 361–371.

Idso, S. B., 1981. Relative rates of evaporative water losses from open and vegetation covered water bodies. Wat. Resour. Bull. 17: 46–48.

Idso, S. B. & M. G. Anderson, 1988. A comparison of two recent studies of transpirational water loss from emergent aquatic macrophytes. Aquat. Bot. 31: 191–195.

Jones, M. B., 1987a. Wetlands. In Baker N. R. & S. P. Long (eds), Photosynthesis in Contrasting Environments. Elsevier, Amsterdam: 103–138.

Jones, M. B., 1987b. The photosynthetic characteristics of papyrus in a tropical swamp. Oecologia 71: 355–359.

Jones, M. B. & F. M. Muthuri, 1985. The canopy structure and microclimate of papyrus (*Cyperus papyrus* L.) swamps. J. Ecol. 73: 481–491.

Jones, M. B. & F. M. Muthuri, 1997. Standing biomass and carbon distribution in a papyrus (*Cyperus papyrus* L.) swamp on Lake Naivasha, Kenya. J. Trop Ecol. 13: 347–356.

Knapp, A. K. & E. Medina, 1999. Success of C_4 photosynthesis in the field: lessons from communities dominated by C_4 plants. In Sage R. F. & R. K. Monson (eds), C_4 Plant Biology. Academic Press, San Diego: 251–283.

Linacre, E. T., B. B. Hicks, G. R. Sainty & G. Grauze, 1970. The evaporation from a swamp. Agric. Meteorol. 7: 375–386.

Long, S. P., 1999. Environmental responses. In Sage R. F. & R. K. Monson (eds), C_4 Plant Biology. Academic Press, San Diego: 215–249.

Moncrieff, J. B., J. M. Massheder, H. DeBruin, J. Elbers, B. Huesunkveld, P. Kabat, S. Scott, H. Soegaard & A. Verhoef, 1997a. A system to measure surface fluxes of momentum, sensible heat, water vapour and carbon dioxide. J. Hydrol. 189: 580–611.

Moncrieff, J., S. Valentini, S. Greco, G. Seufert & P. Ciccioli, 1997b. Trace gas exchange over terrestrial ecosystems: methods and perspectives in micrometeorology. J. exp. Bot. 48: 1133–1142.

Muthuri, F. M., 1985. The Primary Productivity of Papyrus (*Cyperus papyrus* L.) in Relation to Environmental Variables. PhD Thesis. University of Nairobi, Kenya.

Muthuri, F. M., M. B. Jones & S. K. Imbamba, 1989. Primary productivity of papyrus (*Cyperus papyrus* L.) in a tropical swamp – Lake Naivasha, Kenya. Biomass 18: 1–14.

O'Toole, J. C. & V. S. Tomar, 1982. Transpiration, leaf temperature and water potential of rice and barnyard grass in flooded fields. Agric. Meteorol. 26: 285–296.

Parkinson, K. J., 1981. An improved method for measuring soil respiration in the field. J. App. Ecol. 18: 221–228.

Penman, H. L., 1948. Natural evaporation from open water, bare soil and grass. Proc. R. Soc. Lond: Series B. 193: 120–145.

Piedade, M. T. F., W. J. Junk & S. P. Long, 1991. The productivity of the C_4 grass *Echinochloa polystacha* on the Amazon floodplain. Ecology 72: 1456–1463.

Rijks, D. A., 1969. Evaporation from a papyrus swamp. Quart. J. Royal Met. Soc. 95: 643–649.

Schimel, D. S., 1995. Terrestrial ecosystems and the carbon cycle. Global Change Biol. 1: 77–91.

Schuepp, P. H., M. Y. Leclerc, J. I. Macpherson & R. L. Desjardins, 1990. Footprint predictions of scalar fluxes from analytical solutions of the diffusion equation. Boundary Layer Meteorol. 50: 353–373.

Thompson, K. & A. C. Hamilton, 1983. Peatlands and swamps of the African continent. In Gore, A. J. P. (ed.), Ecosystems of the World. Elsevier, Amsterdam: 331–373.

Hydrobiologia **488:** 115–122, 2002.
D.M. Harper, R. Boar, M. Everard & P. Hickley (eds), Lake Naivasha, Kenya.

The dynamics and ecology of exotic tropical species in floating plant mats: Lake Naivasha, Kenya

C. S. Adams[1,*], R. R. Boar[2], D. S. Hubble[3], M. Gikungu[4], D. M. Harper[3], P. Hickley[5] & N. Tarras-Wahlberg[6]
[1]*The Environment Agency, Cobham Road, Ipswich IP3 9JE, U.K.*
[2]*School of Environmental Sciences, University of East Anglia, Norwich NR4 7TJ, U.K.*
[3]*Department of Biology, University of Leicester, Leicester LE1 7RH, U.K.*
[4]*Department of Entomology, National Museums of Kenya, P.O. Box 40658, Nairobi, Kenya*
[5]*The Environment Agency, National Coarse Fish Centre, Hoo Farm Estate, Kidderminster, DY11 7RA, U.K.*
[6]*Kyrkeryds Lillegård, 355 92 Växjö, Sweden*
*(*Author for correspondence)*

Key words: water hyacinth, *Eichhornia crassipes*, *Salvinia molesta*

Abstract

The floating water fern *Salvinia molesta* has occurred in Lake Naivasha since the early 1960s and during this period has obstructed fishing activities and navigation. In recent years the extent of *Salvinia* has declined markedly. Since 1988, a second floating weed, *Eichhornia crassipes* (water hyacinth), has colonised the lake. *Salvinia* formed large mats around the edges, bays and lagoons and had very few plant or invertebrate species associated with it. Mats of *E. crassipes*, however, support a greater abundance and variety of animal and plant species, which shows a relationship with mat size. This seems to have facilitated plant succession but without a zonation typical of the classic hydroseral sequence for the lake.

Introduction

Lake Naivasha is a shallow, endorheic, freshwater lake situated in warm and semi-arid conditions in the eastern Rift Valley of Kenya, lying within an enclosed basin at an altitude of approximately 1890 m (Fig. 1). Rainfall is bimodal with peaks during April and October but shows inter-annual irregularity and an annual evapotranspiration rate that exceeds local rainfall (Harper et al., 1995) as a consequence of long-term climatic changes that affect the catchment (Vincent et al., 1979). The lake as a result of this has a surface area that fluctuates between 100 and 150 km^2, which influences its productivity (Harper, 1991). Its limnology and productivity have also been affected more recently by human intervention. Intensive horticultural development in the last two decades has accelerated papyrus clearance (Boar et al., 1999), drainage and application of fertilizers. This has come on top of exotic species introductions and arrivals between 1926 and 1988. The lake is now considered moderately eutrophic (Kitaka, 2000) because of agricultural developments in its catchment as well as its lakeshore. These changes have affected the aquatic plant species composition and abundance (Harper, 1992) as well as phytoplankton biomass (Hubble & Harper, 2001).

Three main rivers flow into the northern section of Lake Naivasha and transport suspended sediments from upstream erosion during rainfall events. The River Malewa is the largest and flows continually with a watershed area of 1730 km^2 (Gaudet & Melack, 1981). The rivers Gilgil and Karati are seasonal depending on rains and their volumes of flow are much smaller. The area of river inflow was formerly a swamp consisting of river channels flowing into the lake through a marshy area dominated by papyrus (*Cyperus papyrus*). Up to 30 years ago this area was a genuine papyrus swamp (Gaudet, 1977; Gouder et al., 1998) but the area of *C. papyrus* is now only a fragment of its former area (Johnson et al., 1998).

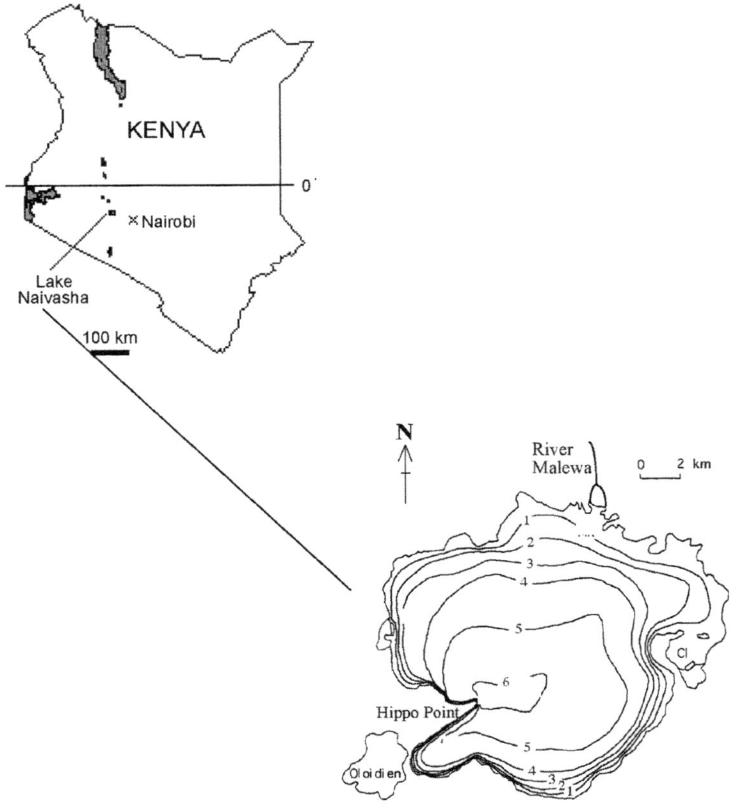

Figure 1. Location map of Lake Naivasha.

In the past 40 years, two highly productive and damaging exotic floating plants have appeared in the lake; 'Kariba weed', the aquatic fern, *Salvinia molesta* D.S. Mitchell [*S. auriculata*] (Salviniaceae) in the late 1960s and the Water Hyacinth, *Eichhornia crassipes* (Mart.) Solms (Pontederiaceae) in 1988. Both these plants formed dense mats, which out-compete less vigorous water plants for both light and space. Both weeds reproduce vegetatively and are distributed to new areas by winds and water currents. *E. crassipes* remains the world's most problematic water weed despite widespread and various approaches to its control (Heard & Winterton, 2000). In the last decade, control at Naivasha has focused upon biological control measures with the introduction of the Salvinia weevil, *Cyrtobagus salvinia*, in the early 1990s, and an Eichhornia weevil (*Neochetina* sp.) in the late 1990s.

S. molesta was abundant up to the end of the 1980s, occupying up to 25% of the lake surface in mobile mats (Johnson et al., 1998). In the 1990s, however, many small floating mats (based on *C. papyrus* and *E. crassipes*) formed when parts of the fringing vegetation broke off and floated due to a combination of wind action and changes in lake level. These mats were temporary in both form and location as they moved with the wind, and may have either split further, coalesced or rejoined the fringing vegetation. Such mats or islands have been reported from a number of wetlands around the world (e.g., Little, 1969), but there has been little ecological study since their initial description (Sahni, 1927 and citations below).

Because of their size and integrity, floating mats offer a colonisation opportunity for other herbaceous plants and shrubs usually found in the littoral zone of lakes. Such plants, if they reproduce by seed, germinate and may complete their life cycles on floating mats. In Lake Kariba, *Salvinia* provided a substratum stable enough for the growth of a number of emergent aquatic and semi-aquatic plants such as *Ludwigia* sp., *Phragmites* sp., *Typha* sp. and *Scirpus* sp. with a total of 40 plant species recorded (Boughey, 1963). Over a period of 10 years, Penfound & Earle (1948) found 63 species: 33 aquatic; 21 wetland and 9 terrestrial, growing on mats of *E. crassipes* in Louisiana in the southern United States. Floating islands may also provide a local habitat for invertebrates and juvenile fish (Oliver & McKaye, 1982; Sazima, 1985; IUCN 1986; Gopal, 1987; Dibble et al., 1996). The invertebrates can be both aquatic species associated with roots (Schramm et al., 1987) or terrestrial species in the aerial vegetation parts (Toft unpublished).

E. crassipes has an ability to root in damp mud (which *S. molesta* cannot do for more than a few days) and so in Naivasha, as in other locations that it has colonized, the littoral zone is overwhelmingly dominated by both floating plants of *E. crassipes* in the shallow water and rooted plants on the shore edge. The dominance has a physical stability, for, as water level changes, rooted plants can float and *vice versa*. It is thus possible that the classic zonation of vascular plants from land to the open water, described for Naivasha by Gaudet (1977), has been be altered. Gaudet classified 108 plant species in a primary successional sequence from lake edge to dry land after a period of naturally low water levels that occurred between 1971 and 1973. The zones were: the seedling zone dominated by *Nymphaea nouchali* seedlings that did not survive further drying; the sedge zone dominated by *Cyperus papyrus*, *C. digitatus* and *C. immensus* and, further inland, the Composite Zone dominated by *Conyza*, *Gnaphalium* and *Sphaeranthus* (Fig. 2).

The present study had the overall objective of examining whether these two exotic species, forming floating mats, were influencing the richness of flora and fauna in Lake Naivasha. This paper describes the changes in the distribution of *E. crassipes* and *S. molesta* which have occurred in the lake since 1988, documents species of plant and invertebrate recorded on them and analyses aspects of the composition of different-sized floating islands of *E. crassipes*.

Methods

The percentage cover of all plants in 4 × 2 m quadrats placed at random in 155 sites on the floating swamp at the edge of the lake was recorded at intervals over the period 1988 to 1998. GPS co-ordinates of each site were recorded on first visit. Sub-sets of the same sites were sampled on later visits. Where *C. papyrus* was present, quadrats were positioned on floating mats at its the lakeward edge. Plants were identified to species as far as possible mainly using Agnew & Agnew (1994) and a composite list compiled.

Samples for invertebrates were collected from individual *E. crassipes* which were taken into a net and later taken apart by hand. Invertebrates were removed by pooter or directly into sample tubes by pipette and preserved in 70% ethanol. A portion of plant was taken back to the laboratory where it was placed into traditional Tullgren funnels, and the extracted fauna collected prior to subsequent microscope examination.

Islands were sampled on a transect stretching from the south-west of the lake at Hippo point to the north, adjacent to the entry of the Malewa river (Fig. 1). Mat size was measured across the largest dimension and at right angles to that and area estimated from this, assuming a rectangle. Plant and animals were counted as 'Recognisable Taxonomic Units' (RTU).

Results

E. crassipes arrived in 1988, where it was recorded as isolated plants in the northern shallows. In 1989, 81% of the 155 sites surveyed around the lake had a percentage cover of more than 75% *S. molesta*. By 1993, when 132 sites were surveyed, only 5% of the sites supported the same high density. However, by this time *E. crassipes* had reached in excess of 75% cover in 63% of sites (Fig. 3). Present-day *S. molesta* cover is negligible and cover by *E. crassipes* has remained relatively stable. It usually forms a narrow fringe 0–5 m wide around much of the lake. Larger mats occur in sheltered bays and inlets.

Fifty-one species of plants were recorded on or in floating mats of *E. crassipes* and *S. molesta* between 1988 and 1998 (Tables 1 and 2), including submerged plants. Figure 4 shows plants apportioned between mat-forming species, grass and sedge species, herbs and shrub species. *Cyperus pectinatus* was the most frequent grass or sedge, and appeared to have an important role in binding the *E. crassipes* mat, facilitating colonisation by other species.

Since 1993, the species richness of herbaceous colonisers of the mats appears to have increased with a peak of 23 species found in August 1997. This was coincident with the first records of colonization of floating mats by the shrubby species *Hibiscus* sp. and *Sesbania* sp. It was thus 5 years after the first appear-

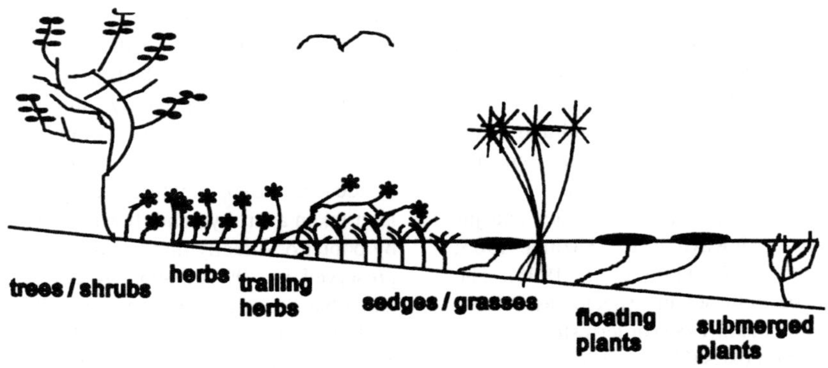

Figure 2. Hydroseral succession of plant species around the shore of Lake Naivasha in the 1970s. After Gaudet (1977).

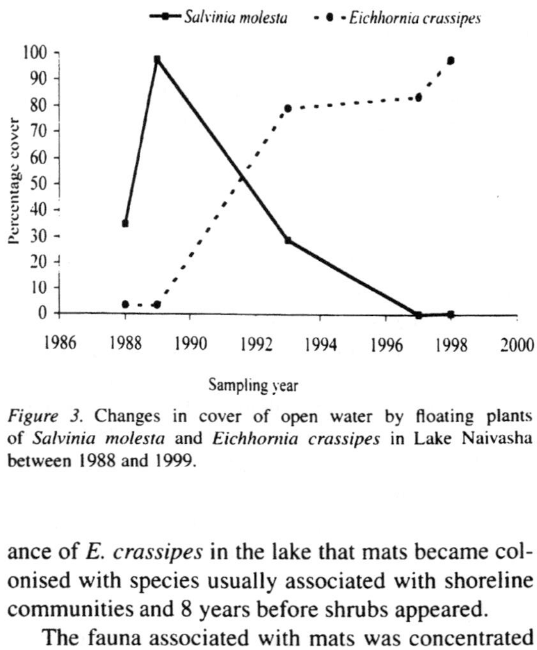

Figure 3. Changes in cover of open water by floating plants of *Salvinia molesta* and *Eichhornia crassipes* in Lake Naivasha between 1988 and 1999.

Table 1. Species list of all plants recorded on floating mats of *Salvinia molesta* and *Eichhornia crassipes* in Lake Naivasha, compiled over the period 1988–1999

Floating species	**Herbaceous species**
Azolla africana	*Basella alba*
Eichhornia crassipes	*Bidens pilosa*
Pistia stratiotes	*Circium arvense*
Salvinia molesta	*Callitriche truncata*
Wolffia arrhiza	*Commelina benghalensis*
	Conyza sp.
Sedges	*Crassocephalum picridifolium*
Cyperus dives (immensus)	*Crotalaria barkae*
C. flavescens	*Crassula schimperi*
C. marginatus	*Diplanche fusca*
C. papyrus	*Enydra fluctuans*
C. pectinatus	*Epilobium hirsutum*
C. rigidifolius	*Gnaphalium luteo-album*
	Hydrocotyle sp.
Grasses	*Ipomoea cairica*
Acroceras zizanioides	*Iris* sp.
Agrostis stolonifera	*Ludwigia stolonifera*
Miscanthidium sp.	*Lythrum rotundifolium*
Pennicetum clandestinum	*Polygonum salicifolium*
Setaria veristicelata	*P. senegalense*
Typha latifolia	*Pycnostachys coerulae*
	P. deflexifolia
	Senecio sp.
Submerged species	*Sphaeranthus napierae*
Najas horrida	*Tagetes minuta*
Nymphaea nouchali	*Veronica* sp.

ance of *E. crassipes* in the lake that mats became colonised with species usually associated with shoreline communities and 8 years before shrubs appeared.

The fauna associated with mats was concentrated in the layer between the leaves and the aquatic roots of the plants. The major groups of invertebrates found in this layer were Oligochaeta (mainly *Alma emini*), Insecta and Arachnida (both aquatic and terrestrial). Within this zone, dead plant material is continuously being broken down and wind-blown dust accreted in a soil-forming process, in which the earthworm *A. emini* appears to be very important. Young individuals of *Procambarus clarkii* and *Micronecta scutellaris* are common in and amongst the roots in *E. crassipes.* In contrast, *S. molesta* mats harbour a very limited fauna, both in species diversity and number of individuals. To date (2001), no individuals of the introduced weevils, *Cyrtobagus salviniae and Neochetina* sp., have been identified in any floating vegetation.

There was an increase in RTU number with increasing island size. The upper limit for plant richness was reached before that for invertebrates, as the rich-

Table 2. Taxonomic list of invertebrates recorded on *Eichhornia crassipes* at Naivasha

Turbellaria	
	Dugesia sp.
Oligochaetaea	
	Alma emeni
	Brachiura sowerbyi
	Potamothrix sp.
Mollusca	
	Bulinus sp.
	Physa acuta
Diplura (terrestrial)	
Aranaea	
	Hydracarina
	Arachnida (terrestrial)
Crustacea	
	Procambarus clarkii
	Ostracoda
	Copepoda
Insecta (terrestrial)	
	Collembola
	Thysanoptera
	Orthoptera
	Dermaptera
	Cicadoidea
	Staphylinidae
	Formicidae
Insecta (aquatic)	
Coleoptera	
	Hydaticus sp.
	Rhantus sp.
	Cybister sp.
	Helochares sp.
	Berosus sp.
	Eochrus sp.
	Canthyrus sp.
	Hydrovatus sp.
	Methles sp.
	Synchortus sp.
	Bidessus sp.
	Helodidae
Hemiptera	
	Micronecta sp.
	Plea sp.
	Mesoveliidae
	Lygaiedae
Diptera	
	Culicidae
	Ceratopogonidae
	Chironomidae
	Tipulidae
	Sciomyzidae
Odonata	
	Enallagma sp.
Ephemeroptera	
	Cioeon sp.
Trichoptera	
	Ecnomus sp.

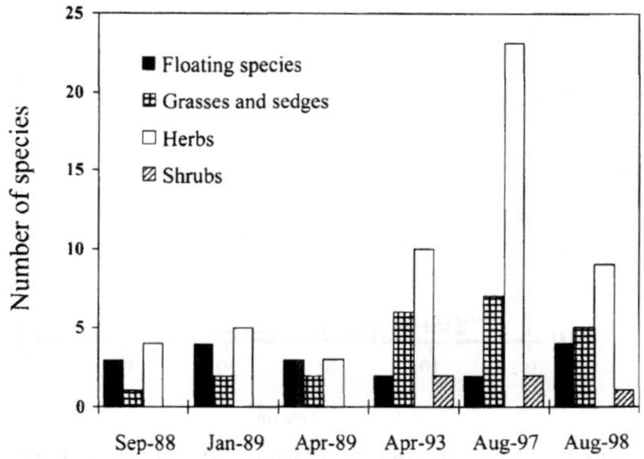

Figure 4. The species richness of groups of plants found growing on floating mats of *Eichhornia crassipes* in Lake Naivasha over the period 1988–1999.

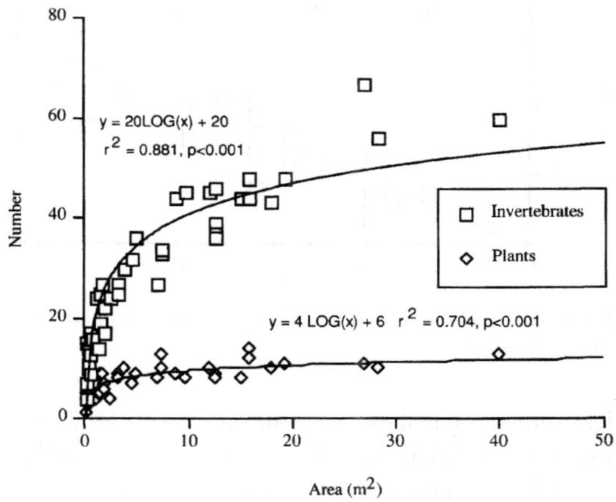

Figure 5. Change in plant and invertebrate species numbers with change in island size.

ness of the latter continued to increase with island size (Fig. 5).

Invertebrate density decreased with island size in a power relationship, with density more or less constant for islands above approximately 10 m^2 in area (Fig. 6). No similar relationship was evident for plants.

As plant species richness increased, so did invertebrate richness (Fig. 7).

Discussion

S. molesta had been first recorded in Lake Naivasha in 1962, inside the lagoons that were common inside the lake's fringing papyrus. Measures to control the spread of the weed involved herbicide applications and phys-

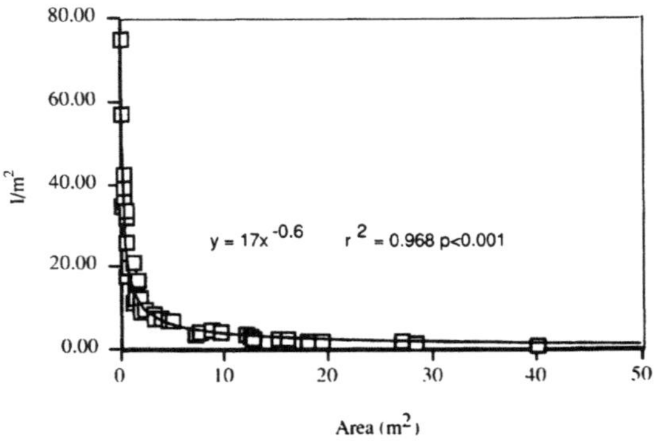

Figure 6. Changes in invertebrate density with increasing island area.

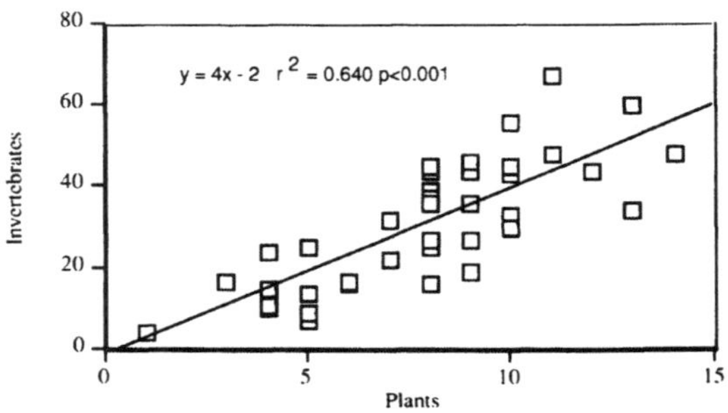

Figure 7. Correlation between numbers of plant and invertebrate species on islands.

ical removal but these were unsuccessful. However, experiments conducted during these times showed that *S. molesta* exposed to the windy and choppy conditions of the open lake quickly perished. However, reproduction in the more sheltered and shaded lagoons resulted in plants being pushed out so that, between 1975 and 1983, floating mats covered large areas of the northern and western open water of the lake (Harper, 1992). When lake level receded in the late 1980s, the sheltered lagoons almost totally disappeared and this resulted in a dramatic decline (Tarras-Wahlberg, 1986; Harper et al., 1990). At present, *S. molesta* is found only in the wetland at the lower reaches of the Gilgil river. In this area, the Gilgil river flows only gently, and the area is sheltered from the exposed conditions of the open lake (Everard et al., this volume). These conditions are similar to those of the former sheltered lagoons and hence are a likely explanation why *S. molesta* has survived only in this part of the lake system. The lack of records of the introduced weevil implies that it has had a limited part in the disappearance of *S. molesta* and that instead the loss of suitable habitat from the lake is a more likely cause of the plant's decline. Additional evidence for this is that no *S. molesta* is to be found in the lower reaches of the Malewa river (except during high water levels of the 1999 'El Nino' rains when extensive lagoons were once again formed in the river mouth at the northern end of the lake).

E. crassipes has colonised the full extent of Lake Naivasha shoreline in only a few years but the rate of its growth is small compared with its invasion and persistence in Lake Victoria, the Sudd and elsewhere throughout the tropics and sub-tropics. Interesting features of *E. crassipes* in Lake Naivasha are that individual plant biomass is low and that the plant has not colonised the open lake in large mats. High exposure to winds and relatively low water temperature may have constrained its spreading. Water hyacinth is at its most productive when maintained at water temperatures of around 28 °C (Bock, 1969). The annual mean water temperature in Lake Naivasha, (at an altitude of 1890 m) is around 21°C (Muthuri et al., 1989) and is thus rather low for tropical waters. A genetic basis for differences between productivity of particular clones of *E. crassipes* may also exist (Geber et al., 1992).

The absence of individuals of the weevil *Neochetina* in any of the *E. crassipes* samples investigated to date, implies that the introduction of the weevil has not (or not yet) significantly affected the spread of the water hyacinth. The largest floating mats occur in the north of the lake around the inflow of the River Malewa, which may advantage the plant with a nutrient supply as well as shelter (prevailing winds for most of the year push floating plants to the north (Litterick et al., 1979)).

The plant species richness of the islands of *E. crassipes*, although not reaching the total recorded by Gaudet (1977) indicate from this study that this floating exotic plant is unlikely to be reducing the biodiversity of semi-aquatic plants in Naivasha. The results also indicate that the islands represent an extension of the swamp and so partially replace its loss to agriculture at declining water levels. The recent evidence of shrub species indicate that the islands are also undergoing successional processes as described on land (Kaul & Zutshi, 1966; Gaudet, 1977). This is aided by the accumulation of suspended organic matter and the activities of wet soil animals such as *A. emini* which encourages further colonization (Trivedy et al., 1978). The nature of this succession though is to dissemble the spatial zonation of plants described by Gaudet (1977) with a temporal succession of species occurring on the mats themselves.

The invertebrate richness cannot be compared directly with other habitats, but the aquatic part of the taxonomic list represents a broad cross-section of the full list for the lake from the 1980s published by Clark et al. (1989). The terrestrial component is not identified adequately for meaningful interpretation but represents aerial dispersal and indicates the availability of arthropod prey for birds such as plovers, ibises, gulls and terns which are common on *E. crassipes* mats (Henderson & Harper, 1992). The increase in richness with island size is likely to represent the increased probability of colonization, whilst the reduced density with island size probably reflects a less suitable environment in the middle of islands (Gopal, 1987; Bailey & Litterick, 1993).

Acknowledgements

This study formed part of the work of the University of Leicester research project at Lake Naivasha, which since 1987 has been authorised by the Office of the President, Government of Kenya under research permit to Dr D. M. Harper no. OP 13/001/12C 46. The project was funded by the Earthwatch Institute, Boston, USA, and Oxford, England. The data collection would not have been possible without the assistance of numerous Earthwatch volunteers. Our sincere thanks go to assistance from colleagues in Kenya, in particular Prof. Ken Mavuti, Jill & Angus Simpson and Velia Carn.

References

Agnew, A. D. Q. & S. Agnew, 1994. Upland Kenya Wild Flowers. 2nd edn. East Africa Natural History Society, P.O. Box 44486, Nairobi, Kenya. 374 pp.

Bailey, R. G. & M. R. Litterick, 1993. The macroinvertebrate fauna of water hyacinth fringes in the Sudd swamps (River Nile, southern Sudan). Hydrobiologia 250: 97–103.

Boar, R. R., D. M. Harper & C. S. Adams, 1999. Biomass allocation in *Cyperus papyrus* in tropical wetland, Lake Naivasha, Kenya. Biotropica 31: 411–421.

Bock, J. H., 1969. Productivity of the water hyacinth *Eichhornia crassipes* (Mart.) Solms. Ecology 50: 460–464.

Boughey, A., 1963. The explosive development of floating weed vegetation on L. Kariba. Adamsonia 3: 49–61.

Clark, F., A. Beeby & P. Kirby, 1989. A study of the macroinvertebrates of Lakes Naivasha, Oloidien and Sonachi, Kenya. Rev. Hydrobiol. Trop. 22: 21–33.

Dibble, E. D., K. J. Killgore & G. O. Dick, 1996. Measurements of plant architecture in seven aquatic plants. J. Freshwat. Ecol. 11: 311–318.

Everard, M., A. Kuria, M. Macharia, J. A. Vale & D. M. Harper, 2002. Aspects of the biodiversity of the rivers in the Lake Naivasha catchment. Hydrobiologia 488 (Dev. Hydrobiol. 168): 43–55.

Gaudet, J. J., 1977. Natural drawdown on Lake Naivasha, Kenya and the formation of Papyrus swamps. Aquat. Bot. 3: 1–47.

Gaudet, J. J. & J. M. Melak, 1981. Major ion chemistry in a tropical African basin. Freshwat. Biol. 11: 309–331.

Gerber, M. A., M. A. Watson & R. Furnish, 1992. Genetic differences in clonal demography in *Eichhornia crassipes*. J. Ecol. 80: 329–341.

Gopal, B., 1987. Aquatic Plant Studies I. Water Hyacinth. Elsevier, New York.

Gouder, A.-C. de B., D. M. Harper, F. Malaisse & J.-J. Symoens, 1998. Dynamique récente et cartographie de la végétation aquatique (1960–1996) du lac Naivasha (Rift Valley, Kenya). Bull. Séanc. Acad. r. Sci. Outre-Mer 44: 373–389.

Harper, D. M., 1991. Primary production in Lake Naivasha, Kenya. Verh. Int. Ver. Limnol. 24: 1112–1116.

Harper, D. M., 1992. The ecological relationships of aquatic plants at Lake Naivasha, Kenya. Hydrobiologia 232: 65–71.

Harper, D. M., C. Adams & K. Mavuti, 1995. The aquatic plant communities of the Lake Naivasha wetland, Kenya: Pattern, dynamics and conservation. Wetlands Ecol. Manage. 3: 111–123.

Harper, D. M., M. Muchiri & K. M. Mavuti, 1990. Ecology and management of Lake Naivasha, Kenya, in relation to climatic change, alien species introductions and agricultural development. Environ. Conserv. 17: 328–335.

Heard, T. A. & S. L. Winterton, 2000. Interactions between nutrient status and weevil herbivory in the biological control of water hyacinth. J. Appl. Ecol. 37: 117–127.

Henderson, I. G. & D. M. Harper, 1992. Bird distribution and habitat structure on Lake Naivasha, Kenya. African J. Ecol. 30: 223–232.

Hubble, D. S. &. D. M. Harper, 2001. Impact of light regimen and self-shading by algal cells on primary productivity in the water column of a shallow tropical lake (Lake Naivasha, Kenya). Lakes Reservoirs: Res. Manage. 6: 143–150.

IUCN, 1986. MAB Information System Biosphere Reserves. Compilation 4. UNESCO, Paris.

Johnson, G. M., D. M. Harper & K. M. Mavuti, 1998. Information for the sustainable management of shallow lakes: Lake Naivasha, Kenya. In Harper, D. M. & A. G. Brown (eds), The Sustainable Management of Tropical Catchments. John Wiley & Sons, Chichester: 259–274.

Kaul, V. & D. P. Zutshi, 1966. Some ecological considerations of floating islands in Srinagar lakes. Proc. Nat. Acad. Sci. India 36B: 273–281.

Kitaka, N., 2000. Phosphorus supply to a shallow tropical lake and its consequences – Lake Naivasha, Kenya. PhD thesis, University of Leicester.

Litterick, M., J. J. Gaudet, J. Kalff & J. M. Melack, 1979. The limnology of an African lake; Lake Naivasha, Kenya. Unpublished manuscript distributed at the International Conference on Tropical Limnology, Nairobi, May 1979. 73 pp.

Little, E. C. S., 1969. The floating islands of Rawa Pening. Proc. Acad. Nat. Sci. 15: 146–153.

Muthuri, F. M., M. B. Jones & S. K. Imbamba, 1989. Primary productivity of papyrus (*Cyperus papyrus*) in a tropical swamp, Lake Naivasha, Kenya. Biomass 18: 1–14.

Oliver, M. K. & K. R. McKaye, 1982. Floating islands: a means of fish dispersal in Lake Malawi, Africa. Copeia 4: 748–754.

Penfound, W. T. & T. T. Earle, 1948. The biology of water hyacinth. Ecol. Monogr. 16: 447–472.

Sahni, B., 1927. A note on the floating islands and vegetation of Khajiar, near Chamba in the N. W. Himalayas. J. Indian Bot. 6: 1–7.

Sazima, I., 1985. Use of water hyacinths as shelter, foraging place, and transport by young piranhas, *Serrasalmus spilopleura*. Environ. Biol. Fishes 12(2): 237–240.

Tarras-Wahlberg, N., 1986. Observations on *Salvinia* and its environment at Lake Naivasha (Kenya). J. East Africa nat. Hist. Soc. Nat. Mus. 76: 189.

Toft, J., unpublished. The effect of the exotic aquatic plant *Eichhornia crassipes* (Water Hyacinth) on the fish/invertebrate food web in the Sacramento/San Joaquin Delta, California.

Trivedy, R. K., K. P. Sharma, P. K. Goel & B. Gopal, 1978. Some ecological observations on floating islands. Hydrobiologia 60(2): 187–190.

Vincent, C., T. Davies & A. Beresford, 1979. Recent changes in the level of Lake Naivasha, Kenya, as an indicator of equatorial westerlies over East Africa. Climate Change 2: 175–189.

Hydrobiologia **488:** 123–128, 2002.
D.M. Harper, R. Boar, M. Everard & P. Hickley (eds), Lake Naivasha, Kenya.

Organochlorine and organophosphorus pesticide concentrations in water, sediment, and selected organisms in Lake Naivasha (Kenya)

S. M. Gitahi[1,*], D. M. Harper[2], S. M. Muchiri[3], M. P. Tole[4] & R. N. Ng'ang'a[5]
[1]*Lake Naivasha Riparian Association, P.O. Box 1011, Naivasha, Kenya,*
[2]*Department of Biology, University of Leicester, Leicester LE1 7RH, U.K.*
[3]*Department of Fisheries, Moi University, Box 3900, Eldoret, Kenya*
[4]*School of Environmental Studies, Moi University, P.O. Box 3900, Eldoret, Kenya*
[5]*Kenya Plant Health Inspectorate Service, P.O. Box 49592, Nairobi, Kenya*
*(*Author for correspondence)*

Key words: organochlorinated and organophosphorus pesticide residues, bioaccumulation, *Procambarus clarkii*, *Micropterus salmoides*

Abstract

Water, sediment, red swamp crayfish (*Procambarus clarkii*) and black bass (*Micropterus salmoides*) from Lake Naivasha were analyzed for selected organochlorine and organophosphorus pesticide residues. The mean *p,p*'-DDT, *o,p*'-DDT and *p,p*'-DDE residue levels recorded in black bass (28.3 ($\pm$ 30.0), 34.2 ($\pm$54.0) and 16.1 ($\pm$16.1) μg kg^{-1}, respectively) and crayfish (4.6 ($\pm$5.1), 3.2 ($\pm$2.8), and 1.4 ($\pm$1.1) μg kg^{-1}, respectively), were higher than previously recorded. This indicated recent usage of technical DDT in the lake's catchment. Levels of *p,p*'-DDT, higher than those of *p,p*'-DDE further emphasized this. Mean lindane, dieldrin, β-endosulfan and aldrin concentrations in black bass were 100.5, 34.6, 21.6 and 16.7 μg kg^{-1}, respectively. The same residues were detected at lower concentrations in crayfish at 2.0, 2.0, 2.0 and 1.9 μg kg^{-1}, respectively. The higher fat content (3.7 $\pm$ 2.7% SD) in black bass (compared to 0.6 $\pm$ 0.3% in crayfish) accounted for the significantly higher residue concentrations in black bass. Organophosphate pesticides were the most commonly used pesticides in the lake's catchment, but none was detected in any of the samples. The results indicate that there is need for further work to identify sources and fate of pesticide contaminants, as well as to improve monitoring of pesticide use throughout the catchment.

Introduction

Pesticide use has resulted in acute and chronic ecological damage, either by direct injury to non-target organisms such as birds and fish, or by indirect effects such as elimination of natural enemies. Pesticides are a singular form of environmental hazard in that they are actually designed to harm living things. Many organochlorine compounds are very persistent in the environment and have a tendency to bioaccumulate significantly through food chains (UNEP, 1993). Organophosphates have several advantages over other types of pesticides, including high acute toxicity to target organisms, but they are not persistent in the environment as are organochlorines, as they decompose to non-toxic products. However, their acute toxicity is of concern.

Aquatic environments are particularly affected because pesticides applied to land-based agriculture systems are eventually carried into water bodies through surface runoff, rivers, and groundwater flow. Therefore, chronic and acute effects of pesticide pollution and bioaccumulation in aquatic flora and fauna are more severe.

Due to the undesirable effects on environmental quality and animal health, the production and usage of organochlorine compounds was banned or severely restricted during the 1970s and 1980s in most developed countries. However, the demand for organochlorine pesticides has been increasing in some developing countries in Africa, Latin America and Asia (Tanabe et al., 1993). There has been widespread use of pesticides in Kenya in the last four decades because agriculture has been the mainstay of Kenya's economy. Organochlorines have been extensively used and particularly, DDT and endosulfan for the control of maize and cotton pests (Munga, 1990). Lindane, dieldrin, aldrin, endrin and heptachlor have also had wide usage in Kenya, which has made their presence ubiquitous in the environment (Wandiga, 1996).

Lake Naivasha, Kenya's second largest freshwater lake, is a valuable national centre for fishing, agriculture and tourism. The lake is also important for biodiversity and received RAMSAR recognition in 1995. But in recent years the lake has faced intensified horticulture/floriculture development for the export market, and this is thought to contribute pesticide pollution into the lake. The purpose of this study was to establish the extent of organochlorine and organophosphate insecticide contamination in lake Naivasha.

Materials and methods

Pesticide use survey

Information on pesticides commonly used in the lake Naivasha basin was obtained through visits to the area Public Health office and horticultural companies, as well as to farmers' shops in different townships in the lake's catchment.

Field sampling

Sampling wascarried out at five selected sites around Lake Naivasha between August 1997 and January 1998. Black bass (*Micropterus salmoides* Lacepede) samples were collected by gill nets, while crayfish (*Procambarus clarkii* Girard) samples were collected from the littoral water hyacinth zone using traps and sweep-nets, and grouped by carapace length. Crayfish muscle was extracted by peeling. The black bass and crayfish muscle samples were weighed and immediately frozen prior to analysis. Sediment samples obtained by use of a core sampler, lake water samples, and water samples from the main river, the Malewa, were also collected. Water samples were stored in glass bottles at 2–4 °C, and sediment at −5 to −10 °C, awaiting analysis. All the samples were transported to the laboratories in coolboxes with freezer packs.

The samples were analysed for the organochlorines lindane (γ-HCH, hexachlorocyclohexane), dieldrin, aldrin, *p,p*'-DDT, *o,p*'-DDT, *p,p*'-DDE and β-endosulfan, as well as for the organophosphates diazinon, malathion and parathion.

Reagents

All solvents (*n*-hexane, acetone, dichloromethane, diethyl ether) as well as other analytical material used (anhydrous sodium chloride, Florisil) were of analytical grade quality (Merck, NJ, USA), double-distilled to ensure purity. Florisil and individual pesticide standards were available at the Kenya Agricultural Research Institute laboratories, Nairobi.

Extraction and cleanup

Extraction and cleanup followed previously described protocol (NRI, 1991). Black bass liver or crayfish muscle was dried using anhydrous sodium sulphate, and individually extracted using 3:1 hexane:acetone. An aliquot of each extract was evaporated to dryness to calculate the tissue fat content. Anhydrous sodium sulphate was added to sediment samples to speed dessication. The samples were then individually soxhlet extracted, using 1:1 acetone:hexane. Water samples were extracted using dichloromethane. At this stage extracts in hexane could be injected for organophosphate quantification. The extracts were cleaned up with Florisil, using 6 and 15% diethyl ether in hexane as the elution mixture. The final pesticide extract was obtained in 5 ml of hexane.

Chromatographic analyses

Gas chromatographic analyses were carried out using Varian 3400 series GLC, fitted with an electron capture detector (ECD) for organochlorines and a nitrogen–phosphorus detector (NPD) for organophosphates. Two μl aliquots of the pesticide extracts and a pesticide standard mixture were injected into the GLC for analysis. A DB-5 capillary column (30 m, 0.2 mm i.d.) was used to separate organochlorine insecticides, while for organophosphates a DB-5 megabore capillary column (15 m, 0.53 mm i.d.) was employed (both J&W Scientific, Folsom, CA, USA). The carrier gas was nitrogen.

Results

Most commonly used pesticides

The most commonly used pesticides (excluding herbicides) in the Lake Naivasha basin were organophosphates, carbamates, and pyrethroids, in that order, with very few organochlorines (Table 1).

Recovered pesticide residues

All the organochlorines were detected in the black bass and crayfish samples in varying concentrations and frequencies of occurrence (Tables 2 and 3). None of the pesticide residues were detected in sediment samples, or in the lake and river water samples. No organophosphates were detected in any of the samples.

Fat content

The mean fat content was 3.73% (± 2.71% SD) and 0.56% (± 0.27% SD) in black bass liver and crayfish muscle, respectively. No correlation was observed between size and the corresponding fat content in either black bass or crayfish ($r^2 < 0.2$).

The influence of size

No correlation between the concentration of contaminants and size was observed in either black bass or crayfish from all stations. The relationship between total weight and lindane concentration is shown for both species in Figure 1.

Discussion

The influence of size

Munga's (1990) study on fish from the Hola Irrigation Scheme strongly indicated that organochlorine pesticide levels were dependent on the amount of fat in fish, caused by concomitant changes in the total lipid content as organisms increase in age. The lack of correlation between residue and size in the present study was probably due to the lack of correlation between fat content and black bass or crayfish size.

Organochlorines

While *p,p*'-DDT is the functional compound used, its conversion in the environment produces the metabolite *p,p*'-DDE (Kanja et al., 1986; Mugambi et al., 1989; Pastor et al., 1996). Environmental conversion of DDT to the more persistent DDE may result in samples containing predominantly DDE. In the present study mean *p,p*'-DDT levels were slightly higher than mean *p,p*'-DDE levels in both black bass and crayfish. This illustrated a fairly recent exposure to functional DDT.

Today the use of DDT in Kenya is banned, except in the control of mosquitoes by public health authorities (PCPB, 1998). DDT has in the past been detected in Kenya in fish from different areas. Koeman et al. (1972) and Greichus et al. (1978) reported very low levels of organochlorine residues in *Tilapia grahami* from Lake Nakuru, while Lincer et al. (1981), in a study of DDT contamination in some Rift Valley lakes, reported DDT residue levels of below 1 ppm in *T. grahami* from Lake Nakuru and in *T. nilotica, Clarias mossambicus, Barbus gregorii* and *Labeo cylindricus* from Lake Baringo. Mitema & Gitau (1989) found DDT to range from 2 μg kg^{-1}(ppb) to 4.51 mg kg^{-1} (ppm) in the fillet of Nile perch from Lake Victoria. Mugachia et al. (1992a,b), in a study on catfish, common carp and tilapia from the Athi River estuary and Masinga Dam, found mean levels of *p,p*'-DDE, *o,p*'-DDT and *p,p*'-DDT of below 1 ppm. Munga (1990) found DDT to be 0.13, 0.92 and 0.38 mg kg^{-1} (ppm) wet mass in the muscle, liver and eggs, respectively, in *Labeo gregorii* from the Hola Irrigation Scheme.

Lincer et al. (1981) reported undetectable to low levels of *p,p*'-DDE in fish from Lake Naivasha. DDT residues were found to be 0.001 and 0.003 mg kg^{-1} in *Tilapia spirulus niger* and *Micropterus salmoides*, respectively. These authors however, found levels of DDE in the white pelican of Naivasha to have doubled since 1970, indicating an increase in organochlorine contamination of the system. A positive bioaccumulation of *p,p*'-DDE in the food-chains of lakes Naivasha, Baringo and Nakuru was found (Lincer et al., *op. cit.*). Frank et al. (1977) studying organochlorine residues in Kenyan birds of prey, reported that most raptors from agricultural areas (including Naivasha) contained residues of DDT metabolites, whereas those from non-agricultural areas did not contain detectable levels of any organochlorines. The most widespread contaminant was *p,p*'-DDE whose concentration ranged from <0.01 to 40 mg kg^{-1}.

The DDT residue levels in the present study are higher than those reported previously in fish from Lake Naivasha. Wandiga (1996) has shown that organochlorine pesticides have a shorter half-life in the tropics than in temperate climates, and suggests that

Table 1. Pesticides most commonly used in intensive horticulture/floriculture around Lake Naivasha and in small-scale agriculture in the lake's catchment

Intensive horticulture/ floriculture			Small-scale shops in catchment		
Trade name	Common name	Family	Trade name	Common name	Family
Lannate	Methomyl	Carbamoyloxime (I)	Triatix	Amitraz	Amidine
Apollo	Clofentezine	Tetrazine (A)	Dimethoate	Dimethoate	Organophosphate
Brigade	Bifenthrin	Pyrethroid (I)	Karate	Lambda-cyhalothrin	Pyrethroid
Mitac	Amitraz	Amidine (I.A)	Diazinon	Diazinon	Organophosphate
Nemacur	Fenamiphos	Organophosphate (N)	Brigade	Bifenthrin	Pyrethroid
Orthene	Acephate	Organophosphate (A.I)	Almatix	Amitraz	Amidine
Pentac	Dienochlor	Organochlorine (A)	Poultry louse powder	Permethrin	Pyrethroid
Thiodan	Endosulfan	Organochlorine (I)	Mboga dust	Fenitrothion	Organophosphate
Cascade	Flufenoxuron	Benzoylurea (I)	Sevin	Carbaryl	Carbamate
Dursban	Chlorpyrifos	Organophosphate (A.I)	Marshall	Carbosulfan	Carbamate
Folimat	Omethoate	Organophosphate (A.I)	Py-grease	Pyrethrin	Pyrethroid
Furadan	Carbofuran	Carbamate (I)	Diazol	Diazinon	Organophosphate
Hostathion	Triazophos	Organophosphate(I.A. N)	Furadan	Carbofuran	Carbamate
Mesurol	Methiocarb	Carbamate (A.I.M)	Taktik	Amitraz	Amidine
Omite	Propargite	Sulfite (A)			
Tedion	Tetradifon	Bridged-diphenyl (A)			
Temik	Aldicarb	Carbamoyloxime (I)			
Vydate	Oxamyl	Carbamoyloxime (I)			
Kelthane	Pirimiphos-methyl	Organophosphate (I)			

I –Insecticide: A – Acaricide: N – Nematicide: M = Molluscicide.

Table 2. Mean, range and percent frequency of occurrence of organochlorine pesticide residue levels (μg kg^{-1}) in blackbass from Lake Naivasha[a]

	o,p'-DDT	*p,p*'-DDT	*p,p*'-DDE	Lindane	β-Endosulfan	Dieldrin	Aldrin
Mean[b]	34.2	28.3	16.1	100.5	21.6	34.6	16.7
(± SD)	(54.0)	(30.0)	(16.1)	(266.0)	(30.5)	(56.0)	(18.8)
Minimum	1.5	2.6	0.5	3.7	1.2	1.8	0.8
Maximum	186.6	100.7	58.2	1595.9	161.8	265.8	70.7
Positive[c]	20	27	23	35	31	25	28
(% frequency)	(57.1%)	(77.1%)	(65.7%)	(100%)	(88.6%)	(71.4%)	(80.0%)

[a] 0.2 μg kg^{-1}= lowest working concentration (below which residues could not be quantified).
[b] Residue levels given on wet weight basis. Mean was calculated for positive samples.
[c] Total samples = 35.

within 3–4 years much of it is already volatilised in tropical soils. This further suggests recent DDT application in the. Naivasha catchment.

Aldrin and dieldrin usage in Kenya is restricted only to termite control in the building industry (PCPB, 1998). Dieldrin is a functional pesticide, but it is also a persistent environmental metabolite of aldrin. Therefore use of aldrin contributes to dieldrin levels in a system (Mugambi et al., 1989). Dieldrin is the more persistent of the two. Therefore the current results indicate that aldrin has been in considerable use in the lake's catchment. The extent of aldrin use in Kenya has

Table 3. Mean, range and frequency of occurrence of organochlorine pesticide residue levels (μg kg^{-1}) in crayfish from Lake Naivasha[a]

	o,p'-DDT	*p,p*'-DDT	*p,p*'-DDE	Lindane	β-Endosulfan	Dieldrin	Aldrin
Mean[b]	3.2	4.6	1.4	2.0	2.0	2.0	1.9
(± SD)	(2.8)	(5.1)	(1.1)	(2.3)	(1.8)	(2.5)	(1.6)
Minimum	0.5	1.1	0.3	0.2	0.3	0.4	0.1
Maximum	11.9	25.0	4.2	10.4	7.5	11.1	6.1
Positive[c]	30	22	34	29	23	22	27
(% frequency)	(85.7%)	(62.9%)	(97.1%)	(82.9%)	(65.7%)	(62.9%)	(77.1%)

[a] 0.2 μg kg^{-1}= lowest working concentration (below which residues could not be quantified).
[b] Residue levels given on wet weight basis. Mean was calculated for positive samples.
[c] Total samples = 35.

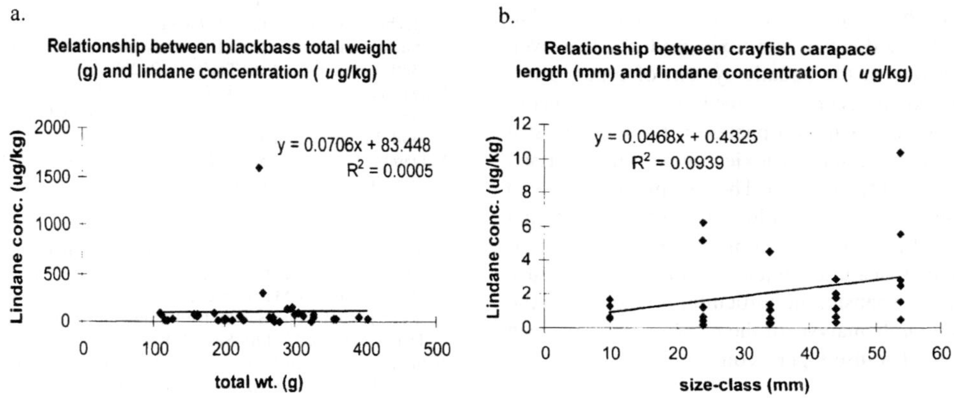

Figure 1. Relationship between size (black bass total weight (a) and crayfish carapace length (b) and lindane concentration (μg kg^{-1}).

probably been much less than the use of DDT, but relatively high aldrin and dieldrin levels have been found in Kenyan human milk as compared to the corresponding levels in other countries (Kanja et al., 1986) and in domestic fowl eggs from central Kenya (Mugambi et al., 1989).

Most of the highest residue concentrations recorded in black bass and crayfish were those of lindane. Therefore lindane may be in significant use in the lake's catchment. Technical HCH is a widely used insecticide in Kenya but it also contains β-HCH as a minor component (Kanja et al., 1986). Lindane in Kenya is restricted to use for seed dressing only. It was observed on the shelves of agricultural stores in the lake's catchment. The relatively biodegradable and less persistent nature of the HCHs, and the fact that lindane is less persistent than β-HCH (Tanabe et al., 1993) can only indicate either continuing, or recent, input into the lake of lindane. The present results are similar to previous findings in Kenya. Low lindane levels of 10–139 ppb were found in mother's milk (Kanja et al., 1986) and chicken eggs (10–100 ppb: Mugambi et al., 1989). Low lindane levels have also been reported in the liver and fillet of common carp, catfish and tilapia from the Tana River system, at 9–140 μg kg^{-1}(Mugachia et al., 1992a,b).

β-Endosulfan levels in the present study are much higher than the levels reported by Munga (1990), of 0.01 and 0.05 μg kg^{-1}respectively, from *Clarias gariepinus* and *Labeo gregorii* in the Tana River. Use of endosulfan (Thiodan) has grown from the 1980s, as an alternative to DDT, especially in the country's cotton growing areas, because cotton pests had developed resistance to DDT (S.O. Wandiga, pers. comm.). It was evident that technical endosulfan was in considerable use in the Lake Naivasha catchment.

Organophosphates

The most commonly used pesticides (excluding herbicides) in the L. Naivasha basin were organophosphates, carbamates, and pyrethroids, in that order, with relatively few organochlorines. The majority of these pesticides are registered in Kenya by the Pest Control Products Board (PCPB, 1998). However, none of the samples registered positive for diazinon, malathion or parathion. This is probably because of their relatively short half-lives, whereby they do not remain active in the environment for long (Lalah and Wandiga, 1996).

Lack of organophosphate residues in samples does not necessarily indicate lack of impact of organophosphate compounds on the environment. They may still impact on the aquatic system in the short term, before they break down to non-toxic products. However, while the fate in the aquatic system of organophosphate compounds is not clear, the most serious concern relating to the organophosphates is that they are still suspected to have serious toxicological impacts upon terrestrial wildlife and man. The vast majority of dead raptors sent to the Ornithology Department National Museums of Kenya, come from the Naivasha region. It is suspected that agrochemicals are the culprit because of the rapid expansion in horticultural practices in the region although analysis of the cadavers has not been carried out (Thomsett, pers. comm).

Acknowledgements

Financial support from both the KWS–Netherlands Wetlands Conservation and Research Training Programme and logistical help from the Earthwatch Lake Naivasha Research Project is gratefully acknowledged. We wish to thank the Kenya Agricultural Research Institute's National Agricultural Research Laboratories, for logistical and technical support.

References

Frank, L. G., R. M. Jackson, J. E. Cooper & M. C. French, 1977. A survey of chlorinated hydrocarbon residues in Kenyan birds of prey. E. Afr. Wildl. J. 15: 295–304.

Greichus, Y. A., A. Greichus, B. D. Aman & J. Hopcraft, 1978. Insecticides, polychlorinated biphenyls and metals in African lake ecosystems III, Lake Nakuru, Kenya. Bull. Environ. Contam. Toxicol. 19: 455.

Kanja, L., J. U. Skaare, I. Nafstad, C. K. Maitai & P. Lökken, 1986. Organochlorine pesticides in human milk from different areas of Kenya 1983–1985. J. Toxicol. Environ. Health 19: 449–464.

Koeman, J. H., J. H. Pennings, J. J. M. De Goeij, P. S. Tjioe, P. M. Olindo & J. Hopcraft, 1972. A preliminary survey of the possible contamination of Lake Nakuru in Kenya with some metals and chlorinated hydrocarbons. J. Appl. Ecol. 9: 411.

Lalah, J. O. & S. O. Wandiga, 1996. The persistence and fate of malathion residues in stored beans (*Phaseolus vulgaris*) and maize (*Zea mays*). Pestic. Sci. 46: 215–220.

Lincer, J. L., D. Zalkind, L. H. Brown, & J. Hopcraft, 1981. Organochlorine residues in Kenya's Rift Valley lakes. J. Appl. Ecol. 9: 411.

Mitema, E. S. & F. K. Gitau, 1990. Organochlorine residues in fish from Lake Victoria in Kenya. Afr. J. Ecol. 28: 234.

Mugachia, J. C., L. Kanja & F. Gitau, 1992a. Organochlorine pesticide residues in fish from Lake Naivasha and Tana River, Kenya. Bull. Environ. Contam. Toxicol. 49: 207–210.

Mugachia, J. C., L. Kanja & T. E. Maitho, 1992b. Organochlorine pesticide residues in estuarine fish from the Athi River, Kenya. Bull. Environ. Contam. Toxicol. 49: 199–206.

Mugambi, J. M., L. Kanja, T. E. Maitho, J. U. Skaare & P. Lokken, 1989. Organochlorine pesticide residues in domestic fowl (*Gallus domesticus*) eggs from central Kenya. J. Sci. Food. Agric. 48: 165–176.

Munga, D., 1990. DDT and endosulfan residues in four tropical fish species. In: Khan, M. R. & H. J. Gijzen (eds), Environmental Pollution and its Management in Eastern Africa. Proceedings of a Symposium organised by the Faculty of Science, University of Dar es Salaam, 11–14th Sept. 1989.

Natural Resources Institute, 1991. NRI Pesticide Management Section. Training Manual for Pesticide Residue Analysis Vol. 2. Natural Resources Institute, U.K.

Pastor, D. J. Boix, V. Fernandez, & J. Albaiges, 1996. Bioaccumulation of organochlorinated contaminants in three estuarine fish species (*Mullus barbatus*, *Mugil cephalus* and *Decentrarcus labrax*). Mar. Pollut. Bull. 32: 257–262.

PCPB, 1998. List of Pest Control Products Provisionally Registered by the Pest Control Products Board for use in Kenya. (Revised June, 1998). PCPB, Nairobi.

Tanabe, S., A. N. Subramanian, A. Ramesh, P. L. Kumaran, N. Miyazaki & R. Tatsukawa, 1993. Persistent organochlorine residues in dolphins from the Bay of Benghal, south India. Mar. Pollut. Bull. 26: 311–316.

Wandiga, S. O., 1996. Organochlorine pesticides: curse or blessing in tropical agriculture? In: Wandiga, S. O. & N. L. Abuodha (eds), Environment and Development in Kenya. Research Papers Presented at Academy Public Lecture Series. Kenya National Academy of Sciences, Nairobi.

UNEP, 1993. United Nations Environment Programme Environmental Data Report 1993–94. Blackwell Publishers, Oxford.

Hydrobiologia **488:** 129–142, 2002.
D.M. Harper, R. Boar, M. Everard & P. Hickley (eds), Lake Naivasha, Kenya.

Feeding of the exotic Louisiana red swamp crayfish, *Procambarus clarkii* (Crustacea, Decapoda), in an African tropical lake: Lake Naivasha, Kenya

Andrew C. Smart[1,4], David M. Harper[1,*], François Malaisse[2], Sophie Schmitz[2], Stephanie Coley[1] & Anne-Christine Gouder de Beauregard[3]
[1]*Department of Biology, University of Leicester, Leicester LE1 7RH, U.K.*
[2]*National Botanic Garden of Belgium, Domaine de Bouchout, B-1860 Meise, Belgium*
[3]*Laboratoire d'Ecologie, FUSAGx, Passage des Déportés 2, B-5030 Gembloux, Belgium*
[4]*Current address: Duchy College, Rosewarne, Camborne, TR14 0AB Cornwall, U.K.*
**Author for correspondence. E-mail: dmh@le.ac.uk*

Key words: submerged aquatic macrophytes, omnivory, freshwater crayfish

Abstract

The Louisiana red swamp crayfish, *Procambarus clarkii*, has had a major impact on aquatic ecosystems in Europe and America where it has been introduced for human consumption. In Lake Naivasha, where it was introduced in 1970, it is suspected of causing the complete disappearance of native floating-leaved and submerged plants by 1982 and again by 1996; recovery of submerged plants occurred in between, concurrent with a decline in *P. clarkii*. Examination of the feeding of *P. clarkii* by microscopic analysis of gut contents collected from the lake confirmed information from both its native and exotic ranges; that it is an omnivore with green plants and insects the predominant items recognised. Feeding experiments confirmed this omnivory, with a preference from herbivory over carnivory. They further showed the importance of native plants as opposed to exotic water hyacinth *Eichhornia crassipes*. It is concluded that *P. clarkii* could well have accounted for the observed elimination of native plant species in Lake Naivasha; the cyclical nature of populations of submerged plant species and of *P. clarkii* in inverse proportion to each other lead us to hypothesise that *P. clarkii* is a 'keystone' species in the lake ecosystem.

Introduction

The feeding behaviour of the Louisiana red swamp crayfish, *Procambarus clarkii* (Girard, 1852), has been widely studied, principally because of concern over its wide distribution outside its native range (the southern United States of America and northern Mexico) (Hobbs et al., 1989; Correia & Costa, 1994). Lowery & Mendes (1977b) noted that the water lily *Nymphaea nouchalii* var. *caerulea* had declined in Naivasha after *P. clarkii* introduction and suggested that it was feeding on this plant, a suggestion supported by Hofkin et al. (1991) following experiments. *P. clarkii* can exist on decomposing plant matter (Sommer & Goldman, 1983), or on detritic plant material and associated 'microbiota' (Morshiri & Goldman, 1969; Mason, 1975; Avault et al., 1983; Wiernicki, 1984; Avault & Brunson, 1990). Adults tend towards herbivory (Mason, 1975; Olsen et al., 1991; Ilhéu & Bernardo, 1993a) while juvenile crayfish are carnivorous or omnivorous. Other authors have shown that it is carnivorous and even cannibalistic (Lorman & Magnuson, 1978; Ilhéu & Bernardo, 1993a). Hofkin et al. (1992), found experimental evidence that *P. clarkii* are active predators of snails. Other experimental studies have identified predation of snail egg masses and a significant negative association between *P. clarkii* presence and snails in the field (Hofkin et al., 1991). Benthic macro-invertebrates with reduced escape reactions (Odonata, Ephemeroptera larvae and planorbid snails) were selected in laboratory trials of various crayfish (Rickett, 1974; Matthews et al., 1993; Ilhéu & Bernardo, 1993a). Schoener (1971) and Ilhéu & Bernardo (1993b) argue that adult crayfish prefer animal food but switch to phyto/detritophageous feeding when predation efficiency is low. *P. clarkii* is known to predate *Biomphalaria pfeifferi* and *B. glabrata*, the molluscan intermediate hosts of human schistosiomiasis in Kenya (Hofkin et al., 1991; Hofkin et al., 1992). Ilhéu & Bernardo (1993a) found that adults in an introduced population in Portugal take invertebrates when vegetation is not readily available and that animal food can make up 85% of the diet. However, *P. clarkii* will eat plant detritus if no animals are present and plant detritus in preference to plant material (Ilheu & Bernardo, 1995). *P. clarkii* therefore has a diverse diet, feeding on green plants, animal or plant detritus, periphyton, benthos, plankton, agriculture by-products or microbially enriched detritus (Huner, 1983). The preferences appear to switch according to availability of food items. Plant detritic material is also the main component of the stomach contents of several other crayfish species, together with aquatic macroinvertebrates, particularly insect larvae (Prins, 1968; Dean, 1969; Rickett, 1974; Mason, 1975; Covich, 1977; Abrahamsson, 1996).

There are no indigenous species of crayfish on mainland Africa, but expanding populations of *P. clarkii* can now be found in Kenya, Uganda and Zambia (Arrignon et al., 1990). Its introduction into new areas has caused the reduction or even disappearance of several indigenous aquatic plants and damaged shores through tunnelling (Sommer & Goldman, 1983). *P. clarkii* was introduced to Lake Naivasha in 1970 when 300 individuals were released into the lake (Oluoch, 1990). It was first harvested in 1975 producing a yield of several hundred metric tonnes per annum until exports ceased in 1983 (Harper et al., 1990). Now approximately 40 metric tonnes are caught annually for local consumption (mainly tourism). It is the main prey item in the diet of black bass (*Micropterus salmoides*) over 260 mm in size (Hickley et al., 1994), which is one of three fin-fish species exploited commercially in the lake.

Field observations made upon Lake Naivasha over the past two decades have suggested that *P. clarkii* has been responsible for the complete disappearance of the floating-leaved and submerged native plant species (Harper, 1991; Harper et al., 1995; Hickley & Harper, 2002). This has happened twice; first in 1982 when they were totally absent from Naivasha but present in Lake Oloidien and then in 1996 when they were absent from both. The submerged plants reappeared, in abundance, between 1987 and 1993 but with only limited spatial or temporal reappearance of the once-widespread *Nymphaea nouchalii* var. *caerulea* (Harper et al., 1995; Gouder et al., 1998). The inverse correlation between *P. clarkii* abundance in the lake (sampled as importance value in the diet of large-mouthed bass *Micropterus salmoides* Lacépède) and the area covered by submerged plant beds (Hickley & Harper, 2002) led us to the belief that the abundance of the former was directly responsible for the decline and disappearance of the latter.

The present paper describes observations and experiments that are part of a wider study, one objective of which is to test the hypothesis that the exotic *P. clarkii* has become a keystone species in this lake because of the consequence for the remainder of the ecosystem of native plant species loss. The null hypothesis tested here is that *P. clarkii* does not have a dietary preference for green plant material.

Study site

Lake Naivasha is an endorheic freshwater lake lying in the eastern part of the Rift Valley about 80 km south of the equator and 100 km north west of Nairobi (0° 45′ S, 36° 20′ E). The lake, at 1890 m altitude, is the highest of the Kenyan Rift Valley lakes (Harper et al., 1990; LNROA, 1993). It has a surface area of between 100 and 150 km^2 caused by fluctuating lake volume that is itself a major influence on the macrophyte and animal communities (Gaudet, 1977; Gaudet, 1979; Vincent et al., 1979; Harper, 1991; Verschuren et al., 2000) (Fig. 1). Some water characteristics vary over time as a consequence, conductivity between 250 and 400 $\mu S\ cm^{-1}$ for example. Short-term changes relate more to climate; surface temperature varies between 18 and 26.5 °C for example but the diurnal variation is greater than the seasonal (unpublished data). The water is rich in carbonates, with Na^{+} and Ca^{2+} as major cations; mean pH value is 8.5 (Litterick et al., 1979; Gaudet & Melack, 1981; Burgis & Mavuti, 1987; Harper et al., 1995). Its water, fish and ecosystem form important resources for local people, horticulture and tourism (Harper, 1992; Harper et al., 1995). As well as *P. clarkii*, the lake harbours several exotic species with a significant impact (Harper et al., 1990). Two exotic floating plants arrived unintentionally; the water fern *Salvinia molesta* in the late 1960s and the water hyacinth, *Eichhornia crassipes*, in 1988 (Harper et al., 1995). Observations had shown that *P. clarkii* used *E. crassipes* plants extensively for shelter.

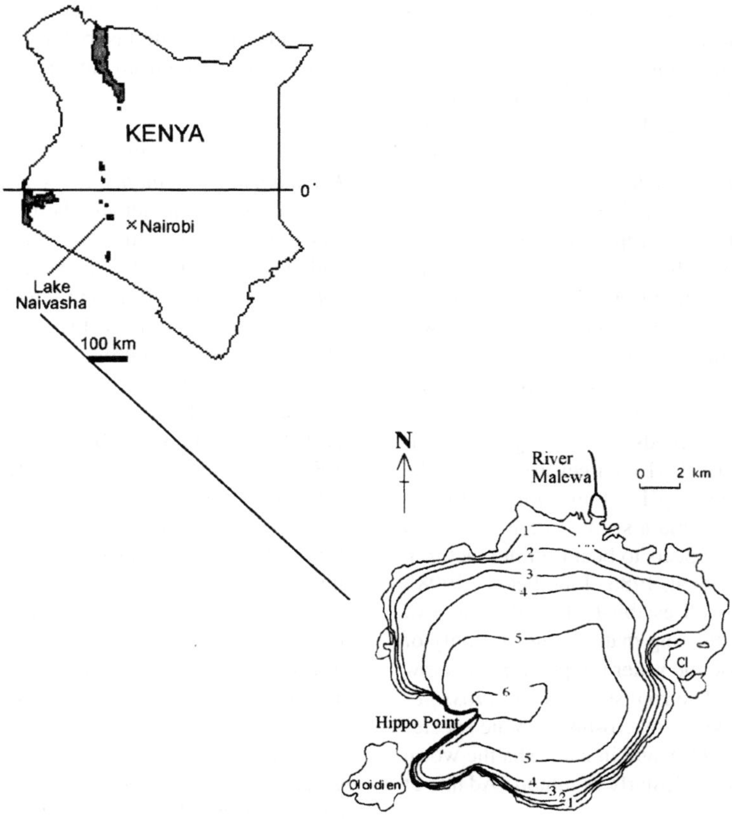

Figure 1. Location Map of Lake Naivasha, Kenya. After Hickley et al. (2002).

Ecology of the experimental animal and plants

Louisiana red swamp crayfish

P. clarkii has a mean size of 100 mm and large individuals can reach a length of 200 mm (Arrignon et al., 1990; Arrignon, 1996). Sexual maturity is generally reached at 11 months (Oluoch, 1990) and seems to be dependent on water levels (Guttiérez-Yurrita & Montes, 1999). Life span at low latitudes does not exceed 3 years but can reach 5 years in higher latitudes. *P. clarkii* is extremely tolerant of poor water quality (Arrignon et al., 1990): oxygen >3 ppm; alkalinity >50 ppm (in CaCo3); pH of 6.5–8.5; salinity <15% and temperature of 22–25 °C. Open water with light penetration of less than 40 cm appear to be the preferred habitat; it copes well with alternatively inundated and dry areas through burrowing (inundation periods allow the proliferation of the macrophyte components of its diet and many of its predators are eliminated during the dry periods). Among macrophytes identified in Louisiana as preferential habitat of *P. clarkii* are *E. crassipes* and several *Cyperus* spp. (Arrignon et al., 1990), also present in Lake Naivasha. At Naivasha, *P. clarkii* is an important component of the diet of *M. salmoides* (Hickley et al., 1994; Hickley & Harper, 2002) and heron, egrets, ibis, fish eagle, marsh mongoose and otter are above-water predators (pers. obs.).

Eichhornia crassipes

Water hyacinth, *Eichhornia crassipes* (Mart.) Solms, is an indigenous monocotyledonous species of tropical South America (François, 1969) that has become a widespread exotic pest of tropical zones (Blackburn et al., 1969; MacDonald & Wolverton, 1979; De Parcevaux, 1980). An attractive plant in flower, it has been widely cultivated as an ornamental pond plant, which has facilitated its spread. *E. crassipes* is semi-amphibious and usually floats but can also survive rooted in wet margins. The main root, which may be several meters long, supports dense (up to 70 cm^{-1}) lateral roots. The plant has rosette characteristic leaves and a lilac white inflorescence (François, 1969; Bruynseels & Lawalrée, 1981). The fruit is a dehiscent capsule containing 3–364 seeds. The total production per plant varies from a few seeds to more than 5000. Seeds sink to the bottom of water when mature and remain viable for at least 15 years (Blackburn et al., 1969). *E. crassipes* lives in stagnant freshwaters or in slow currents; is a megatherm species with its optimum growth in a temperature between 25 and 30 °C. The plant can not survive more than 5 weeks when the temperature exceeds 34 °C and growth is almost stopped below 10 °C (De Parcevaux, 1980). The species tolerates extreme pH within a range of 4–10 although its growth optimum seems to correspond to 7. A pH value of 4.2 appears to be a limiting factor in Africa (Berg, 1959; Gopal, 1986; Tabuti et al.; 1998). The species is now well established on the shore of Naivasha and forms floating mats. Plants are only of moderate size because the water temperature is below optimum for growth. The root system of *E. crassipes* shelters a large number of macro-invertebrate species (Bailey & Litterick, 1993), and in Naivasha the worm *Alma emini* Mich is particularly common (Adams et al., 2002).

Najas horrida

Najas horrida A. Br. Ex Magn. syn. *Najas pectinata* auct. Afr. Non (Parl.) Magnus is a tropical submerged annual aquatic herb with roots. Thin, ramified stems carry simple sessile leaves (3.2–20.0 mm long). Isolated flowers are located at the stems' node, and produce achenes (Triest & Symoens, 1983). It is frequently found in deep waters (in Naivasha from 0.8 to 2.2 m depth, with an average of 1.36 m), anchored on hard bottom or rocks (Gouder et al., 1998).

Potamogeton octandrus

Potamogeton octandrus Poiret is an aquatic herb with underground rhizomes and roots, with an entirely or nearly submerged habit. Thin, ramified stems support floating and submerged leaves (20–45 mm long, 1 mm width). Paired green flowers produce small fruits (2.0 mm long, 1.5 mm width) (Lisowski et al., 1978). A rare plant in Kenya, it is considered a pioneer species in shallow locations (0.60 m depth in average, from 0.40 to 0.90 m). In Naivasha its appearance and spread is sudden and fast but it declines or disappears under competition from other plant species (Gouder et al., 1998).

Potamogeton pectinatus

Potamogeton pectinatus L. is a cosmopolitan submerged perennial aquatic herb, with underground roots and ramified rhizomes. The stems, cylindrical and thin, reach up to 3 m long and carry threadlike and submerged leaves (4–12 cm length, 1–2 mm width), which vary according to the depth (Lisowski et al., 1978). The spike inflorescences produce ovoid seeds (3–4 mm) up to 11.500 per square metre in some areas of Lake Naivasha. It occurs here at an average depth of 1.20 m (amplitude: 0.65–1.70 m, Gouder et al., 1998). *P. pectinatus* is widespread in alkaline waters (pH 7.0–9.0 in average) and has a large temperature and a tolerance to polluted and turbid waters (Kantrud, 1990). It is adapted to currents and water level fluctuations but it is not resistant to wave action. This species is well adapted in Naivasha as water movements are a common feature, as well as a fluctuating transparency. It was present with other species in the adjacent Lake Oloidien in the 1980s but the latter lake became increasingly saline to the present day as a consequence of evaporation of water maintained by groundwater input from Naivasha (the two lakes had been connected by surface water only between 1979 and 1981). Species died out as salinity increased and *P. pectinatus* was the last to disappear at a conductivity of 1800 $\mu S\ cm^{-1}$ in 1996.

Potamogeton schweinfurthii

Potamogeton schweinfurthii A. Benn is a submerged perennial aquatic herb, with underground rhizomes and roots. The stems, up to three metres long, supports broad leaves (11–25 cm length, 9–30 mm width), the upper sometimes floating (Lisowski et al., 1978). The spike inflorescences, up to 10 cm long, are held above the water. It is a dynamic species with mainly vegetative reproduction through stolons (runners) but can produce 500–12 000 seeds m^{-2} (Gouder et al., 1998).

Methods

Collection of P. clarkii *from the lake*

Large crayfish (carapace length greater than or equal to 40 mm) were removed from the lake using traps while medium (carapace length 20–40 mm) and small crayfish (carapace length below 20 mm) were collected from floating *E. crassipes* using methods described in Harper et al. (2002).

Examination of gut contents

Animals were placed in a refrigerator to slow down their metabolism and then killed by freezing. They were removed from the freezer and allowed to thaw before dissection. Juvenile animals were dissected by removing a dorsal section of the carapace exposing the stomach. The pyloric and cardiac sections were removed intact by severing the gut, then reaching beneath the stomach with forceps to sever the oesophagus. Connective tissue generally broke away as the stomachs were removed. Adult animals were dissected by removing a section of the lateral body wall and lifting to expose the pyloric and cardiac stomachs. Stomachs were removed by severing the connection to the gut and the oesophagus and cutting connective tissue as the two parts were lifted away. Attempts to lift adult stomachs directly from a dorsal incision frequently led to rupture and loss of material. Stomachs were preserved in 70% Industrial Methylated Spirits or 4% Formalin for later examination. Examination was made on a Leica Zoom Stereo-Microscope at ×5 to ×30 magnification. Stomach contents were emptied into a Petri-dish and sorted to separate vegetation and detritus from identifiable animal fragments. Stomachs were inverted to ensure all material was examined. Five slides were made of samples of material from each gut to recognise oligochaete chaetae. One hundred and seventy eight guts were examined in this way.

Experiments

All experiments were undertaken in a laboratory located on the south-west shore of Lake Naivasha. Water, plant and animal material used in the experiments were all taken from the lake. *P. clarkii* were kept in opaque basins or tanks (9 l; 1 l or 300 ml; depending on size category) with a refuge tube and a piece of plastic mesh for cover. Animals were starved for 24 h before the beginning of the experiments. Where possible, the lighting regime followed natural conditions but occasionally lights were on for a short time after sunset. Regularly oxygenated lake water was used which was renewed every 2 or 3 days to minimise variations of oxygen and ensure clean and odourless water. In laboratory conditions the physical and chemical parameters of water were: 24 °C, 2.5–4.0 mg.l^{-1} O_2, 0.35 mS cm^{-1} conductivity.

Ex situ *herbivory of water hyacinth*

In the laboratory, 32 crayfish (8 blocks) were fed four treatments of water hyacinth: leaves, floats (petioles), roots (containing *Alma eminii*), roots where *Alma* was removed. One crayfish was placed in each basin and received one of the four food types. Food in the system was weighed every 3 days over a period of 24 days by removing the crayfish; filtering the water and removing any plant material for weighing; then replacing all plant material and crayfish in the basins.

Ex situ *herbivory of submerged plant species*

Crayfish of each size category were presented with a known wet weight of a submerged plant species over a set period of time. The quantity of macrophyte remaining after the time period was measured and recorded. Trials were set up using two or three crayfish of each size category for either 24 h or 5 days (without disturbance, measuring *Najas horrida*, *Potamogeton pectinatus*, *P. schweinfurthii* at the end of the trial); and using one crayfish of large and medium size categories over 19 days (measuring the quantity of *Najas horrida*, *Potamogeton octandrus*, *P. pectinatus*, *P. schweinfurthii* remaining every 2 or 3 days). The initial trials (with no disturbance) were replicated twice, while the latter (19 day) experiment was set up in 8 blocks with 4 treatments and controls.

In situ *herbivory of submerged plant species*

Two *P. clarkii* (>40 mm) or 4 crayfish (20–40 mm) were enclosed in a metal cage of approximately 30 cm × 30 cm × 30 cm, with a 2 mm mesh on all surfaces, along with measured quantities of macrophyte. The cage was then submerged in the lake to a depth of 1–2 m, for 7 days. At the end of this period, the plant material was removed and wet weighed to determine the quantity removed by the crayfish.

Ex-situ *herbivory of mixed submerged macrophytes*

Known quantities of submerged macrophytes were presented to one, two or three *P. clarkii*, in the laboratory, for a period of 5 days. The quantity of plant material remaining at the end of the trial was wet weighed and the quantity presumed eaten determined.

Table 1. The percentage occurrence of identifiable food items in crayfish stomachs. Data separated by carapace length (mm) and year of examination. n. d. = species not determinable

	1992		1994		1999	
	<35 mm	>35 mm	<35 mm	>35 mm	<35mm	>35 mm
Total examined	12	8	30	40	38	20
Najas	33	50	33	27	0	0
Green plant, n.d.	0	0	15	70	68	30
Plant detritus	0	25	3	0	8	35
Micronecta sp.	83	37	27	36	16	10
Chironomidae	17	12	13	17	5	0
Ecnomus sp.	17	0	0	0	5	0
Alma emenii	0	0	0	5	0	0
Coleoptera	0	0	0	2	3	0
Mosquito larvae	0	0	0	7	0	0
P. clarkii	0	0	0	5	3	0
Gastropoda	0	0	0	2	0	5

Ex situ *carnivory of macro-invertebrates and food preferences*

P. clarkii of each size category were presented with known numbers of the same or two different types of macro-invertebrate over 24 h. Each crayfish was in a tank with a refuge and plastic mesh to act as an inedible substratum. Numbers (or length in the case of *A. eminii*) of animals remaining at the end of the trial were recorded.

Ex situ *herbivory in the presence of animals and carrion*

This was to establish the crayfish preferences between animal food (worms and fish) and vegetable food (submerged plants). Various densities and size categories of *P. clarkii* were placed into containers with macrophytes and *A. eminii* or macrophytes and fish carrion for a period of 3 days. The quantity of macrophyte eaten was compared with and without animal material.

Results

Analysis of gut contents

The food items were predominantly plant material, either as large amounts of detritus or small unidentified fragments (Table 1, Fig. 2). Approximately 60% of adults had empty stomachs in 1999 (when there were no submerged plants in the lake) compared to 1% in 1994 and 13% in 1992 (when there were). No evidence of oligochaete prey was found from examination of microscope slides. The three main dietary items were *Najas* or unidentified green plants, *Micronecta* and chironomid larvae; the two animal groups being those that were in high densities (Clark et al., 1989).

There were significant differences between adult and juvenile food items (χ^2, $p < 0.001$ (Yates' correction)), apart from the invertebrates in 1994, which were equally distributed between adults and juveniles (Fig. 3). The overall dietary strategy of individuals (omnivory, carnivory and herbivory) between each year recorded was significantly different for adults (χ^2, $p < 0.001$ between 1992 and 1994; and 1992 and 1999) but not juvenile diets in 1994 and 1999 (Fig. 4).

Ex situ *herbivory of water hyacinth*

During the 24 days when *P. clarkii* was offered only *E. crassipes* plant items, leaves (Fig. 5) were significantly removed compared to losses in the controls due to decay (Newman–Keuls test, $p < 0.05$). Bite marks were made on the decomposing edge of the leaves during the experiment. No significant differences existed between the treatment and control for petioles (Fig. 5). Significant differences between control and crayfish treatments appeared after 12 days in both experiments with roots (Fig. 5). Examination of the changes with time of the four alternates (Fig. 6) show that the leaf biomass eaten is greater but that this is only small (approx. 0.15 g crayfish^{-1} d^{-1}).

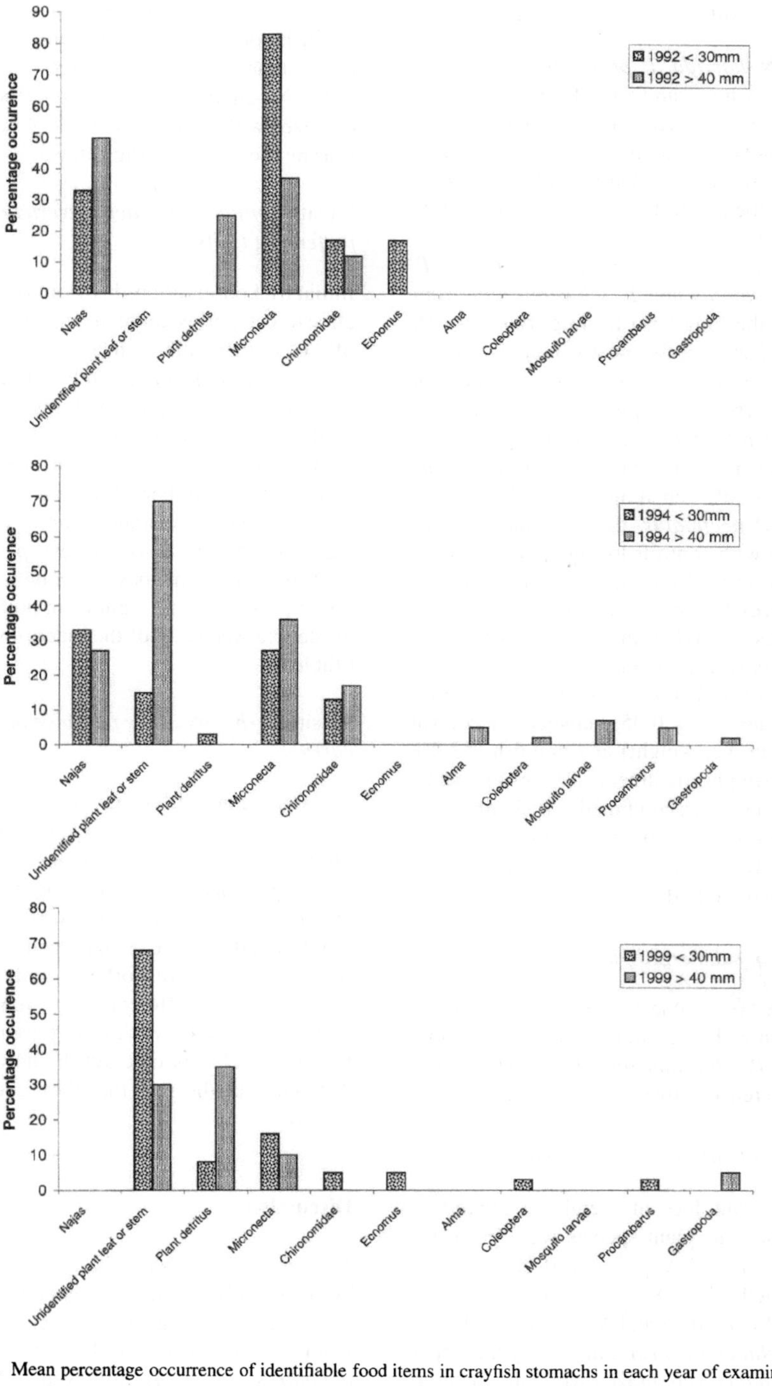

Figure 2. Mean percentage occurrence of identifiable food items in crayfish stomachs in each year of examination.

Ex situ *herbivory of submerged plant species*

The removal of *N. horrida, P. pectinatus* and *P. schweinfurthii* after 24 h (mean from 10 replicates) and 5 days (mean from 2 replicates) was not significant (χ^2 analysis). However, in all treatments, *P. pectinatus* was the macrophyte eaten the least, while *N. horrida* was consumed to the greatest extent in four of the five treatments (Fig. 6).

However, crayfish began eating *N. horrida, P. octandrus, P. pectinatus* and *P. schweinfurthii* from the beginning of the long term feeding trial (Fig. 7). The juvenile crayfish were the most voracious with an apparent preference for *N. horrida* (eaten in 10 days), then *P. octandrus* (16 days), *P. pectinatus* (5% remaining) and *P. schweinfurthii* (19% remaining after 19 days). Adult crayfish consumed less, with *P. octandrus* eaten most rapidly (6% remaining after 19 days). In all cases except *P. pectinatus*, the plant material deteriorated rapidly with controls losing up to 40% over the period of the trial. Newman–Kuels test showed that the differences between reduction of plant wet weight for all sizes of crayfish and the control were not significant for *N. horrida, P. octandrus* and *P. schweinfurthii. P. pectinatus* did show a significant reduction (Newman–Kuels test; $p < 0.05$) between the control (7.9% reduction in wet weight) and crayfish (52.2% reduction in wet weight) treatments. The mean quantity of all four plant species combined was significantly lower than the natural loss of the control plants for adults (Newman–Kuels test; $p < 0.05$) and juveniles ($p < 0.01$, Dunnett method).

In situ *herbivory of submerged plant species*

Over 7 days in the lake (Table 2) up to 10 g of plant per crayfish was eaten with an adult preference for *N. horrida* compared to *P. schweinfurthii*, although there was no statistical difference between juveniles and adults.

Ex-situ *herbivory of mixed submerged macrophytes*

There was significant loss of weight compared to controls when the four plant species were given together to crayfish (ANOVA, $p < 0.001$ all sizes and densities combined) (Fig. 8). The crayfish showed preferences for *P. octandrus* and *N. horrida*, followed by *P. schweinfurthii* and *P. pectinatus*. Only the upper leaves of *P. pectinatus* were grazed while the material from *P. schweinfurthii* consumed included decomposing leaves. An increase in the number of crayfish per experiment increased the plant removal, with single crayfish consuming a significantly smaller proportion (55%) than two (70%) or three (75%) crayfish ($p < 0.01$). No significant interaction was found for different sizes with both size classes of crayfish apparently sharing the same feeding habits.

Ex situ *carnivory of macro-invertebrates and food preference trails*

Initial trial observations had confirmed that crayfish of all size categories can consume all of the prey items offered to them although large crayfish have difficulty catching *Micronecta* sp. and do not crush and digest the shell of gastropod molluscs to the same extent as the smaller size categories. The results (Table 3) confirm the observations with almost total removal of prey. The number of *Micronecta* sp. and gastropod shells taken was significantly lower by the adult crayfish (ANOVA & Tukey Test; $p < 0.001$).

When experiments were conducted with two different prey items, no significant preference (ANOVA) existed between any of the macro-invertebrates tested (Table 4).

Ex situ *herbivory in the presence of animals and carrion*

No significant difference in consumption occurred when worms and submerged plant material were offered together (Fig. 9). However, consumption of submerged macrophytes was significantly reduced in the presence of fish (*Tilapia* sp.) carrion ($p < 0.001$). Single crayfish gave a significant reduction of plant weight when it was offered with either worms or carrion but no difference was found when two or three individual crayfish were present ($p < 0.05$). Different size classes of crayfish showed no significant difference in the rate that they consumed plants or worms.

Discussion

P. clarkii at Lake Naivasha is opportunistic, and may even switch between diets depending on the most optimal feeding strategy at the time. In *in-situ* and *ex-situ* feeding experiments, it showed a preference for *P. octandrus* and *Najas*, followed by *P. schweinfurthii* and *P. pectinatus*. Crayfish removed from beds of submerged macrophytes dominated by a single species for

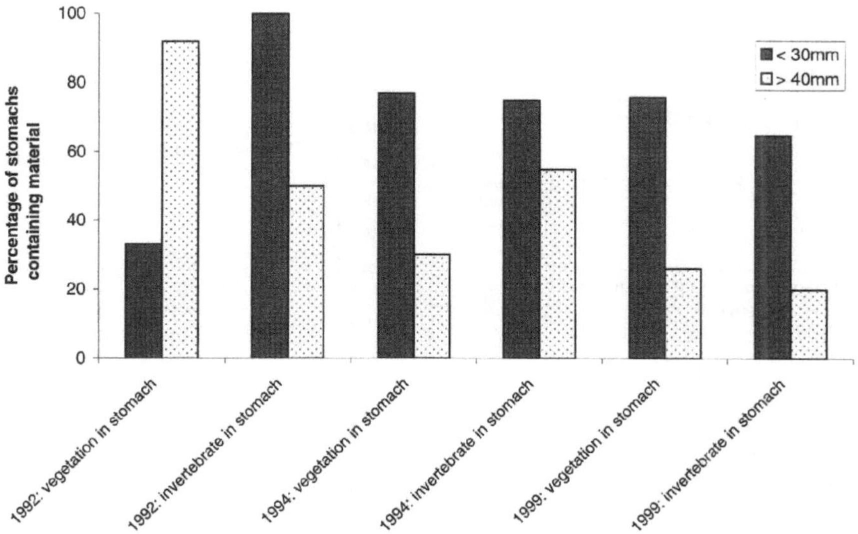

Figure 3. Mean percentage of animal and plant food items, in adult and juvenile crayfish in the threeyears of collection.

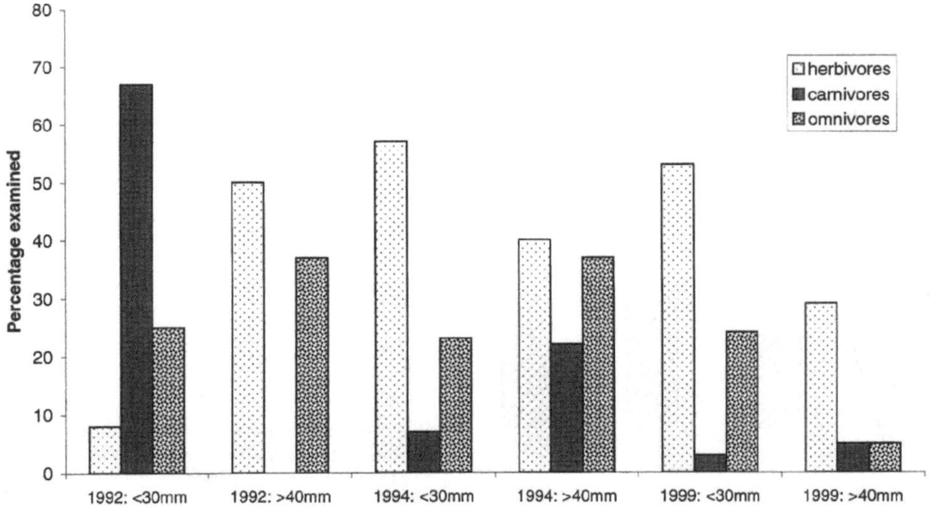

Figure 4. Mean percentage of *P. clarkii* individuals, which were herbivores, carnivores or omnivores in each year ofcollection.

experiments may have been habituated to feeding on that species and that may have affected results of these trials. However, the crayfish impact on the pioneer species, *P. octandrus*, could well influence the colonisation strategy of submerged vegetation and may impede the reappearance of that species. *P. clarkii's* major way of feeding is by cutting (observations in the experiments) and so macrophytes are unlikely to remain connected to their root systems in the presence of crayfish. Harper et al. (1990) and Hofkin et al. (1991) implied that the impact of this feeding strategy strongly implicated the crayfish as the main cause of the decline in abundance of *Nymphaea nouchalii* var. *caerulea* in Naivasha. *P. clarkii* showed little preference for different animal species in experimental conditions, eating whatever they came across.

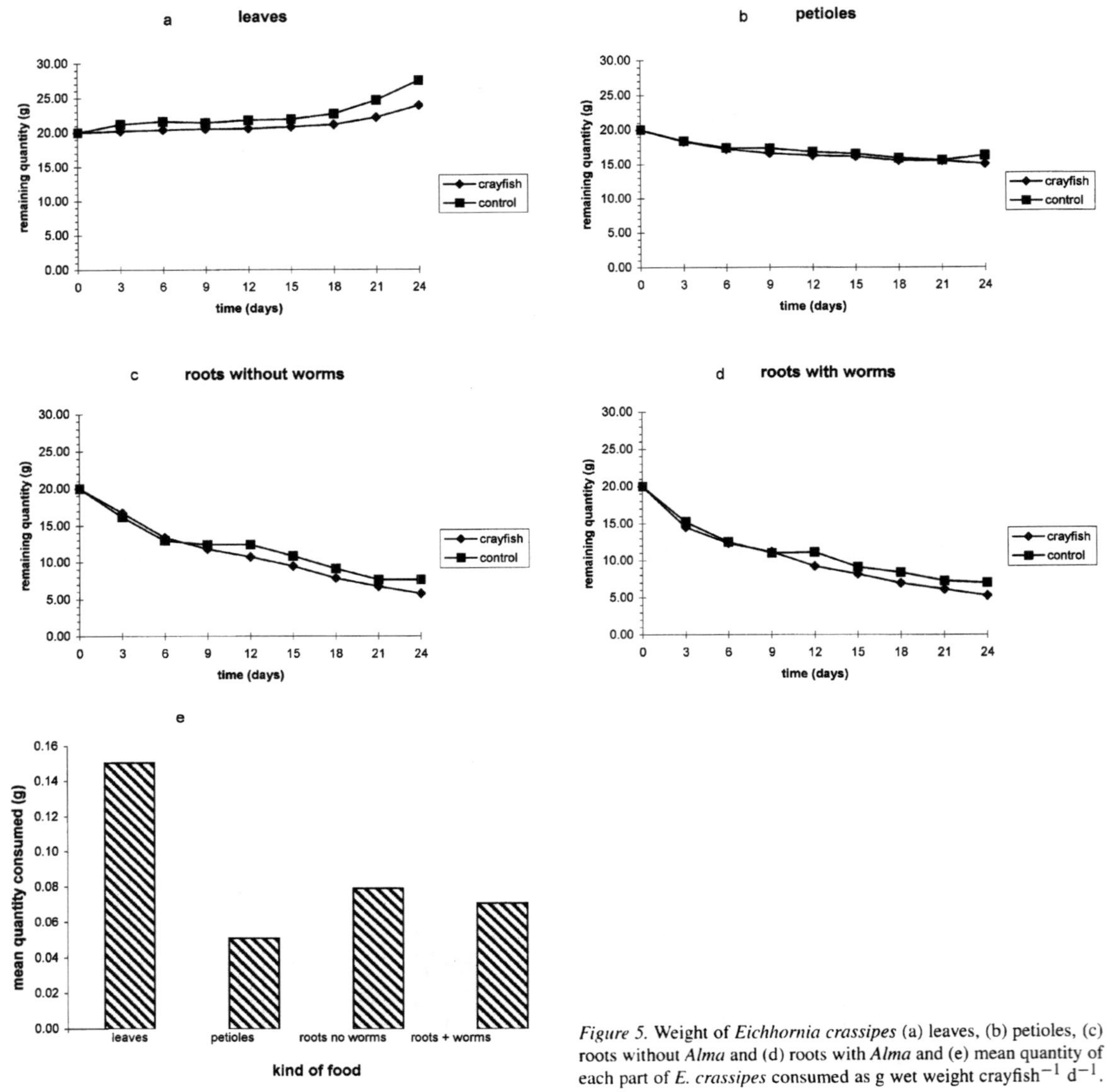

Figure 5. Weight of *Eichhornia crassipes* (a) leaves, (b) petioles, (c) roots without *Alma* and (d) roots with *Alma* and (e) mean quantity of each part of *E. crassipes* consumed as g wet weight crayfish^{-1} d^{-1}.

Highly mobile species such as *Micronecta* sp., were less easily consumed by adult *P. clarkii*. There did not appear to be a preference for fish carrion over macro-invertebrates in the laboratory.

When the submerged plant populations collapsed in Naivasha (as at the beginning of the 1980s and in the mid-late 1990s) *P. clarkii* was left with only detritus and a reduced availability of invertebrates (because of the absence of submerged plant habitat) together with *E. crassipes* in the latter period. These results show that no part of *E. crassipes* can compensate for the native plant food loss and suggest that the collapse of the *P. clarkii* population, which occurred following the first disappearances of submerged plants, may be a proximate consequence of food shortage. Food shortage may not be the ultimate cause of death, as predation on weakened individuals unprotected by the visual refuge of macrophyte beds in the lake or in the shallows is much more likely. The number of predators known to feed on crayfish in Naivasha is

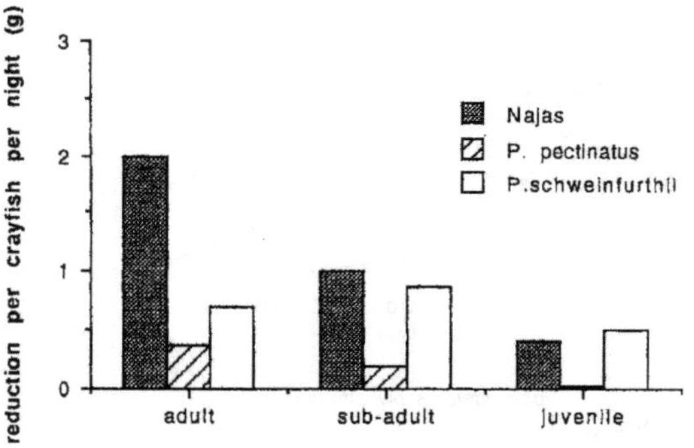

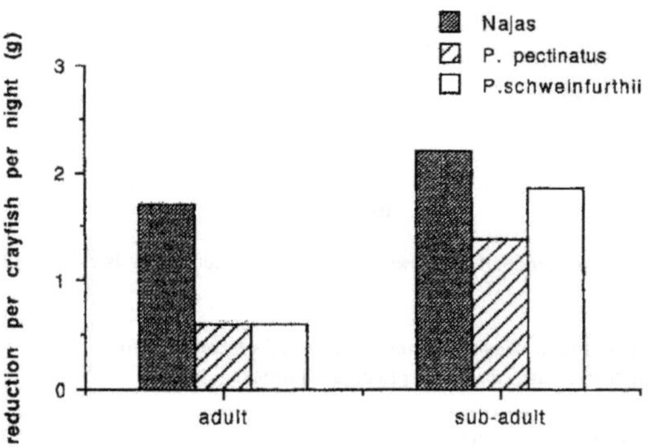

Figure 6. Reduction of submerged macrophyte wet weight by *P. clarkii* during 5 day feeding trials expressed as g wet weight crayfish^{-1}d^{-1}.

Table 2. Reduction in macrophyte wet weight crayfish^{-1} d^{-1} over 7 days feeding in the lake

Plant species	*N. horrida*	*P. schweinfurthii*
P. clarkii (carapace length)		
>40 mm	10.7	5.9
<40 mm	2.5	2.6

high (Fig. 10). Moreover, *P. clarkii* has in the past suffered from muscle-wasting disease, first recorded by Lowery & Mendes (1977a). They found this reduced tail weights at Naivasha to 10% of normal (Lowery & Mendes, 1977b) while Mikkola (1979) estimated reductions to 37% (male) and 27% (female). Lindquist & Mikkola (1979) identified the causes of this disease to be accumulation of material on gills, bacterial growth that leads to hypoxia and starvation.

Table 3. Mean number of potential prey items removed by *P. clarkii* where *n* (prey items) = 5 in each container with a single crayfish (unless otherwise indicated). Each crayfish size category was replicated ten times.

	Crayfish carapace size		
	<20 mm	20–40 mm	>40 mm
Oligochaeta	4.4	5	5
Chironomid larvae	5	5	9.7 (n = 10)
Micronecta	4.5	4.7	1.8 (n = 10)
Gastropods (tissue)	4.9	4.6	4.8
Gastropods (shells)	4.1	4.6	2.3
Alma (12–15 mm long)	No data	84% length	90% length
Ostracoda	5	5	5

The dramatic impact of crayfish species, both native and exotic, on lentic ecosystems, is well known. Mathews et al. (1992, 1993) demonstrated by a combination of fieldwork, before and after crayfish elimination through plague, and enclosures into which crayfish were added, that the native *Austropotamobius pallipes* was important in regulating both lake vegetation by grazing and community structure by predation on otherwise-dominant invertebrate species such as *Gammarus* or chironomid larvae. Lodge et al. (1994) demonstrated by enclosure experiments that the exotic omnivorous crayfish *Orconectes rusticus,* controlled the abundance of macrophytes, snails as well as periphyton in the littoral zone of lakes in the northern United States. As a consequence of studies such as these, together with the detrimental effects on biodiversity that exotic introductions have, Lodge et al. (1998) have suggested that management of exotic species should be focussed on preventing introduction, because elimination is almost impossible once introduction has occurred. *P. clarkii* is the most introduced crayfish in the world (Correia & Ferreia, 1995) which is spreading widely in Kenya (Lowery & Mendes, 1977a). In many agricultural areas within its introduced range, it reaches pest status because its burrowing behaviour causes irrigation banks to leak and earth dams to collapse (Correia & Ferreia, 1995). Despite its agricultural damage risk as well as its known ecological disruption it has potential benefits as a biological control of snail species which are the intermediate host of schistosomiasis (Hofkin et al., 1991). In Kenya, at least, the benefits have to be evaluated against the disbenefits if any management strategy is considered.

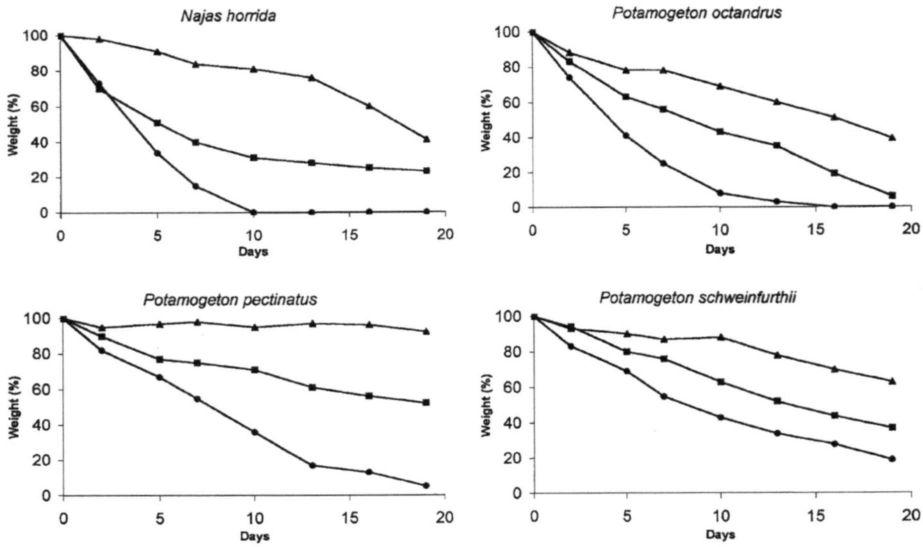

Figure 7. Reduction in wet weight of four submerged macrophyte species in the presence of *P. clarkii* during 19 day feeding trials expressed as a percentage of the original weight.

Table 4. Number of potential prey items removed by *P. clarkii* where initial density of prey items was 5 of each of two different taxa in each container with a single crayfish. Each crayfish size category was replicated 10 times

	Crayfish carapace size					
	<20 mm		20–40 mm		>40 mm	
Prey species	Species A	Species B	Species A	Species B	Species A	Species B
Micronecta & Ostracoda	4.5	4.6	5	5		
Chironomid larvae & Ostracoda	5	5	5	5		
Oligochaeta & Ostracoda	5	5	5	5		
Micronecta & Gastropod	4.7	4.9				
Alma eminii & Snails					4.7	3.5

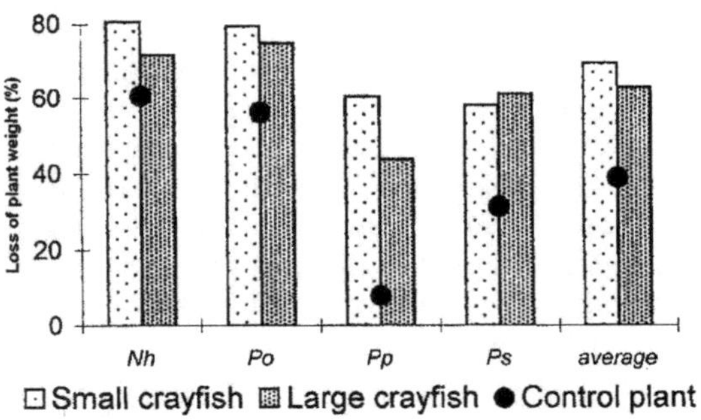

Figure 8. Total loss of plant wet weight as a percentage of the original weight compared with control loss for adult and juvenile crayfish.

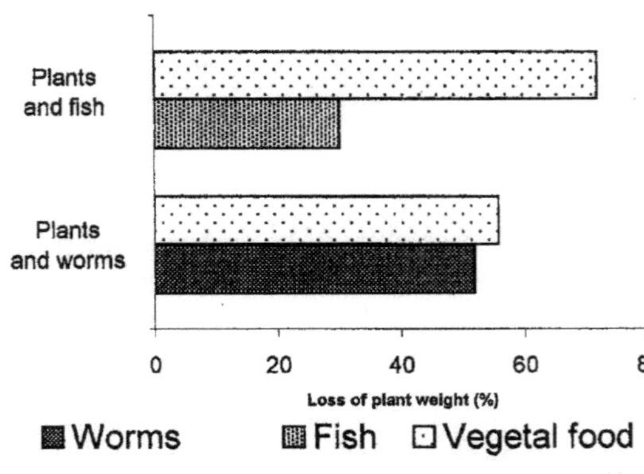

Figure 9. Reduction of submerged plant weight with and without the presence of animal food.

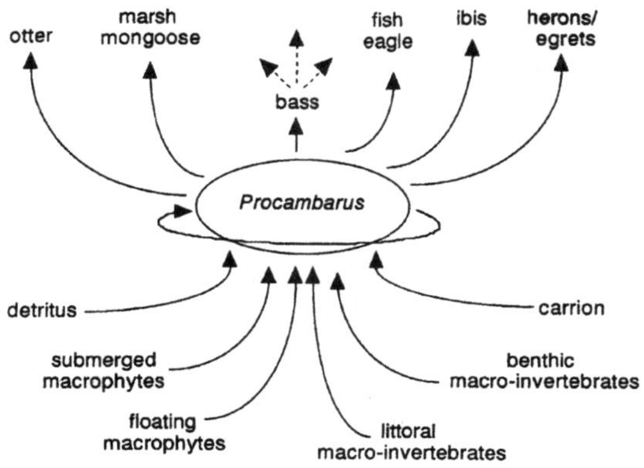

Figure 10. The position of *P. clarkii* in the Naivasha food web.

Acknowledgements

This study formed past of the work of the University of Leicester research project at Lake Naivasha, which since 1987 has been authorised by the Office of the President, Government of Kenya under research permit to Dr D. M. Harper no. OP 13/001/12C 46. The project was funded by the Earthwatch Institute, Boston U.S.A. and Oxford, England. The data collection would not have been possible without the assistance of numerous Earthwatch volunteers between 1992 and 1998. Our sincere thanks go to the numerous colleagues in Kenya in particular Kamau Mbogo, Jill & Angus Simpson and Velia Carn.

References

Abrahamsson, S. A. A., 1996. Dynamics of an isolated population of the crayfish *Asatsu astacus*. Oikos 17: 96–107.

Adams, C. S., R. R. Boar, D. S. Hubble, M. Gikungu, D. M. Harper, P. Hickley & N. Tarras-Wahlberg, 2002. The dynamics and ecology of exotic tropical species in floating plant mats: Lake Naivasha, Kenya. Hydrobiologia 488 (Dev. Hydrobiol. 168): 115–122.

Arrignon, J. C. V., 1996. Produire et Vendre de l'écrevisse. La Pisciculture Française d'eau vive et d'Étang Saumâtre et Marine 126 (n° spécial): 36 pp.

Arrignon, J. C. V., J. V. Huner & P. J. Laurent, 1990. L'écrevisse Rouge des Marais. Maisonneuve et Larose, Paris: 87 pp.

Avault, J. W. & M. W. Brunson, 1990. Crawfish forage and feeding systems. Rev. Aquat. Sci. 3: 1–10.

Avault, J. W., J. S. Forester & R. P. Romaire, 1977. Length–weight relationships of two commercially important crayfishes of the genus *Procambarus*. Freshwater Crayfish 3: 463–470.

Avault, J. W., R. P. Romaire & M. R. Miltner, 1983. Feeds and forages for red swamp crawfish, *Procambarus clarkii*: 15 years reasearch at Louisiana State University reviewed. Freshwater Crayfish: 362–369.

Bailey, R. G. & M. R. Litterick, 1993. The macroinvertebrate fauna of water hyacinth fringes in the Sudd swamps (River Nile, southern Sudan). Hydrobiologia 250: 97–103.

Berg, A., 1959. Analyse des conditions impropres au développement de la jacinthe d'eau *Eichornia crassipes* (Mart.) Solms dans certaines rivières de la cuvette congolaise. Bulletin agricole du Congo Belge et du Ruanda-Urundi L(2): 365–393.

Blackburn, R. D., L. G. Holm & L. W. Weldon, 1969. Aquatic weeds. Sciences 166: 699–709.

Bruynseels, G. & A. Lawalrée, 1981. Flore d'Afrique centrale (Zaïre, Rwanda, Burundi), Spermatophytes, Pontederiaceae. Jardin Botanique National de Belgique, Meise: 11 pp.

Burgis, M. J. & K. M. Mavuti, 1987. The Gregory Rift. In Burgis, M. J. & J.-J. Symoens (eds), African Wetlands and Shallow Water Bodies. Directory. Coll. "Travaux et Documents", O.R.S.T.O.M., Paris 211: 331–358.

Clark, F. L., A. Beeby & P. Kirby, 1989. A study of the macroinvertebrates of Lakes Naivasha, Oloidien and Sonachi, Kenya. Rev. d'Hydrobiol. Tropic. 22: 21–33.

Correia, A. M. & A. C. Costa, 1994. Introduction of the red swamp crayfish *Procambarus clarkii* (Crustacea, Decapoda) in São Miguel, Azores, Portugal. *Arquipélago* 12: 67–73.

Correia, A. M. & O. Ferreira, 1995 Burrowing behaviour of the introduced red swamp crayfish *Procambarus clarkii* (Decapoda: Cambaridae) in Portugal. J. crust. Biol. 15: 248–257.

Covich, A. P., 1977. How do crayfish respond to plants and mollusca as alternate food resources? Freshwater Crayfish 3: 165–179.

De Parcevaux, M. S., 1980. Possibilités d'utilisation de la Jacinthe d'eau en France. Colloque CENECA: 4 pp.

Dean, J. L., 1969. Biology of the crayfish *Orconectes causeyi* and its use for control of aquatic weeds in trout lakes. Technical Papers, United States Bureau of Sport Fishenes and Wildlife 24: 1–15.

François, J., 1969. Recherches expérimentales sur l'écologie de la jacinthe d'eau *Eichornia crassipes* (Mart.) Solms. Thèse de doctorat, Faculté des Sciences Agronomiques, Gembloux, tome 1: 334 pp.

Gaudet, J. J. & J. M. Melack, 1981. Major ion chemistry in a tropical lake basin. Freshwat. Biol. 11: 309–333.

Gaudet, J. J., 1977. Natural drawndown on Lake Naivasha, Kenya and the formation of papyrus swamps. Aquat. Bot. 3: 1–47.

Gaudet, J. J., 1979. Seasonal changes in nutrients in a tropical swamp: North Swamp, Lake Naivasha. J. Ecol. 67: 953–981.

Gopal, B., 1986. Water Hyacinth: Biology, Ecology and Management. Elsevier Science Publishers, Amsterdam: 471 pp.

Gouder, A.-C. de B., D. M. Harper, F. Malaisse & J.-J. Symoens, 1998. Dynamique récente et cartographie de la végétation aquatique (1960–1996) du lac Naivasha (Rift Valley, Kenya). Bull. Séanc. Acad. r. Sci. Outre-Mer 44: 373–389.

Gutiérrez-Yurrita, P. J. & C. Montes, 1999. Bioenergetics and phenology of reproduction of the introduced red swamp crayfish, *Procambarus clarkii*, in Doñana National Park, Spain, and implications for species management. Freshwat. Biol. 42: 561–574.

Harper, D. M., 1991. Primary production in Lake Naivasha, Kenya. Verh. int. Ver. Limnol. 24: 1112–1116.

Harper, D. M., 1992. The ecological relationships of aquatic plants at Lake Naivasha, Kenya. Hydrobiologia 232: 61–65.

Harper, D. M., C. Adams & K. Mavuti, 1995. The aquatic plant communities at the Lake Naivasha wetland, Kenya: pattern, dynamics and conservation. Wetlands Ecology and Management 3: 111–123.

Harper, D. M., K. M. Mavuti & S. M. Muchiri, 1990. Ecology and Management of Lake Naivasha, Kenya, in Relation to Climatic Change, Alien Species' Introductions and Agricultural Development. Environmental Conservation 17: 328–336.

Hickley, P. & D. M. Harper, 2002. Fish population and habitat changes in the artificially stocked fishery of Lake Naivasha,

Kenya. In Cowx, I. (eds), Management and Ecology of Lake and Reservoir Fisheries. Fishing Books News, Blackwell, Oxford.

Hickley, P., R. North, S. M. Muchiri & D. M. Harper. 1994. The diet of the largemouth bass *Micropterus salmoides* in Lake Naivasha, Kenya. J. Fish Biol. 44: 607–619.

Hickley, P., R. Bailey, D. M. Harper, R. Kundu, M. Muchiri, R. North & A. Taylor, 2002. The status and future of the Lake Naivasha fishery, Kenya. Hydrobiologia 488 (Dev. Hydrobiol. 168): 181–190.

Hobbs, H. H., J. P. Jass & J. V. Huner, 1989. A review of global crayfish introductions with particular emphasis on two North American species (Decapoda, Cambaridae). Crustaceana 56: 299–316.

Hofkin, B. V., D. K. Koech, J. Ouma & E. S. Loker, 1991. The North American crayfish *Procambarus clarkii* and the biological control of schitosome transmitting snails in Kenya: laboratory and field investigations. Biological Control 1: 183–187.

Hofkin, B. V., G. M. Mkoji, D. K. Koech & E. S. Loker, 1991. Control of schistosome-transmitting snails in Kenya by the North American crayfish *Procambarus clarkii*. Am. J. Trop. Med. Hyg. 45: 339–344.

Hofkin, P. V., D. M. Hofinger, D. K. Koech & E. S. Loker, 1992. Predation of *Biomphalaria* and non-target molluscs by the crayfish *Procambarus clarkii*: implications for the biological control of schitosomiasis. Ann. trop. Med. Parasit. 86: 663–670.

Huner, J. V., 1983. *Procambarus* in North America and elsewhere. Chap. 9. In Holdich, D. M. & R. S. Lowery (eds), Freshwater Crayfish. Biology, Management, and Exploitation. Croom Helm, London & Sydney and Timber Press, Portland, Oregon: 239–261.

Ilhéu, M. & J. M. Bernardo, 1993a. Aspects of trophic ecology of red swamp crayfish (*Procambarus clarkii* Girard) in Alentejo, South of Portugal. Actas VI Congreso Español de Limnología, Granada: 417–423.

Ilhéu, M. & J. M. Bernardo, 1993b. Experimental evaluation of food preference of red swamp crayfish, *Procambarus clarkii*: vegetal versus animal. Freshwater Crayfish 9: 359–364.

Ilhéu, M. & J. M. Bernardo, 1995. Trophic ecology of red swamp crayfish *Procambarus clarkii* (Girard) preferences and digestibility of plant foods. Freshwater Crayfish 10: 132–139.

Lindqvist O. V. & H. Mikkola, 1979. On the etiology of the muscle wasting disease in *Procambarus clarkii* in Kenya. Freshwater Crayfish 4: 363–372.

Lisowski, S., F. Malaisse, J.-J. Symoens & J. Van de Velden, 1978. Flore d'Afrique Centrale (Zaïre, Rwanda, Burundi), Spermatophytes, Potamogetonaceae. Jardin Botanique National de Belgique, Meise: 12 pp.

Litterick, M. R., J. J. Gaudet, J. Kalff & J. M. Melack, 1979. The limnology of an African lake, Lake Naivasha, Kenya. Workshop on African Limnology, SIL UNEP, Nairobi: 61 pp.

LNROA, 1993. A Three Phase Environmental Impact Study of Recent Developments around Lake Naivasha (phase I). John Goldson Associates, Nairobi: 124 pp.

Lodge, D. M., M. W. Kershner & J. E. Aloi, 1994. Effects of an omnivorous crayfish (*Orconectes rusticus*) on a freshwater littoral food web. Ecology 75: 1265–1281.

Lodge, D. M., R. A. Stein, K. M. Brown, A. P. Covich, C. Bronmark, J. E. Garvey & S. P. Klosiewski, 1998. Predicting impact of freshwater exotic species on native biodiversity: challenges in spatial scaling. Aust. J. Ecol. 23.

Lorman, J. & J. J. Magnuson, 1978. The role of crayfish in aquatic ecosystems. Fisheries 3: 8–10.

Lowery, R. S. & A. J. Mendes, 1977a. *Procambarus clarkii* in Lake Naivasha and its effects on established and potential fisheries. Aquaculture 11: 111–121.

Lowery, R. S. & A. J. Mendes, 1977b. The biology of *Procambarus clarkii* in Lake Naivasha, Kenya; with a note on its distribution. Freshwater Crayfish 3: 203–210.

MacDonald, R. C. & B. C. Wolverton, 1979. The water hyacinth: from prolific pest to potential provider. Ambio 8: 2–9.

Mason, J. C., 1975. Crayfish production in a small woodland stream. Freshwater Crayfish 2: 449–479.

Mathews, M. & J. D. Reynolds, 1992. Ecological impact of crayfish plague in Ireland. Hydrobiologia 234: 1–6.

Matthews, M. A., J. D. Reynolds & M. J. Keatinge, 1993. Macrophyte reduction and benthic community alteration by the crayfish *Austrapotamobius pallipes* (Lereboullet). Freshwater Crayfish 9: 289–299.

Mikkola, H., 1979. Ecological and social problems in the use of the crayfish *Procambarus clarkii* in Kenya. Freshwater Crayfish 4: 198–205.

Morshiri, G. A. & C. K. Goldman, 1969. Estimation of assimilation efficiency in crayfish *Pacilastacus leniusculus*. Arch. Hydrobiol. 68: 298–306.

Nyström, P. & J. A. Strand, 1996. Grazing by a native and an exotic crayfish on aquatic macrophytes. Freshwat. Biol. 36: 673–682.

Olsen, T. M., D. M. Lodge, G. M. Capelli & R. J. Houlihan, 1991. Mechanism of impact of an introduced crayfish (*Orconectes rusticus*) on littoral congeners, snails, and macrophytes. Can. J. Fish. aquat. Sci. 48: 1853–1861.

Oluoch, A. O., 1990. Breeding biology of the Louisiana red swamp crayfish *Procambarus clarkii* Girard in Lake Naivasha, Kenya. Hydrobiologia 208: 85–92.

Prins, R., 1968. Comparative ecology of the crayfish *Orconectes rusticus* and *Cambarus tenebrosus* in Doe Run, Meade County, Kentucky. Int. Rev. ges. Hydrobiol. 53: 667–714.

Rickett, J. D., 1974. Trophic relationship involving crayfish of the genus *Orconectes* in experimental ponds. Progressive Fish-Culturist 36: 207–211.

Schoener, T. W., 1971. Theory of feeding strategies. Ann. Rev. Ecol. Syst. 2: 369–403.

Sommer, T. R. & C. R. Goldman, 1983. The crayfish *Procambarus clarkii* from California ricefields: ecology, problems and potential for harvest. Freshwater Crayfish 5: 418–428.

Tabuti, J. R. S., H. Oryem-Origa, G. M. Mutumba & J. T. Kashambuzi, 1998. Partitioning of Soluble Calcium in the Water Hyacinth (*Eichornia crassipes* [Martius] Solms) in Lake Victoria, Uganda. Bot. Acta 111: 297–302.

Triest, L. & J.-J. Symoens, 1983. Flore d'Afrique Centrale (Zaïre-Rwanda-Burundi), Spermatophytes: Najadaceae. Jardin Botanique National de Belgique, Meise: 9 pp.

Triest, L., 1987. A Revision of the Genus *Najas* L. (Najadaceae) in Africa and Surroundings Islands. A.R.S.O.M., Classe des Sciences Naturelles et Médicales, Mémoires in-8°, Nouvelle Série, Tome 21, fasc. 4, Bruxelles: 88 pp.

Triest, L., 1989. Flora of Tropical East Africa, Najadaceae. A.A. Balkema, Rotterdam, Brookfield: 9 pp.

Troupin, G., 1988. Flore du Rwanda, Spermatophytes. Volume IV, Agence de Coopération Culturelle et Technique, Annales Série in-8°-Sciences Économiques, no 16, 1987. Musée Royal de l'Afrique centrale, Tervuren: 651 pp.

Verschuren, D., J. Tibby, K. Sabbe & N. Roberts, 2000. Effects of depth, salinity, and substrate on the invertebrate community of a fluctuating tropical lake. Ecology 81: 164–182.

Vincent, C., T. Davies & A. Beresford, 1979. Recent changes in the level of Lake Naivasha, Kenya, as an indicator of equatorial westerlies over East Africa. Climatic Changes 2: 175–189.

Wiernicki, C., 1984. Assimilation efficiency by *Procambarus clarkii* fed Elodea (Egera densa) and its products of decomposition. Aquaculture 36: 203–215.

Hydrobiologia **488:** 143–151, 2002.
D.M. Harper, R. Boar, M. Everard & P. Hickley (eds), Lake Naivasha, Kenya.

Distribution and abundance of the Louisiana red swamp crayfish *Procambarus clarkii* Girard at Lake Naivasha, Kenya between 1987 and 1999

David M. Harper[1,*], Andrew C. Smart[1,6], Stephanie Coley[1], Sophie Schmitz[2], Anne-Christine Gouder de Beauregard[2], Rick North[3], Chris Adams[4], Paul Obade[5] & Mbogo Kamau[5]
[1]*Department of Biology, University of Leicester, Leicester LE1 7RH, U.K.*
[2]*Laboratoire d'Ecologie, FUSAGx, Passage des Déportés 2, B-5030 Gembloux, Belgium*
[3]*Environment Agency, Midlands Region, Welshpool Road, Shrewsbury, Shropshire, U.K.*
[4]*Environment Agency, Anglian Region, Ipswich, Suffolk, U.K.*
[5]*Kenya Marine and Fisheries Research Institute, Box 837, Naivasha, Kenya*
[6]*Current address: Duchy College, Rosewarne, Camborne, Cornwall TR14 0AB, U.K.*
**Author for correspondence. E-mail: dmh@le.ac.uk*

Key words: length–weight relationship, population structure, density, food–web interactions, *Eichhornia crassipes*

Abstract

The abundance and habitat distribution of *Procambarus clarkii*, an exotic herbivore at Lake Naivasha, Kenya, was studied between 1987 and 1999. Its growth and morphometrics were similar to other locations in the world where the species have been studied. *P. clarkii* was found primarily amongst the floating, exotic vegetation where it achieved densities of over 500 m^{-2} by the end of the study. It is suggested that these densities became possible because the habitat offers protection against predation and an adequate food supply of bacterial-enriched detritus exists in the absence of green plant material. Both adults and juveniles forage on the lake bed at night.

Introduction

Crayfish are well known, on the one hand, to be important members of aquatic food webs (Mathews et al., 1993; Lodge et al., 1994) and on the other hand, to create havoc as exotic species in modified food webs (Lodge et al., 1998). The Lousiana crayfish, *Procambarus clarkii* Girard is now the dominant invertebrate in the littoral habitats of Lake Naivasha. It was introduced to the lake in 1970 when 300 individuals were released and transferred to various sites around the margins (Oluoch, 1990). It was first harvested in 1975, initially as an export crop. In latter years it has been mainly for local consumption, yielding a few thousand kg per annum.

Lowery & Mendes (1977a) identified densities of up to four adults m^{-2} in 1973. They, and then Olouch (1990), found that the Naivasha population showed continuous egg production, with 60% of males sexually active throughout the year, unlike the U.S. populations where reproduction is seasonal. In the U.S., Burras et al. (1995) found that *P. clarkii* produced tertiary burrows (a 'pipe' with end chamber) of median depth 3.3 m; median diameter 6.4 cm and median volume of 4000 cm^3. In Naivasha, burrows of indeterminate shape and volume appeared at low lake levels (Lowery & Mendes, 1977a). Males guard females in their burrows when the latter are 'berried' (Olouch, 1990). Adults mature at a larger size (males 40, females 42 mm; Oluoch 1990) than either its native Louisiana (males 31 mm, females 32 mm; Penn, 1943) or other exotic locations such as the Azores (males 36.4 mm, females 36.2 mm; Correia & Costa 1994). Other studies have shown temperature and food availability to influence size at maturity (Huner & Romaine, 1978) and Naivasha, although equatorial, is less-warm than much of the southern U.S. in summer, because of its altitude (1890 m).

The distribution of crayfish around Naivasha was first shown by Lowery & Mendes (1977b) to be concentrated on the east and north east of the lake. Factors influencing distribution were suggested to include the fishing effort (concentrated near Naivasha); and availability of vegetation and detrital food (more in the shallow areas). Changes in the ecology of Lake Naivasha over the last 30 years have been dramatic. The native aquatic vegetation has been subjected to pressures from fluctuating lake levels and the exotic floating species *Salvinia molesta* and *Eichhornia crassipes* (Harper et al., 1995; Gouder et al., 1998). *P. clarkii* is now considered the major factor that reduced submerged and floating plant species. Studies of diet choice lend support to this belief (Smart et al., 2002).

The maximum commercial exploitation of *P. clarkii* was recorded in 1983 (Harper et al., 1995). An 8-week intensive survey of aquatic plants of the lake at that time (December 1982–January 1983) found no floating-leaved native or any submerged plants in the main basin or Crescent Island lagoon (Rich & Harper 1984). In the adjacent Oloidien lake (which was then (under high water level; Becht & Harper, 2002) connected through *Cyperus papyrus* swamp to the main lake but whose conductivity was three times higher (ca. 800 μS cm^{-1}), a clear zonation of submerged plants occurred between 0.5 and 3 m depth (Harper & Rich, 1984). *P. clarkii* was noted during the plant survey to be very abundant in the main lake and absent from Oloidien, but regrettably its density was not then systematically recorded.

This inverse distribution of *P. clarkii*, and aquatic plants, together with evidence of crayfish diets in their natural range (e.g., Penn, 1943), subsequently led to the hypothesis that, in Lake Naivasha, a high density of *P. clarkii* was responsible for the elimination of floating-leaved and submerged aquatic plants. The area covered by plants was mapped each year (Harper et al., 1995; Gouder et al., 1998); and this showed that the area occupied by submerged plants increased from zero at the beginning of 1983 to 20 km^2 in 1987, then decreased almost to zero again by 1995. An inverse correlation between plant area in the lake and crayfish abundance was highly significant when the measure of abundance used was the proportion of crayfish in the diets of large-mouthed bass, *Micropterus salmoides* (Hickley & Harper, 2002).

The aquatic plants involved in this dynamic inverse relationship during the 1990s were submerged species with very little contribution from the single native floating-leaved species, *Nymphaea nouchalii* var. *caerulea*. This latter had been an abundant, characteristic species at Lake Naivasha throughout the 20th Century, regardless of lake level. A bank of viable seeds has built up in lake sediments, which results in rapid germination when soils are newly flooded by water rises following earlier heavy rains. The last record of extensive natural beds of *N. nouchalii* var. *caerulea* was in September 1974, when a hot air balloon was filmed dragging its basket through lily beds in *C. papyrus*-fringed lagoons (Root & Root, 1976). The underside of the lily leaves was 'dripping' with *P. clarkii*. By September 1977 *N. nouchalii* var. *caerulea* had completely disappeared from the lake (J. Root, pers comm.) and between 1988 and 2000 isolated clumps were only recorded occasionally in small areas and for a short time. The longest period of occurrence was inside Crescent Island lagoon, where a continuous thin belt (2 m wide, which corresponded to depths of 0.5 and 2 m, approximately) existed between 1987 and 1992.

A second hypothesis was formulated; that *P. clarkii* had eliminated the *N. nouchalii* var. *caerulea* in Lake Naivasha in the late 1970s and it only reappeared from germination briefly when and where *P. clarkii* was not able to graze it. No published records of *N. nouchalii* var. *caerulea* in Naivasha since 1977 exist and there are only five brief occasions when the species has been observed: (i) Crescent Island lagoon, 1987–92, where no crayfish were recorded until they rafted in on wind-blown *E. crassipes* floating mats; (ii & iii) in May 1988 and again November 1997 when rapid water level rise flooded new shallow lagoons on northern and eastern edges of the lake for a few months; (iv) a small bed (1 m × 5 m) was recorded in 1992 on the south east shoreline in front of horticultural land, where the *C. papyrus* had recently been cleared and the floating exotic vegetation was absent; and (v) for a period of about a year in 1993–4 when a stranded *C. papyrus/E. crassipes* island in the western edge of the lake (vicinity of Korongo Farm) created a 1–2-m deep lagoon about 200 m long off the fixed fringing *C. papyrus*, within which submerged native submerged and floating-leaved aquatic plants germinated. *N. nouchalii* var. *caerulea* has also existed continuously to the present day, adjacent to the lake in one horticultural irrigation pond maintained behind the *C. papyrus* fringe to provide a constant head of water for pumps. The virtual absence of *N. nouchalii* var. *caerulea* from the lake (except in unpredictable areas and periods) meant that testing the hypothesis was limited to laboratory experiments (Smart et al.,

2002). These supported the hypothesis, as have other published experiments (Hofkin et al., 1991).

A deduction from the hypotheses is that the abundance of crayfish, as grazers on native aquatic plants (Smart et al, 2002), is inversely proportional to the abundance of native plants. The study examines the density of crayfish in the decade 1987 to 1999 and aspects of its spatial distribution in the lake as part of the testing of this hypothesis.

Study site

Lake Naivasha (0° 45′ S, 36° 20′ E) lies in the Eastern Rift Valley at 1890 m above sea level, approximately 100 km north west of Nairobi. The lake is freshwater with two rivers draining into the north of the lake and a subterranean inflow and outflow (Ase, 1987; Becht & Harper, 2002). The recent biological history of the lake is reviewed by Harper et al. (1990). *P. clarkii* now occurs in the main lake (approximately 100 m^2 area) and in the Crescent Island lagoon. It has not been recorded since 1982 in Oloidien Lake, which is slightly saline with an alkalinity by the late 1990s of around 2000 $\mu S\ cm^{-1}$ (Fig. 1).

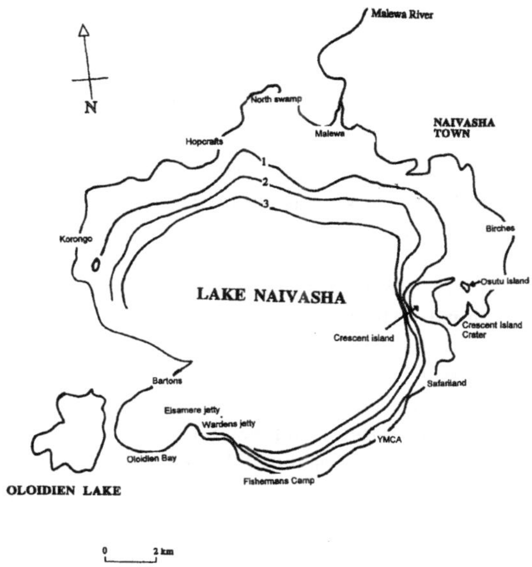

Figure 1. Map of Lake Naivasha showing depth contours (metres) and sample sites. After Hickley et al. (2002).

Methods

Four methods of sampling for crayfish have been used to estimate population abundance during April and August each year. The first was trapping from the sediment, which is not a density measure (and has additional sampling problems which were not addressed (Westman et al., 1978; Brown & Brewiss, 1978). The other three were used for density measures in the shallow littoral zone of floating exotic vegetation species, from a minimum of five locations, covering the four compass quadrants of the lake each year. These different methods were developed of necessity as the vegetation changed over the decade and the previous method became impractical.

1. *Traps.* In water deeper than 1 m, or on the lakeward side of the fringing plant mat, *P. clarkii* was sampled using baited traps of the same design used by commercial fishermen on the lake. A cylindrical wire frame of approximately 50 cm length and 20 cm diameter was covered with plastic 1 cm mesh, with inward-facing funnel openings at either end and a rectangular 20 × 10 flap opening, held shut by a catch. The trap was baited, usually with fish and then sunk, to lie on the bottom for a fixed length of time, usually for half an hour. *P. clarkii* entered by crawling in at either funnel end, was unable to crawl out, and when the trap was raised were removed by hand through the rectangular opening for processing. Density measures were not possible with this trapping and although mark-recapture was tried several times with batteries of traps, at no time were enough marked individuals re-caught. The mesh size of these traps did not reliably hold juveniles. Overnight trapping was carried out in 1992 and 1994 using micromesh fish traps, which did.
2. *Cage.* A wire mesh cube cage of 1 m^3 volume, open at the top and bottom, was used between 1987 and 1990. The floating vegetation of the littoral during this time consisted only of scattered *S. molesta* (and a few other aquatic species; Adams et al., this volume). The vegetation offered no physical impediment to sampling and the cage was dropped to the lake bed in depths of up to 1 m, randomly from a boat. The invertebrate contents of the 1 m^3 water volume enclosed and its sediment were then removed by successive sweeps with a standard macroinvertebrate hand net until no further individuals could be found. This collected *P.*

clarkii from both the sediment and the floating vegetation.

3. *Hand net.* The floating vegetation developed from the early 1990s into a more substantial physical habitat as a result of the spread in the lake of *E. crassipes* (first seen in 1988). It became apparent that *P. clarkii* was burrowing into the mat. *S. molesta* was the dominant species in 1991, with increasing density (of initially small plants) of *E. crassipes* from then onwards, (Adams et al., this volume). The vegetation was not tightly packed and so it was most easily sampled by using a hand net (0.0625 m^2) from beside the boat, inserting it vertically under the plant mat, turning the net to a horizontal position and then raising it through the mat. The vegetation thus collected was placed in a domestic bowl in the boat and carefully examined by hand to remove all crayfish. The sediment below was scraped with the net for approximately 0.0625 m^2 and any crayfish added. The location of net samples was randomised by the selection of compass bearings from the boat centre. The hand net was also used to collect crayfish from the submerged rhizomes on the outside of floating clumps of *C. papyrus*; here the net was scraped vertically upwards for 25 cm to give a similar-sized sample.
4. *Quadrat.* The floating vegetation after 1995 progressively became composed of *E. crassipes* plants (with other swamp colonist species of minor importance to the physical structure). Individual plants were larger than a single hand net and matted together so that it became no longer practical to collect a net sample of 1 m^2. A wooden square quadrat of 1 m^2 was placed on the vegetation surface (at random bearings on the side of the boat driven into the middle of the fringing mat), and then all plants within the quadrat were lifted by hand into a bowl held adjacent, holding the hand net under to collect any *P. clarkii* which fell from the plants. The sediment below was also scraped with the net for approximately 1 m^2 and any crayfish added to the bowl.

The density figures obtained from methods 2–4 are used here to give a continuous picture of change over the decade, even though it has never been possible to compare them directly. Both types of sample method – littoral density and catch per unit effort (catch per trap per hour) – were used independently to compare locations within the lake at the same time, or the whole lake at different times.

The length of the carapace of all animals collected as above was measured using callipers or mm ruler. Four hundred animals during 1994 were weighed (using a 0–30- or 0–100-g spring balance) and measured (using callipers) to establish the relationship between weight, carapace length, full length and total length (defined in Oluoch, 1990). A number of animals across the sampled size ranges were killed and the tail was removed and weighed to examine the incidence of crayfish muscle-wasting disease (Lowery & Mendes, 1977b).

Results

There was a significant linear relationship ($r^2 = 0.97$) between carapace length and total length for *P. clarkii* (Fig. 2) with a relationship TL (total length) = 2.15 + 1.91CL (carapace length). No apparent variation existed between males and females. A strong relationship ($r^2 = 0.76$) existed between carapace length and weight (W) (Fig. 3), as $\log W = -4.45 + 2.93 \log \mathrm{CL}$ (Fig. 3).

There was a greater variation in weight in animals over 40 mm carapace length, indicating the probability that a number of animals are suffering from 'tail-wasting' disease. Figure 4 suggests that animals over 20 g in weight (approximately 40 mm carapace length) may be suffering from the disease.

The density of *P. clarkii* in littoral floating vegetation increased between 1987 and 1996 but with quite wide confidence limits and without statistical significance (Fig. 5). The densities in this habitat were quite high nevertheless, ranging from 6 to 77 m^{-2}. The majority of these were always juveniles (Fig. 6) with the mean carapace length varying between 8 and 18 mm during this time (Fig. 7). In 1996 the proportion of adults jumped to over one-third of the population in

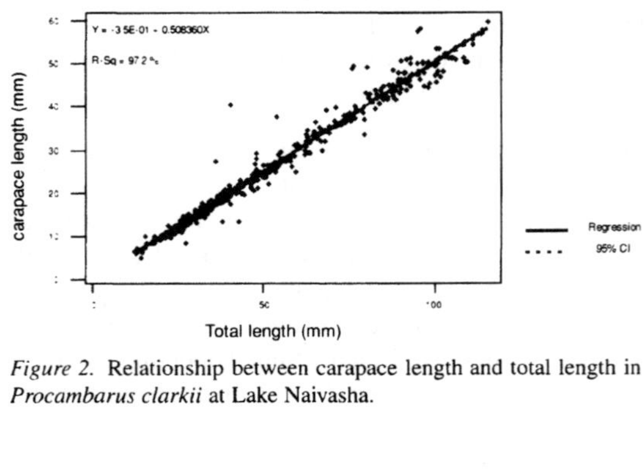

Figure 2. Relationship between carapace length and total length in *Procambarus clarkii* at Lake Naivasha.

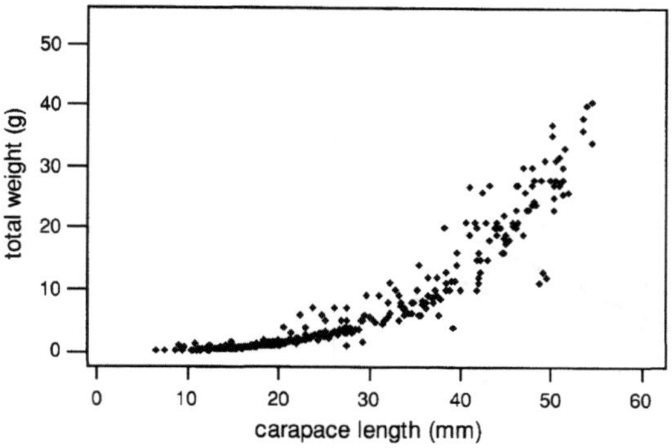

Figure 3. Relationship between carapace length and total weight in *Procambarus clarkii* at Lake Naivasha.

the littoral vegetation (mean size 30±14 mm) but in the following 2 years fell back within the limits of the preceding years and could not be explained by any concurrent environmental measurements. In 1999 the population density increased dramatically reaching over 500 individuals m^{-2} in early 2000; the majority juvenile (mean size 14±5.6 mm with only 2% of the population sexually mature in April 2000).

The abundance of trapped crayfish from the sediments each year to 1996 never exceed three individuals $trap^{-1}$ h^{-1} with no statistical difference between years. The catch in 1999 and early 2000, was three times higher than previously (9.5 and 11, respectively, individuals $trap^{-1}$ h^{-1}).

Adults dominated all the trap returns in daytime throughout the 1990s, whilst juveniles dominated the floating vegetation. A comparison of returns from traps during daytime (1992 mean size 45±5 mm; 1994 mean size 46±11 m) confirmed this ((ANOVA and Tukey test $p < 0.001$; F=189.7; Fig. 8). Trapping overnight captured a full size range of individuals (mean size 1992 35±9 and 1994 32±12 mm).

The vegetation in 1992 consisted of both *S. molesta* and *E. crassipes*, containing a similar density of individuals. These were significantly larger in *E. crassipes* (mean carapace length 22 mm) than in *S. molesta* (mean carapace length 13.2 mm) (ANOVA; $p < 0.01$). Comparison was not possible after 1994, as *S. molesta* became an insignificant part of the vegetation. The only other species from then on to the present contributing, structurally to the littoral vegetation was *C. papyrus*. Where it occurred stranded on the lakebed, its rhizomes supported abundant juvenile *P. clarkii*, marginally but not significantly smaller than *E. crassipes* (e.g., 11.9 mm compared to 14.7 mm in April 1995) Where the *C. papyrus* formed a continuous floating fringe at the edge of the lake, however (usually on the landward edge of a thick *E. crassipes* mat), few *P. clarkii* could be found. *C. papyrus* islands (a dominant landscape feature of the lake up to the early 1980s) were rare with few *P. clarkii* clinging to them.

A significantly higher density of crayfish was found in vegetation around the southern quarter of the lake (Kruskal–Wallis $p <$ df 18) but the catch in traps showed the opposite pattern (Kruskal–Wallis, $p < 0.99$, df 16) with higher catches in the north and west (Table 1). Crayfish trapped from the sediment surface were found in significantly higher numbers at depths between 1 and 2 m with fewer (and many zero catches) in either shallower or deeper water (7.7±2.1 individuals $trap^{-1}$ h^{-1} compared to 1.6±0.8 and 2±1.2, $n = 42$) (Table 2).

Discussion

P. clarkii grows well in Naivasha (corresponding to growth found elsewhere, such as in Portugal and Spain (Correia, 1993; Romaine et al., 1977)) but its economic value is influenced to an unknown extent by tail-wasting disease, which had been reported shortly after the introduction of the species (Lowery & Mendes, 1977a) with tail weights reduced to 10% of normal due to the disease. This is not extreme; Mikkola (1979) estimated reductions to 37% (male) and 27% (female) of normal tail weight. and suggested the accumulation of bacterial growth on gills leading to hypoxia and starvation to be its cause (Lindquist & Mikkola, 1979).

It has, for the last decade, existed primarily in floating exotic vegetation at the edge of the lake where it has achieved a higher density than recorded in other locations outside artificial culture (Huner, 2002). The evidence presented here indicates that adults forage on the lake bed during the day and juveniles also do at night. This suggests that *E. crassipes* does not provide adequate food (confirmed by Smart et al., 2002) but does provide adequate shelter from predation. *P. clarkii* is an important component of the diet of *M. salmoides* in this lake (Hickley & Harper, 2002) and is also fed upon by cormorants (Childress et al., 2002) fish eagles (Harper et al., 2002) as well as other wading birds such as ibises and mammals such as marsh mongoose (pers. obs.). The refuge provided by *E. crassipes* is probably essential to the

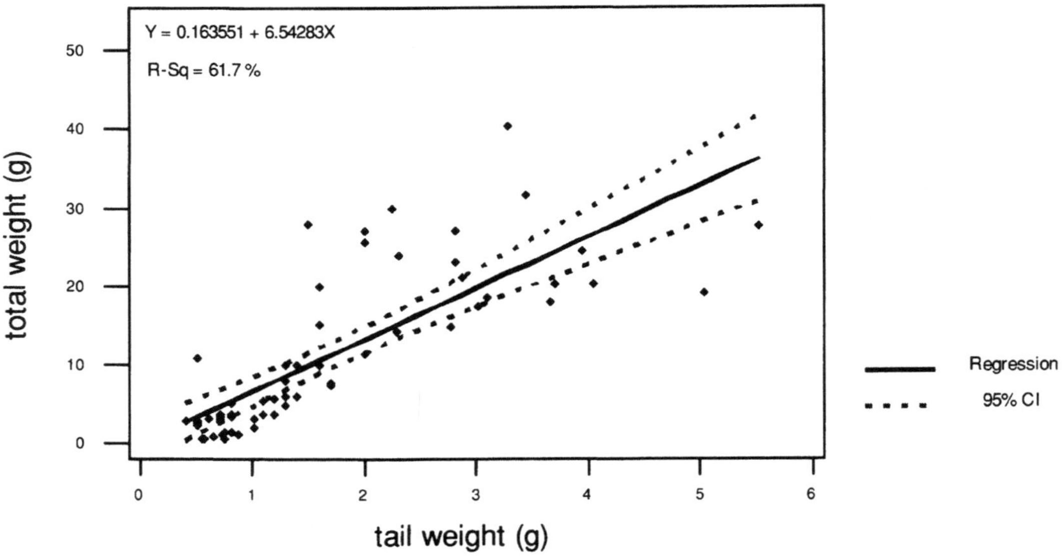

Figure 4. Relationship between tail weight and total weight of *Procambarus clarkii* at Lake Naivasha.

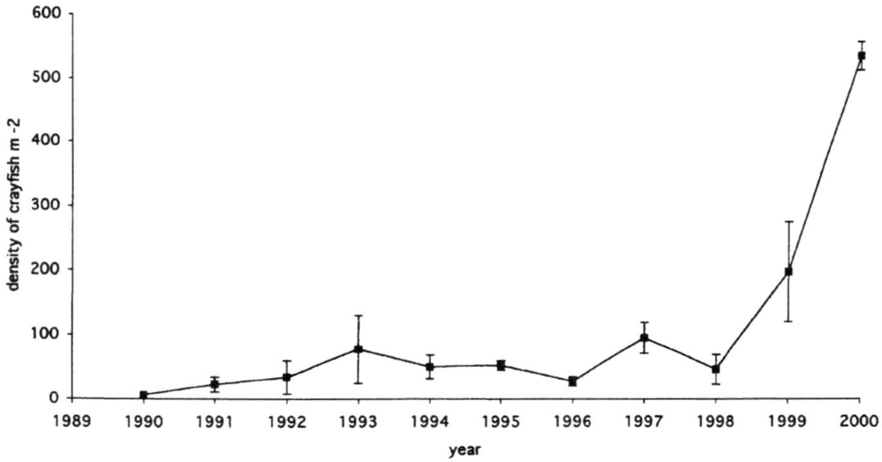

Figure 5. Density of *Procambarus clarkii* in littoral vegetation at Lake Naivasha, 1990–2000.

survival of *P. clarkii*. A prediction was made earlier (Harper et al., 1995) that the food chain in Naivasha of aquatic plants–*P. clarkii*–*M. salmoides* was inherently unstable, with the plants and the crayfish each in a cycle of about 10 years duration between maximum and minimum density. The plant cycle commenced with germination from the seed bank in the sediment when *P. clarkii* declined as a consequence of absence of food plus predation, which last happened around 1985. If the prediction had been correct, then the abundance of plants recorded in 1987 (Harper, 1992) would have been followed by their elimination shortly afterwards, closely followed by a crash in *P. clarkii* and the re-appearance of plants 5 years later. This did not happen however; the plants were eliminated in the early 1990s (Gouder et al., 1998) as predicted but *P. clarkii* remained abundant and prevented plant re-appearance for the remainder of that decade. The one large difference in the lake in the 1990s is the presence of *E. crassipes*. It has been recorded else-

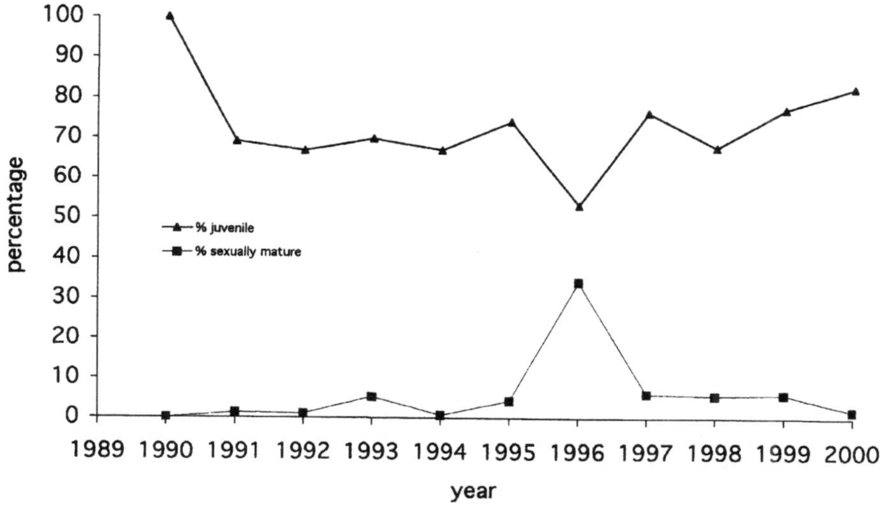

Figure 6. Percentage of *Procambarus clarkii* in littoral vegetation at Lake Naivasha composed of juvenile and sexually mature adults, 1990–2000.

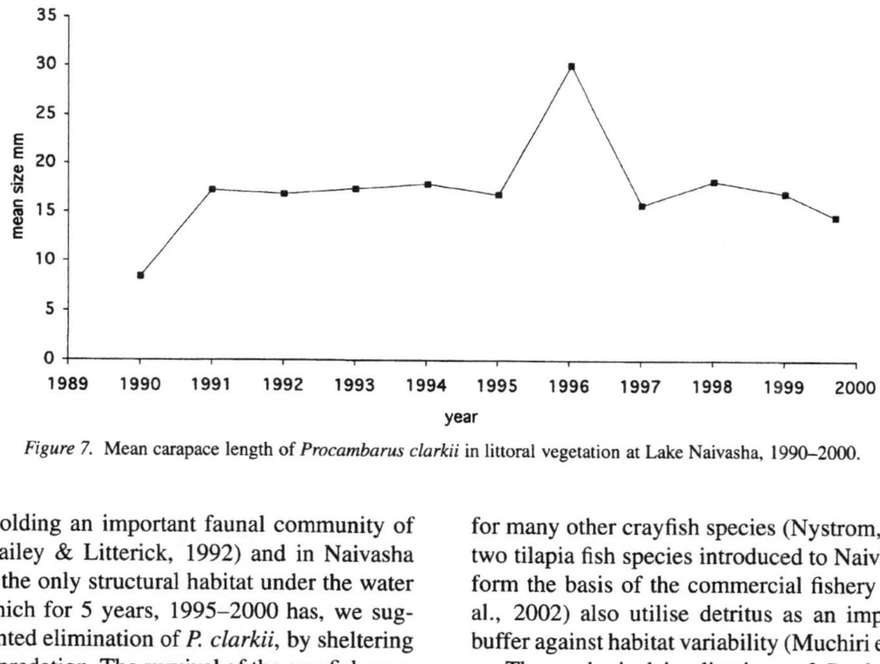

Figure 7. Mean carapace length of *Procambarus clarkii* in littoral vegetation at Lake Naivasha, 1990–2000.

where as holding an important faunal community of its own (Bailey & Litterick, 1992) and in Naivasha it provides the only structural habitat under the water surface, which for 5 years, 1995–2000 has, we suggest, prevented elimination of *P. clarkii*, by sheltering them from predation. The survival of the crayfish population requires not only shelter from predation but also an adequate food supply. This could have been provided by an abundance of bacteria-rich detritus on the lake bed, by the decomposition of an abundant supply of *S. molesta* and then *E. crassipes* plants. Whilst this may not be the most energy-rich, or even preferred food (Ilhéu & Bernardo, 1993) it is an important food for many other crayfish species (Nystrom, 2002). The two tilapia fish species introduced to Naivasha, which form the basis of the commercial fishery (Hickley et al., 2002) also utilise detritus as an important food buffer against habitat variability (Muchiri et al., 1995).

The ecological implications of *P. clarkii* in lake Naivasha are the elimination of aquatic plants; in that sense the effects can be predicted from the knowledge of lakes that have similarly lost their aquatic plant base as a consequence of eutrophication, compounded by the possibility that invertebrate species may also be directly eliminated by *P. clarkii* (Nystrom, 2002). This example of Lake Naivasha thus provides a counterbal-

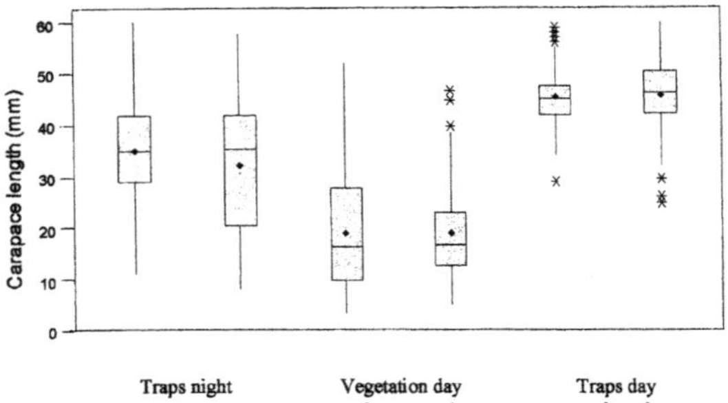

Figure 8. Mean carapace size of *Procambarus clarkii* from daytime traps, nighttime traps and daytime littoral vegetation at Lake Naivasha, 1992 and 1994.

Table 1. Density of *P. clarkii* in the four compass quadrant of littoral vegetation at Lake Naivasha in 1995

	West	North	East	South
Density (m^{-2})	3.30	6.75	5.80	15.00
S.D.	0.89	6.91	5.27	7.18
Indiv $trap^{-1}$ h^{-1}	2.65	3.21	1.31	1.84
S.D.	3.04	3.91	1.20	1.67

Table 2. Catches in traps from different depths at Lake Naivasha, 1997

Depth (m)	Indiv. $trap^{-1}$ h^{-1}	S.D.	*n*
0–1	1.64	1.78	14
1–2	7.65	5.70	20
2 +	1.75	1.98	8

ance to suggestions that *P. clarkii* should be introduced as a biological control agent for the intermediary host molluscs of Schistosomiasis (Hofkin et al., 1991) and furthermore suggests, that *P. clarkii* is a 'keystone' species in this aquatic ecosystem.

Acknowledgements

This study formed past of the work of the University of Leicester research project at Lake Naivasha, which since 1987 has been authorised by the Office of the President, Government of Kenya under research permit to Dr D. M. Harper no. OP 13/001/12C 46. The project was funded by the Earthwatch Institute, Boston USA and Oxford, England. The data collection would not have been possible without the assistance of numerous Earthwatch volunteers. Our sincere thanks go to the numerous colleagues in Kenya in particular Jill & Angus Simpson and Velia Carn.

References

Adams, C. S., R. R. Boar, D. S. Hubble, M. Gikungu, D. M. Harper, P. Hickley & N. Tarras-Wahlberg, 2002. The dynamics and ecology of exotic tropical species in floating plant mats: Lake Naivasha, Kenya. Hydrobiologia 488 (Dev. Hydrobiol. 168): 115–122.

Ase L. E., 1987. A note on the water budget of Lake Naivasha – especially the role of *Salvina molesta* Mitch and *Cyperus papyrus* L. Geograf. Ann. 69A: 415–29.

Bailey, R. G. & M. R. Litterick, 1993. The macroinvertebrate fauna of water hyacinth fringes in the Sudd swamps (River Nile, southern Sudan). Hydrobiologia 250: 97–103.

Becht, R. & D. M. Harper, 2002. Towards an understanding of human impact upon the hydrology of Lake Naivasha, Kenya, Hydrobiologia 488 (Dev. Hydrobiol. 168): 1–11.

Brown D. J. & J. M. Brewiss, 1978. A critical look at trapping as a method of sampling a population of *Austropotamobius pallipes* (Lereboullet) in a mark and recapture study. Freshwat. Crayfish 4: 159–164.

Burras L., G. Blakewood, T. Richard & J. V. Huner, 1995. Laboratory observations on burrowing in different soils by commercially important procambaris crayfish. Freshwater Crayfish 10: 427–434.

Childress, R. B., L. A. Bennun & D. M. Harper, 2002. Population changes in sympatric Great and Long-tailed Cormorants (*Phalacrocorax carbo* and *P. africanus*): the effects of niche overlap or environmental change? Hydrobiologia 488 (Dev. Hydriobiol. 168): 163–170.

Clark, F., A. Beeby & P. Kirby, 1989. A study of the macroinvertebrates of Lakes Naivasha, Oloidien and Sonachi, Kenya. Rev. Hydrobiol. Trop. 22(1): 21–33.

Correia, A. M., 1993. Length–weight relationships for two populations of red swamp crayfish, *Procambarus clarkii* (Decapoda, Cambaridae) from Portugal. Freshwat. Crayfish 9: 442–450.

Correia, A. M. & A. C. Costa, 1994. Introduction of the red swamp crayfish *Procambarus clarkii* (Crustacea: Decapoda) in Sano Miguel, Azores, Portugal Arquipélago. Life mar. Sci. 12A: 67–73.

Harper D. M., 1992. The ecological relationships of aquatic plants at Lake Naivasha, Kenya. Hydrobiologia 232: 65–71.

Harper, D. M., K. M. Mavuti & S. M. Muchiri, 1990. Ecology and management of Lake Naivasha, Kenya, in relation to climatic change, alien species' introductions and agricultural development. Environ. Conserv. 17(4): 328–336.

Harper, D. M., C. Adams & K. Mavuti, 1995. The aquatic plant communities at the Lake Naivasha wetland, Kenya: pattern, dynamics and conservation. Wetlands Ecol. Manage. 3(2): 111–123.

Hickley P., R. Bailey, D. M. Harper, R. Kundu, M. Muchiri, R. North & A. Taylor, 2002. The status and future of the Lake Naivasha fishery, Kenya. Hydrobiologia 488 (Dev. Hydrobiol. 168): 181–190.

Hofkin, B., D. K. Koech, J. Ouma & E. S. Loker, 1991. The North American crayfish *Procambarus clarkii* and the biological control of schistosome-transmitting snails in Kenya: laboratory and field investigations. Biol. Control 1: 183–187.

Huner, J. V., 2002. *Procambarus*. In Holdich, D. (eds), Biology of Freshwater Crayfish. Blackwell Science, Oxford: 541–584.

Huner, J. V. & R. P. Romaine, 1978. Size at maturity as a means of comparing populations of *Procambarus clarkii* (Girard) (Crustacea, Decapoda) from different habitats. Freshwat. Crayfish 4: 53–64.

Ilhéu, M. & J. M. Bernardo, 1993. Trophic ecology of red swamp crayfish *Procambarus clarkii* (Girard) preferences and digestibility of plant foods. Freshwater Crayfish 10: 132–139.

Lindqvist O. V. & H. Mikkola 1979. On the etiology of the muscle wasting disease in *Procambarus clarkii* in Kenya. Freshwat. Crayfish 4: 363–372.

Lodge, D. M., M. W. Kershner, J. E. Aloi & A. P. Covich, 1994. Effects of an omnivorous crayfish (*Orconectes rusticus*) on a freshwater littoral food web. Ecology 75: 1265–1281.

Lodge, D. M., R. A. Stein, K. M. Brown, A. P. Covich, C. Bronmark, J. Garvey & S. P. Klosiewski, 1998. Predicting impact of freshwater exotic species on native biodiversity: challenges in spatial scaling. Aust. J. Ecol. 23.

Lowery, R. S. & A. J. Mendes, 1977a. The biology of *Procambarus clarkii* in Lake Naivasha, Kenya; with a note on its distribution. Freshwat. Crayfish 3: 203–210.

Lowery, R. S. & A. J. Mendes, 1977b. *Procambarus clarkii* in Lake Naivasha and its effects on established and potential fisheries. Aquaculture 11: 111–121.

Madgwick, F. J., 1999. Strategies for conservation management of lakes. Hydrobiologia 395/396: 309–323.

Mathews, M. A., J. D. Reynolds & M. J. Keating, 1993. Macrophyte reduction and benthic community alteration by the crayfish *Austropotamobius pallipes* (Lereboullet). Freshwat. Crayfish 19: 289–299.

Mikkola, H., 1979. Ecological and social problems in the use of the crayfish *Procambarus clarkii* in Kenya. Freshwat. Crayfish 4: 198–205.

Muchiri, M., P. J. B. Hart & D. Harper, 1995. The persistence of two introduced tilapia species in Lake Naivasha, Kenya, in the face of environmental variability and fishing pressure. In Pitcher, T. J. & P. J. B. Hart (eds), The Impact of Species Changes in African Lakes. Chapman & Hall, London. 299–319.

Nystron, P., 2002. Ecology. In Holdich, D. (eds), Biology of Freshwater Crayfish. Blackwell Science, Oxford. 192–235.

Oluoch, A. O., 1990. Breeding biology of the Louisiana red swamp crayfish *Procambarus clarkii* Girard in Lake Naivasha, Kenya. Hydrobiologia 208: 85–92.

Penn, G. H., 1943. A study of the life history of the Louisiana red-crawfish, *Cambarus clarkii* Girard. Ecology 24(1): 1–18.

Rich, T. & D. M. Harper, 1987. Aquatic macrophytes of Lake Naivasha. In Harper, D. M. (eds), Studies on the Lake Naivasha Ecosystem, 1982–84. University of Leicester. 73–100.

Romaine, R. P., J. S. Forester & J. W. Avault, 1977. Length–weight relationship of two commercially important crayfishes of the genus *Procambarus*. Freshwat. Crayfish 3: 463–470.

Root, A. & J. Root, 1976. Safari by Balloon. Survival, Anglia T.V., Norwich.

Smart A. C., D. M. Harper, F. Malaisse, S. Schmitz, S. Coley & A.-C. Gouder de Beauregard, 2002. Feeding of the exotic Louisiana red swamp crayfish, *Procambarus clarkii* (Crustacea, Decapoda), in an African tropical lake: Lake Naivasha, Kenya. Hydrobiologia 488 (Dev. Hydrobiol. 168): 129–142.

Westman, K., O. Sumari & M. Pursiainen, 1978. Electric fishing in sampling crayfish. Freshwat. Crayfish 4: 252–255.

Hydrobiologia **488:** 153–161, 2002.
D.M. Harper, R. Boar, M. Everard & P. Hickley (eds), Lake Naivasha, Kenya.

Population structure and secondary productivity of *Limnodrilus hoffmeisteri* (Claparede) and *Branchiura sowerbyi* Beddard in the profundal zone of Lake Naivasha, Kenya

Phil Raburu[1], Kenneth M. Mavuti[3], David M. Harper[2] & Frank L. Clark[2]
[1]*Department of Fisheries, Moi University, Box 3900, Eldoret, Kenya*
[2]*Department of Zoology, University of Nairobi, Box 30197, Nairobi, Kenya*
[3]*Department of Biology, University of Leicester, Leicester LE1 7RH, U.K.*

Key words: benthic invertebrates, tropical lake, biomass, P/B ratios

Abstract

Lake Naivasha has been well studied since the 1930s but attempts to understand its ecological functioning have had to wait until enough was known about its structure. The energetics of the lake has only been studied to date at primary producer level. Following the identification of the invertebrate components of the littoral and profundal benthos, this study was initiated. The absence of native fish species in Lake Naivasha, combined with a fishery based only on three, introduced species, added an applied dimension to the work. The introduction of additional fish species which will utilize unexploited ecological niches has been suggested. The benthic invertebrates form one such niche. Two oligochaetes dominated the community, *Branchiura sowerbyi* Beddard and *Limnodrilus hoffmeisteri* Claparade. The former had a productivity of 7.4 g m^{-2} ann^{-1} (as dry weight), the latter 0.6. These figures are not particularly high and do not support the introduction of a new fish species on their own.

Introduction

There is a paucity of fish species in Lake Naivasha, with only one native species, believed to have become extinct in the 1960s and five exotic species introduced between 1926 and 1965 (Harper et al., 1990; Hickley et al., this volume). The cause is believed to be the instability of the inflows, which led to the lake becoming dry for much of the 19th Century (Becht & Harper, 2002). The importance of the lake as a commercial fishery since 1959, based on these few species, led Mavuti (1990) to recommend introduction of zooplanktivorous, commercially important fish to exploit uncropped zooplankton because no fish permanently occupy the limnetic zone leading to zooplankton production directly entering the decomposer food chain. Hickley et al. (2002) recommend a benthic-feeding fish such as *Mormyrus* spp. because no benthivore was present. It was known that the benthic invertebrates consisted of two oligochaetes – *Limnodrilus hoffmeisteri* Claparade and *Branchiura sowerbyi* Beddard – together with several species of chironomids with the former predominating in abundance (Clarke et al., 1989).

A knowledge of the life cycle of tubificids is a precursor to the determination of their secondary production. Taxonomic problems, however, have led to scarcity of data on many life cycles. *L. hoffmeisteri* is one of the most widespread and abundant tubificid species in the world (Kennedy, 1965). The life cycle of this species has received appreciable attention in temperate regions (Kennedy, 1966; Ladle, 1971; Aston, 1973; Poddubnaya, 1980) but not in the tropics. Success in establishing the life cycle of *L. hoffmeisteri* has largely been facilitated by culturing the worms in the laboratory (Kennedy, 1966; Aston, 1973). Information on the life cycle of *B. sowerbyi* on the other hand is scarce (Carrol & Dorris, 1972). An account has been given by Casellato (1984), based on cultured populations.

This study focused on the life cycle and production of these two tubificids, *L. hoffmeisteri* and *B. sowerbyi* as the former is the most numerically abundant species while the latter contributes the highest benthic biomass in Lake Naivasha (Raburu, 1991). The specific objectives of the study were to establish the population structure and to estimate the productivity of the two species.

Study area and methods

Lake Naivasha at an altitude of 1890 m a.s.l. is the highest and largest freshwater lake of the Eastern Rift Valley of Kenya. The lake is located at latitude 0°45′ and longitude 36° 20′ E. The main lake (150 km^2) is a more or less circular, endorrheic, basin while two other water bodies, Crescent Island lagoon and Oloidien lake (5.5 km^2) are connected at high water levels, separate at low. Crescent Island lagoon is the deepest part of the lake. The main lake was the only basin sampled in this study as this is the focus of the commercial fishery.

Monthly sampling of the macrobenthos was carried out along four transects off the eastern shore (Safariland SL), the southern (Hippo Point HP), the western (Mennell's Lagoon ML), and the northern (North Swamp (NS), from September, 1989, to August, 1990. Each transect had five sampling stations 500 m apart (Fig. 1). Grab samples were taken in triplicate at each sampling station with a 15 × 15 cm Ekman grab which closes by means of a messenger. The top was covered with a fine (335 μm) stainless steel gauge to prevent animals in the top few centimeters of sediment from escaping as the grab penetrates the soft bottom substrates.

The macro-invertebrates were sieved from the sediment through a 0.3 mm mesh sieve. Sorting was carried out by hand on fresh samples when organisms had their natural colours and were mobile, which contributed to quicker and more efficient sorting. After sieving they were identified, counted and weighed and their lengths measured while still alive, before preservation in 4% formaldehyde. In Naivasha, *L. hofmeisteri* does not occur in association with any other species of the genus *Limnodrilus* and it was presumed that all immature specimens belonged to the same species (Clark et al., 1989).

The fresh weight was recorded when specimens were still alive. After sorting into different species, excess water was removed by placing them onto a filter paper. The weight was recorded to the nearest 0.01 g using an electronic balance. Biomass determinations were derived from dry weight but, since the same specimens whose wet weights were taken (especially *L. hoffmeisteri*) had to be mounted on slides for further analysis, a conversion factor was established from the regression of dry weight on wet weight (Ladle, 1971; Gieve, 1975) of 50 samples of tubificid worms collected from the lake. The dry weight of the worms was obtained by drying them to a constant weight in a vacuum at 60 °C for 36 h.

The population structure of *L. hoffmeisteri* was established by grouping the worms into two categories. These were (a) immature individuals not possessing a penis tube on the 9th segment and (b) mature individuals possessing penis tube on the 9th segment (Brinkhurst, 1966). Other workers (Kennedy, 1966; Aston, 1973) managed to group *L. Hoffmeisteri* into immature, breeding and mature worms. This was not possible in this study because immature and breeding worms are not easy to distinguish either by chaetal characteristics or presence of spermatophores (Ladle, 1971). Classification of *L. hoffmeisteri* is further complicated by the fact that sexually mature worms resorb the penis tube after reproduction (Kennedy, 1966; Ladle, 1971; Aston, 1973; Poddubnaya, 1980) and revert to an immature condition which cannot be differentiated from a truly juvenile worm. In this study, the width of the 9th segment which contains the penis tube was used as an index of size obviating problems with breakage. Worms were mounted in dimethyl hydantin formaldehyde (DMHF) under 22 × 50 cm coverslips. After clearing for 2 days, the worms were observed under a compound microscope (magnification 24×) to locate and measure the width of the 9th segment using a graduated micrometer eyepiece. Worms were grouped into size classes 0.11–0.16, 0.17–0.21, 0.22–0.26, 0.27–0.31, 0.32–0.36, 0.38–0.42, 0.43–0.47 and 0.48–0.52 mm.

The weight of worms in each size class was calculated from the equation obtained from the relationship between the width of the 9th segment and the mass of *L. hoffmeisteri* in different size classes: weight = 0.0239 + 0.161 width.

The number of *B. sowerbyi* cocoons in grab samples were counted every month from September, 1989 to August, 1990. From May to October 1990, the lengths of live *B. sowerbyi* were measured to the nearest 0.1 mm. This was done by placing the worms in a wet groove drilled in the middle of a plastic ruler and the length taken when the worms were relaxed. They were grouped into four length classes: 0.1–3.0cm, 3.1–6.0 cm, 6.1–9.0cm and 9.1–12+ cm. Monthly variation in the number of cocoons and length frequency of *B. sowerbyi* were recorded. The

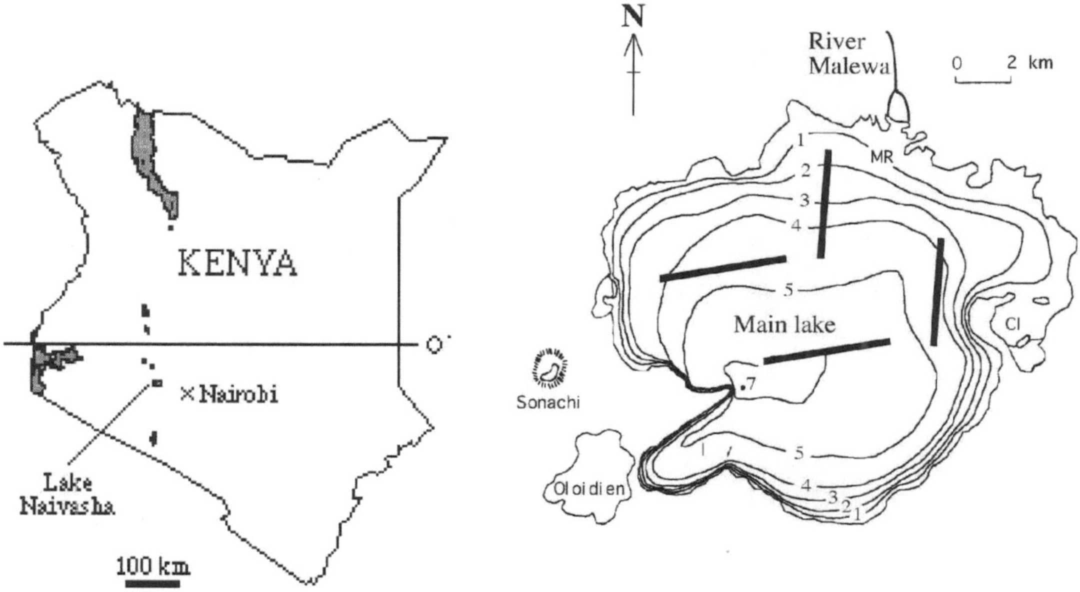

Figure 1. Location of sample transects in Lake Naivasha.

worms in each size class were counted and weighed alive using an electronic balance to the nearest 0.01 gram. Dry weights of tubificid worms could not be obtained on every sampling occasion because worms were mounted for further analysis. A conversion factor was therefore established from the regression of the dry weight on wet weight of both species:

$$\text{Dry weight} = 0.005 + 0.17705\,(\pm 0.0041),$$
$$\text{wet weight } (r = 0.987,\ P < 0.05).$$

The regression coefficient gave the percentage dry weight of wet weight to be 17.7%.

The first two size classes of *L. hoffmeisteri* (0.11–0.16 and 0.17–0.21 mm) were excluded in the production estimate as they were under represented probably due to the mesh size of sieve used. Productivity of *L. hoffmeisteri* and *B. sowerbyi* were estimated using a modification of Hynes method to give a direct estimate of the number of individuals growing into a particular size category (Nj) (Hynes & Coleman, 1968; Menzie, 1980). The equations used were:

$$N_j = in_j \times \text{Pe}/\text{Pa} \times 365/\text{CPI},$$

where i = the size categories into which the organisms are placed; Pe = estimated proportion of life cycle spent in a particular length class; Pa = the actual proportion of life cycle spent in a particular length class; n = the mean number of individuals in the size class; CPI = the cohort production interval in days from hatching to attainment of the largest aquatic size.

N_j is then substituted into the equation estimate production:

$$\text{P} = (N_j - N_{j+1}) \times (W_j W_{j+1})^{1/2},$$

where P = the annual production (g dry weight/m^2/yr). W_j = the mean weight of an individual in the jth size category (g). $(W_j W_j + 1)^{1/2}$ = the geometric mean weight between two size classes (g).

Results

The total annual mean benthic biomass of all macroinvertebrates was 4.07 g dry weight m^{-2} with *Branchiura sowerbyi* accounting for about 58% (2.37 g dry weight m^{-2}), *L. hoffmeisteri*, 27% (1.10 g dry weight m^{-2}). Chironomids made up 8% and other oligochaetes and microturbellarian worms, respectively, 4% and 3% (Raburu, 1991).

Immature worms dominated the population of *L. hoffmeisteri* throughout 12 months (Fig. 2). The lowest proportion of immature in the population composed (57%) occurred in October 1989, a month when the mature worms were breeding. The proportion of immature worms in the population of *L. hoffmeisteri* in

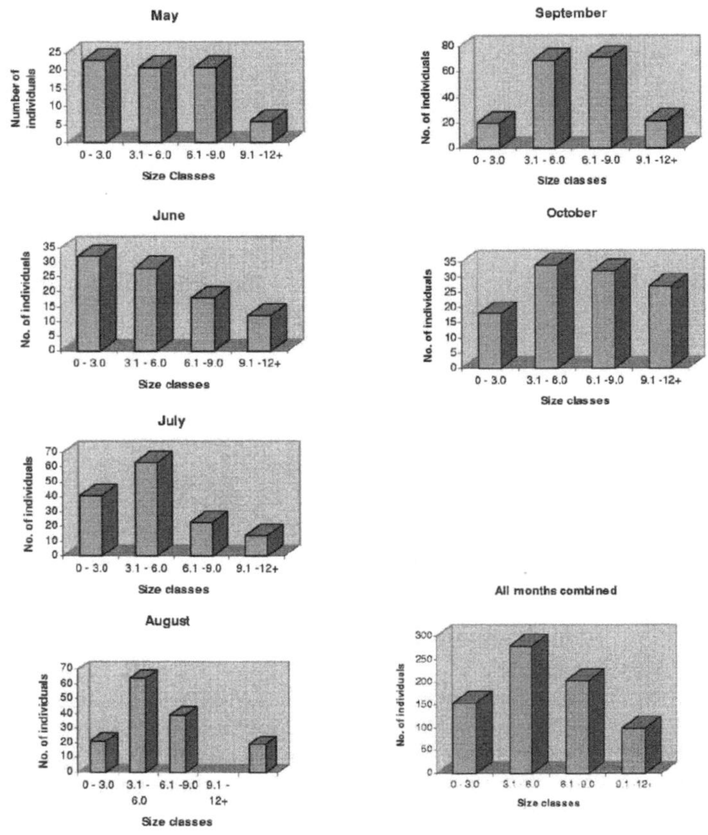

Figure 2. Abundance of *Limnodrilus hoffmeisteri* in Lake Naivasha

the lake then increased consistently from November 1989 to July 1990 when the ratio of immature to mature worms was 17:3. Reproduction of *L. hoffmeisteri* in Lake Naivasha was thus continuous with a peak in October. The smallest size-class was low throughout the study period, probably indicating insufficient collection. Size class 0.22–0.26 mm was most numerous from September 1989 to January 1990, 0.27–0.31 mm in February and March 1990 and 0.22–0.26 mm again from April to August 1990. When monthly observations on the width of the 9th segment of *L. hoffmeisteri* were pooled, the 0.22–0.26 mm size-class had the highest number of individuals followed by the 0.27–0.31 mm and 0.17–0.26 size-classes.

Cocoons of *B. sowerbyi* occurred throughout the study (Fig. 3), indicating continuous breeding, more abundant between January and March reaching a peak in March dropping therafter. Variation in length frequencies of *B. sowerbyi* from May 1990 to October 1990 shows the population was dominated by juvenile worms throughout the year, unlike *L. hoffmeisteri*, where, there was a progression in abundance from one size group to the next.

Productivity

The annual productivity of *L. hoffmeisteri* is shown in Table 1. Since there is one peak in cocoon production a year, the time taken by a young worm to develop to the largest aquatic size was assumed to be one year (CPI). Time taken by the worms to grow from one size class to the next was assumed to be one month. From these assumptions, the parameters Pe/Pa and 365/CPI, were 1.5 and 1.0, respectively. The annual productivity of *L. hoffmeisteri* was 0.65 g dry weight m^{-2} ann^{-1} and the annual production to mean biomass ratio (P/B) of 8 per year.

Table 2 shows that the calculation of the productivity of *B. sowerbyi* was 7.43 g dry weight m^{-2} ann^{-1}

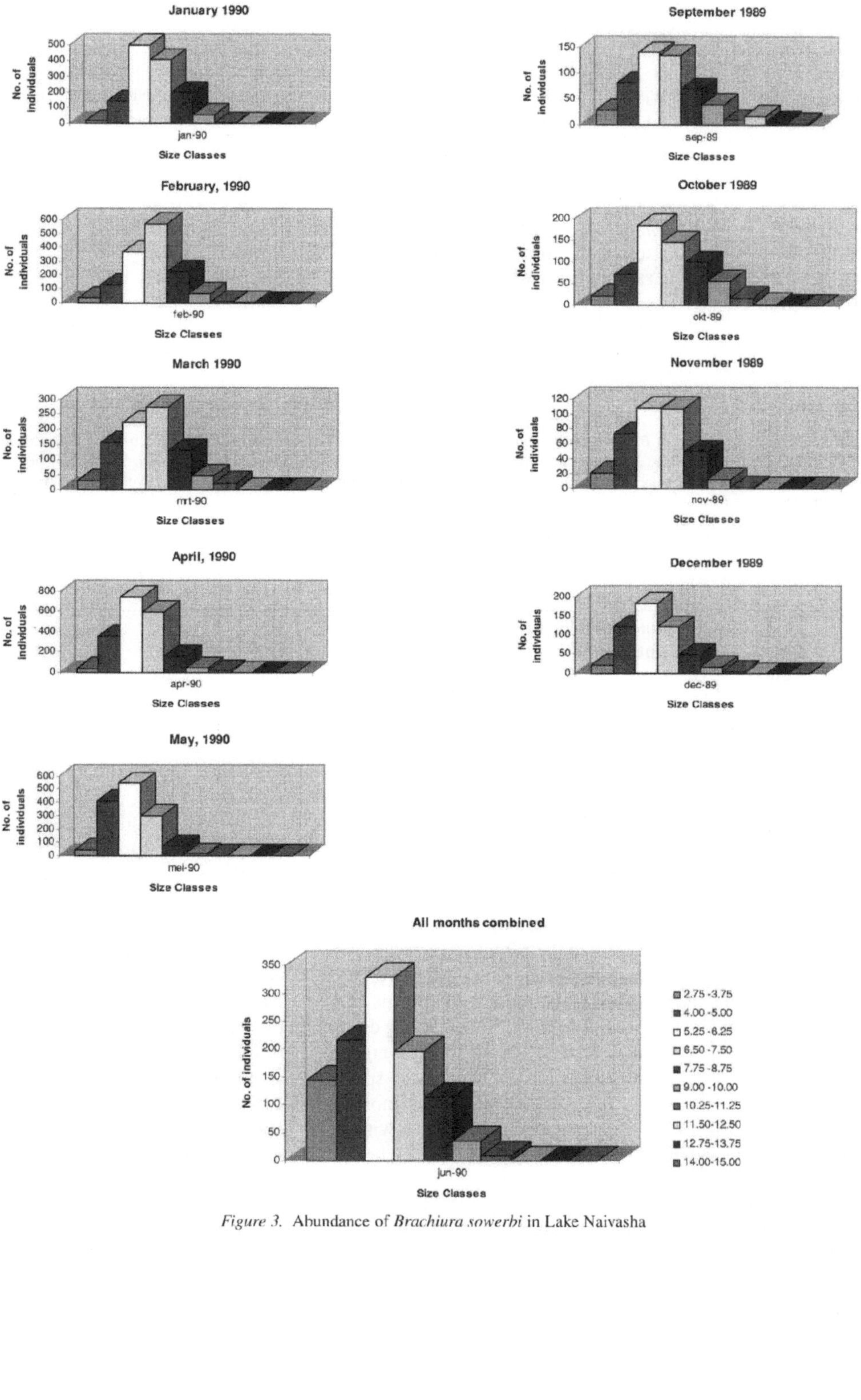

Figure 3. Abundance of *Brachiura sowerbi* in Lake Naivasha

Table 1. The secondary production of *L. hoffmeisteri* by Hynes' method. Column 1 is the total number of worms/m^{-2}, 2 is the mean number of individuals in the size-class, 3 is the mean weight of an individuals in the size category, 4 is biomass, 5 is the number of individuals that develop into a particular size category during the year, 6 is the number of individuals lost between successive size category, 7 is the geometric mean weight between two size classes, 8 is the production

Size group Width (mm)	1 N No m^{-2}	2 $\bar{n}$	3 Wj (g)	4 B (g m^{-2})	5 Nj	6 (Nj–Nj+1)	7 (WjWj+1)– (g)	8 (Nj–Nj+1)×(WjWj+1)– g dry wt m^{-2}yr^{-1}
0.11–0.16	549	46	8.17×10^{-5}	0.045	552	–1788	4.36×10^{-5}	(–0.078)
0.17–0.21	2338	195	2.33×10^{-5}	0.054	2340	–1560	1.76×10^{-5}	(–0.028)
0.22–0.26	4684	390	1.35×10^{-5}	0.062	4680	660	1.54×10^{-5}	0.010
0.27–0.31	4015	335	1.76×10^{-5}	0.070	4020	2472	3.00×10^{-5}	0.074
0.32–0.36	1543	129	5.10×10^{-5}	0.079	1548	1080	9.79×10^{-5}	0.106
0.38–0.42	471	39	1.88×10^{-4}	0.088	468	336	3.79×10^{4}	0.127
0.43–0.47	126	11	7.65×10^{-4}	0.096	132	108	1.95×10^{3}	0.211
0.48–0.52	21	2	4.97×10^{-3}	0.104	24	24	4.97×10^{-3}	0.119
			BB = 0.08				TOTAL	= 0.647

Table 2. The secondary production of *B. sowerbyi* by Hynes' method. Column 1 is the total number of worms m^{-2}, 2 is the mean number of individuals in the size-class, 3 is the mean weight of an individuals in the size category, 4 is biomass, 5 is the number of individuals that develop into a particular size category during the year, 6 is the number of individuals lost between successive size category, 7 is the geometric mean weight between two size classes, 8 is the production

Size group Lengths (cm)	1 2 no m^{-2}	2 $\bar{n}$	3 Wj (g)	4 B (g m^{-2})	5 Nj	6 (Nj–Nj+1)	7 (WjWj+1)– (g)	8 (Nj–Nj+1)×(WjWj+1)– g dry wt - m^{-2}yr^{-1}
0.1–3.0	155	26	2.067×10^{-3}	0.32	138	–138	0.0041	(–0.566)
3.1–6.0	274	46	8.012×10^{-3}	2.20	276	66	0.0089	0.587
6.1–9.0	210	35	9.991×10^{-3}	2.10	210	108	0.0130	1.404
9.1–12+	100	17	1.691×10^{-2}	1.69	102	102	0.0169	1.724
			B = 4.79				TOTAL	= 3.715 × 2
								= 7.43

B = Mean biomass.

and a P/B ratio of 5. In the calculation of productivity of *B. sowerbyi*, Pe/Pa and 365/CPI values used were 0.5 and 1.0, respectively. The worms were assumed to develop to the largest aquatic size in 1 year and they took 2 months to grow from one size class to the next.

The total production of dominating tubificid oligochaetes *B. sowerbyi* and *L. hoffmeisteri* is therefore 8.08 g dry weight m^{-2} ann^{-1}.

Discussion

Comparing the biomass of benthic macroinvertebrate communities in shallow tropical African lakes is not easy due to scanty data on benthic biomass, seasonally or over a number of years. It often depends on the dominating species, which are usually either oligochaetes, dipteran larvae (*Chaoborus* or chironomid) or molluscs. In Lake Chad, the mean benthic biomass for the whole lake was 3.7 g dry weight m^{-2} being dominated by molluscs (Leveque et al., 1983). In Lake George, a very low mean benthic biomass of 0.74 g dry weight m^{-2} was recorded during the dry seasons (Burgis et al., 1973), which rose during wet periods to 9.76 g dry weight m^{-2} in the mid-lake area and 11.31 g dry weight m^{-2} inshore (Darlington, 1977). The dominant species were dipteran larvae (*Chaoborus* and chironomids). In Lake Chilwa, a shallow endorrheic lake with levels experiencing annual fluctuations, the mean benthic biomass during the 1967 dry season was 0.07 g dry weight m^{-2} and was dominated by chironomids (McLachlan & McLachlan, 1969). In Lake Mcllwaine, Zimbabwe, Munro (1966) estimated that

Table 3. Biomass and benthic production in other lakes compared with Lake Naivasha

Lake	Biomass ($g\ m^{-2}$ dry weight)	Productivity ($g\ m^{-2}$ dry wt ann^{-1})	Dominant taxa	Author(s)
Chad 13° N	3.7	–	Molluscs and Insects	Leveque et al. (1983)
	0.257	14.76	Total Molluscs	"
	0.126	–	Total Oligochaete	"
	3.503	–	Total Insecta	"
George 0°	1.2	–	*Chaoborus* and Chironomids	Burgis et al. (1973)
	0.74	–	" " "	Darlington, (1977)
Chilwa 15° S	0.07–0.32	–	Chironomids, Coleoptera and Diptera	McLachlan & McLachlan (1969)
McIlwaine 17° S	3	–	Insects and Oligochaetes	Munro (1966)
	5.0	–	Chironomids	Marshall (1978)
	18.8	–	Oligochaete	"
Kariba	0.02–0.1	–	Chironomids	McLachlan & McLachlan (1969)
Sibaya 27° S	2.1235	0.1534 $g\ m^{-2}d^{-1}$	*Grandidierella lignorum*	
	3.6158	0.281 $g\ m^{-2}d^{-1}$	*Apseudus digitalis* & *Potamon Sidneyi* (2 month study)	Hart (1979)
Naivasha 0° 45′ S	1.095	0.65	*L. hoffmeisteri*	Present study
	2.365	7.43	*B. sowerbyi*	"
Leman (France)		3.33	*L. hoffmeisteri*	Lafont (1987)
Rybinsk Reservoir (Russia)		6–43 (wet weight)	*L. hoffmeisteri*	Poddubnaya (1980)
Donghu (China)		5.7–33.5 (wet weight)	*B. sowerbyi*	Liang (1984)
Sierra Nevada Peatlands		58–74 (wet weight)	Oligochaeta	Erman & Erman (1975)
Lake Esrom (Denmark)	16.5	92.6 Kcal m^{-2} ann^{-1}	Chironomids and *Chaoborus* sp.	Jonasson (1972)
		1.5 " "	*Ilyodrilus hammoniensis*	
		6	*Pisidium casertanum*	

the benthic biomass was 3 g dry weight m^{-2} at a time when the benthos was dominated by insects and oligochaetes. Later, Marshall (1978) found the mean benthic biomass of chironomids and oligochaetes to be 5.0 and 18.8 g dry weight/m^2, respectively, which are extremely high values compared to other African lakes.

Lake Naivasha is morphometrically similar to Lakes McIlwaine, Chad, Chilwa, and George. The benthic biomass in the profundal region of Lake Naivasha falls within the range recorded in other African lakes with higher standing crop than many lakes. In both Lake Naivasha and Lake McIlwaine, the tubificid *B. sowerbyi* contributes about 60% of profundal benthic biomass.

Data on benthic productivity is almost non-existent in African lakes except for Lakes Sibaya and Chad. The productivity of *Caridina nilotica* in Lake Sibaya was 37.5 g dry weight m^{-2} ann^{-1} (Hart, 1979) while the productivity of the entire molluscan assemblage was 14.76 g dry weight m^{-2} ann^{-1} in Lake Chad (Leveque et al., 1983). In Lake Naivasha, the productivity, considering the kind of species studied, was high (Table 3).

Secondary production in benthic communities is normally limited by habitat complexity, biological interactions, temperature and the quantity and quality of food (Benke, 1984). The relative importance among these factors differs among different benthic environments and in their influence on different species in any given community. According to Butler (1982), major factors limiting production of a species are often the same as those limiting life cycle patterns. Temperature does not limit standing stock biomass (Morgan et al., 1980) since high biomasses can be found in very cold and in warm environments. However, it can influence biomass turnover in an environment where there is no stress and abundant food supply (Benke, 1984). In Lake Naivasha, temporal variation in sediment temperature was minimal compared to that in temperate lakes with distinct seasonal fluctuations. Production of benthic communities is therefore unlikely to be limited by temperature. Biological effects (such as competition and predation) on the production dynamics of benthic species is often very complex, and the information about it is scarce (Morgan et al., 1980). The biomass and turnover rates can be limited by the quality and quantity of food. In Lake Naivasha, life cycle patterns and biomass peaks were synchronized with seasonality of rainfall normally associated with injection of fresh food from allochthonous and autochthonous nutrient inputs stimulating primary productivity of phytoplankton. Quality and quantity of food therefore appears to be the major factor limiting the production of benthic macroinverterbrates in Lake Naivasha.

Acknowledgements

This study formed past of the work of the University of Leicester research project at Lake Naivasha, which since 1987 has been authorised by the Office of the President, Government of Kenya under research permit to Dr D. M. Harper no. OP 13/001/12C 46. The project was funded by the Earthwatch Institute, Boston USA and Oxford, England. The data collection would not have been possible without the assistance of numerous Earthwatch volunteers. Our sincere thanks go to the numerous colleagues in Kenya in particular Jill & Angus Simpson.

References

Aston, R. J., 1973. Field and experimental studies on the effect of power station effluent on Tubificidae (Oligochaeta, Annelida). Hydrobiologia 42: 225–242.

Becht, R. & Harper, D. M., 2002. Towards an understanding of human impact upon the hydrology of Lake Naivasha, Kenya. Hydrobiologia 488 (Dev. Hydrobiol. 168): 1–11.

Benke, A. C., 1984. Secondary Production of Aquatic Insects. In Resh, V. H. & D. M. Rosenberg (eds), The Ecology of Aquatic Insects. Praeger Publishers, New York: 625 pp.

Brinkhurst, R. O., 1963. A Guide for the Identification of British Aquatic Oligochaeta. Freshwater Biological Association. Scientific Publication No. 22: 52 pp.

Brinkhurst, R. O., 1966. A Contribution Towards a Revision of the Aquatic Oligochaeta of Africa. Zoologica African 2 (2): 131–166.

Burgis, J. M., J. P. E. C. Darlington, I. G. Dunn, G. G. Ganf, J. J. Gwaba & L. M. McGowan, 1973. The Biomass and Distribution of Organisms in Lake George, Uganda. Proc. R. Soc. Lond. B. 184: 271–289.

Butler, M. G., 1982. Production Dynamics of some Artic *Chironomus* larvae. Limnol. Oceanogr. 27: 728–736.

Carrol, H. J. & T. T. Dorris, 1972. The Life History of *Branchiura sowerbyi*. Am. midl. Nat. 87: 413–422.

Casellato, S., 1984. Life Cycle and Karyology of *Branchiura sowerbyi* Beddard (Oligochaeta, Tubificidae). Hydrobiologia 115: 65–69.

Clark, F. L., A. Beeby & P. Kirby, 1989. A study of the macroinvertbrates of lakes Naivasha, Oloidien and Sonachi, Kenya. Rev. Hydrobiol. Trop. 22: 21–33.

Darlington, S. P. E. C., 1977. Temporal and Spatial Variation in Benthic Invertebrate Fauna of Lake George, Uganda. J. Zool. Lond. 181: 95–11.

Erman, D. C. & N. A. Erman, 1975. Macroinvertebrate composition and production in some Sierra Nevada minerotrophic peatlands. Ecology 56: 591–603.

Gieve, O., 1975. Population structure, food relations and ecological role of marine oligochaetes with special reference to microbenthic species. Mar. Biol. 31: 139–156.

Harper, D. M., K. M. Mavuti & S. Muchiri, 1990. Ecology and management of Lake Naivasha Kenya, in relation to climate change, alien species introductions and agricultural development. Environmental Conservation 17: 328–336.

Hart, R. C., 1979. The Invertebrate Communities: The Zooplankton, Zoobenthos and Littoral Fauna. In Allanson, B. R. (ed.), Lake Sibaya. Monographie Biologiae, 36. Dr W. Junk Publishers, The Hague: 108–161.

Hickley, P., R. Bailey, D. M. Harper, R. Kundu, M. Muchiri, R. North & A. Taylor, 2002. The status and future of the Lake Naivasha fishery, Kenya. Hydrobiologia 488 (Dev. Hydrobiol. 168): 181–190.

Hiltunen, J. K., 1967. Some oligochaetes from Lake Michigan. Trans. am. Microsc. Soc. 86: 433–454.

Hubble, D. & D. M. Harper, 2001a. Impact of light regimen and self-shading by algal cells on primary productivity in the water column of a shallow tropical lake (Lake Naivasha, Kenya). Lakes & Reserves, Research and management 6: 143–150.

Hubble, D. & D. M. Harper, 2001b. What defines a healthy lake? Evidence from Lake Naivasha, Kenya. Aquatic Ecosystem Health & Management 4: 243–250.

Hynes, H. B. N. & J. M. Coleman, 1968. A Simple Method of Assessing the Annual Production of Stream Benthos. Limnol. Oceanogr. 8: 569–573.

Jonasson, P. M., 1972. Ecology and production of the profundal benthos in relation to phytoplankton in Lake Esrom. Oikos 14 (suppl): 1–148.

Kennedy, C. R., 1965. The Distribution and Habitat of *Limnodrilus* Claparede (Oligochaeta: Tubificidae). Oikos *16*: 26–38.

Kennedy, C. R., 1966. The Life History of *Limnodrilus hoffmeisteri* Clap. (Oligochaeta: Tubificidae) and its Adaptive Significance. Oikos 17: 158–168.

Ladle, M., 1971. The Biology of Oligochaeta from Dorset Chalk Streams. Freshwat. Biol. 1: 83–97.

Lafont, M., 1987. Production of Tubificidae in the littoral zone of Lake Leman near Tnonon-les-Bains: A methodological approach. Hydrobiologia 155: 179–187.

Leveque, C., 1979. Biological productivity of Lake Chad. Document at UNEP Workshop on African Limnology, Nairobi, 16–23 December 1979, unpublished.

Leveque, C., C. Dejoux & L. Lauzanne, 1983. The Benthic Fauna: Ecology Biomass and Communities. In Carmouze, J. P. et al., (eds), Lake Chad. W. Junk, the Hague: 233–272.

Liang, V. I., 1984. Annual production of *Branchiura sowerbyi* (Oligochaeta: Tubificidae) in the Donghu Lake, Wuhan, China. Chin. J. Oceanol. Limnol. 2: 102–108.

Marshall, B. E., 1978. Aspects of the ecology of benthic fauna in Lake McIlwaine, Rhodesia. Freshwat. Biol. 8: 241–249.
Mavuti, K. M., 1990. Ecology and role of zooplankton in the fishery of Lake Naivasha. Hydrobiologia, 208: 131–140.
McLachlan, A. J. & S. M. McLachlan, 1969. The Bottom Fauna and Sediments in a Drying Phase of a Saline African Lake (L. Chilwa, Malawi). Hydrobiologia 34: 401–413.
McLachlan, A. J. 1979. Decline and recovery of benthic invertebrate communities. In Kalk, M., A. J. McLachlan & C. Howard-Williams (eds), Lake Chilwa. Studies of Change in a Tropical Ecosystem. Dr W. Junk Publishers, The Hague: 143–160.
Menzie, C. A., 1980. A Note on the Hynes Method of Estimating Secondary Production. Limnol. Oceanogr. 25 (4): 770–773.
Morgan, N. C., T. Backiel, G. Bretschko, A. Duncan, A. Hillbricht-Ilkowska, Z. Kajak, J. F. Kitchell, P. Larsson, C. Leveque, A. Nauwerk, F. Scheimer & J. E. Thorpe, 1980. Secondary Production. In Le Cren E. D. & R. H. LoweMcConnel (eds), The Function of Freshwater Ecosystems. Cambridge University Press, Cambridge, England: 558 pp.
Muchiri, S. M., 1990. The feeding ecology of Tilapia and the fishery of Lake Naivasha. Unpublished Ph.D. Thesis, University of Leicester: 233 pp.
Munro, J. L., 1966. A limnological survey of Lake Mcllwaine, Rhodesia. Hydrobiologia 28: 281–308.
Poddubnaya, T. L., 1980. Life cycles of mass species of tubificidae (Oligochaeta). In Brinkhurst, R. & A Cook (eds), Aquatic Oligochaete Biology, 1st Internat,. Symposium, Sydney.
Poddubnaya, T. L., 1984. Parthenogenesis in Tubificidae. Hydrobiologia 115: 97–99.
Raburu, P. O., 1991. Studies on the Profundal Benthic Macroinvertebrate Productivity in Lake Naivasha, Kenya. Unpublished M.Sc. Thesis, University of Nairobi. 192 pp.

Hydrobiologia **488:** 163–170, 2002.
D.M. Harper, R. Boar, M. Everard & P. Hickley (eds), Lake Naivasha, Kenya.

Population changes in sympatric Great and Long-tailed Cormorants (*Phalacrocorax carbo* and *P. africanus*): the effects of niche overlap or environmental change?

R. Brooks Childress[1,2,*], Leon A. Bennun[1] & David M. Harper[2]
[1]*Ornithology Department, National Museums of Kenya, P.O. Box 40658, Nairobi, Kenya*
[2]*Department of Biology, University of Leicester, Leicester, LE1 7RH, U.K.*
[*]*Author for correspondence: Wildfowl & Wetlands Trust, Slimbridge, Glos. GL2 7BT, U.K.*
E-mail: brooks.childress@wwt.org.uk

Key words: Phalacrocorax carbo, Phalacrocorax africanus, population changes, niche overlap, sympatric species, resource competition, ecological separation

Abstract

Between January 1993 and January 1995, the number of Great Cormorants (*Phalacrocorax carbo*) using Lake Naivasha, Kenya (00° 45′ S, 36° 20′ E) for foraging and resting increased 56%, while the number of sympatric Long-tailed Cormorants (*Phalacrocorax africanus*) decreased 64%. In 1995 and 1996, we documented habitat changes and conducted monthly population and resource-use surveys of the two species in an attempt to discover the most likely reasons for these changes. The increase in Great Cormorants was probably the result of immigration from nearby Lake Nakuru due to extreme water level reductions there. Lake Naivasha also experienced falling water levels and transparency during this period, but these changes were not as severe and are not considered likely reasons for the decline in Long-tailed Cormorant numbers. Despite some probable dietary overlap, the two species were well separated in terms of foraging locations, foraging methods, resting habitats and breeding timing. The decline in Long-tailed Cormorant numbers may be connected with increased disturbance by fishermen along the lake littoral, this species' primary feeding location.

Introduction

The ranges of the Great and Long-tailed Cormorants substantially overlap in sub-Saharan Africa (Johnsgard, 1993). At Lake Naivasha (00° 45′ S, 36° 20′ E) in the southern part of Kenya's Rift Valley, the two species are sympatric. Between January 1993 and January 1995, the number of Great Cormorants using the lake for foraging and resting increased by 56%, while the number of Long-tailed Cormorants decreased by 64% (Table 1; Nasirwa & Bennun, 1994, 1995). These population changes continued through 1996.

The competitive exclusion principle postulates that ecologically similar species using the same set of limited resources cannot coexist in equilibrium (reviewed in Wiens, 1989). Both prey variety and suitable resting habitat are limited at Lake Naivasha. This lake is unusual for a tropical freshwater lake in having only three abundant fish species (Harper et al., 1990) to support the 45 species in its avian fish-eating guild (Hartley, 1984). However, while the limited number of prey species might indicate a high degree of diet overlap, previous studies have shown that ecologically similar species which exhibit a high degree of overlap on one niche dimension (e.g., prey species), often are able to coexist due to compensating differences on other dimensions such as prey size, or foraging locations or methods (Lack, 1945; Schoener, 1974; Brandl & Utschick, 1985). Further, it has also been shown that two similar sympatric species with a large difference in size, such as the Great and Long-tailed Cormorants (Brown et al., 1982), are not likely to utilise the same resources to any important extent (e.g., Lack, 1971; Furness & Barrett, 1985).

Before 1993, there was a breeding population of approximately 3000 Great Cormorants at Lake Nakuru (Bennun, 1992; Nasirwa, pers. comm.), an alkaline lake about 55 km north-west of Lake Naivasha. The years 1993 and 1994 were relatively dry in this section of the Rift Valley (Nasirwa & Bennun, 1994) and Lake Nakuru evaporated almost completely. Due to the low water level and resulting high alkalinity, fish stocks virtually disappeared, followed by the fish-eating birds (Bennun & Nasirwa, 2000). The Great Cormorant population there declined from 3238 in 1992, to 578 in 1993, and to zero in 1994.

In 1995, Great Cormorants established a new breeding colony on Lake Oloidien in the Lake Naivasha ecosystem. Lake Oloidien (00° 49′ S, 36° 16′ E) is a small (5.5 km^2) alkaline satellite of Lake Naivasha (ca. 130 km^2). This colony, which contained 875 nests in its first year, increased to 925 nests in 1996. No birds in the Lake Nakuru population were marked, and it is therefore not certain where they dispersed. However, it seems probable that the increase in Great Cormorants using Lake Naivasha and the establishment of the new breeding colony were the result of immigration from Lake Nakuru. This does not, however, explain the coincidental decrease in the number of Long-tailed Cormorants using Lake Naivasha. Here we assess two hypotheses, not mutually exclusive, to explain this decline:

- There is extensive niche overlap between the two sympatric species, suitable prey and resting resources are limited, and the Great Cormorant is more successful in competing for these limited resources.
- Declines in lake water level and transparency at Lake Naivasha during the 1990s had a greater negative effect on the Long-tailed Cormorant's ability to meet its needs than on that of the Great Cormorant.

Methods

Between September 1995 and August 1996, we conducted monthly population and resource-use surveys of the two cormorant species on lakes Naivasha and Oloidien. For these surveys, the littoral of the two lakes was divided into 16 sections (Fig. 1), the sections containing approximately equal lengths of shoreline. To facilitate historical analysis, the sections were the same as those used by the Department of Ornithology, National Museums of Kenya for its annual waterbird counts.

Each survey of Lake Naivasha consisted of circumnavigating the lake in an inflatable dinghy, keeping approximately 100 m offshore. Due to the size of the lake (approximately 130 km^2), the high winds and rough waters that often occurred in the afternoons, and the daily movement patterns of the cormorants, the surveys were all conducted between about 07:00 and 13:00 h. Surveys usually took place over two successive days with six or seven sections being completed each day. Each survey was begun in a different section, and the direction of travel was altered to vary the survey times for each section throughout the year. Water transparency was recorded with a Secchi disc at the beginning, middle, and end of each section. Because of its smaller size, Lake Oloidien could be surveyed accurately with ×10 binoculars and a ×20–60 telescope from an elevated point of land along the southern shore. In order to maintain time-of-day consistency, the Oloidien surveys were always conducted mid-morning on the day immediately following the Lake Naivasha survey.

During the surveys, each cormorant sighted was recorded, along with its activity (foraging, resting, or flying), foraging location (lake section and estimated distance from shore), and, if appropriate, resting habitat (mud shore, floating water hyacinth, sedges, rocks, man-made structures). The numbers of birds in flocks (foraging, resting, or flying) were counted with the aid of ×10 binoculars and a tally counter, and recorded along with the foraging location or resting habitat of the flock. The number of individual birds in each flock was counted twice. If the counts were within 5% of one another, the mean of the two counts was recorded. If the counts were more than five percent apart, the flock was counted a third time, and the mean of the three counts was recorded.

In order to obtain an indication of the Great Cormorant diet an analysis was made of the species and size of dropped and regurgitated fish collected under the colony. Between the ages of 5 and 8 weeks Great Cormorant nestlings will frequently regurgitate whatever is in their oesophagus when threatened (RBC, unpubl. data). For a period of approximately 8 weeks in both 1995 and 1996, while the main body of nestlings passed through this age category, all dropped and regurgitated fish were collected, measured, and weighed ($n = 190$). Fish identification was straightforward, since the fish species variety in Lake Naivasha is limited and well known, consisting

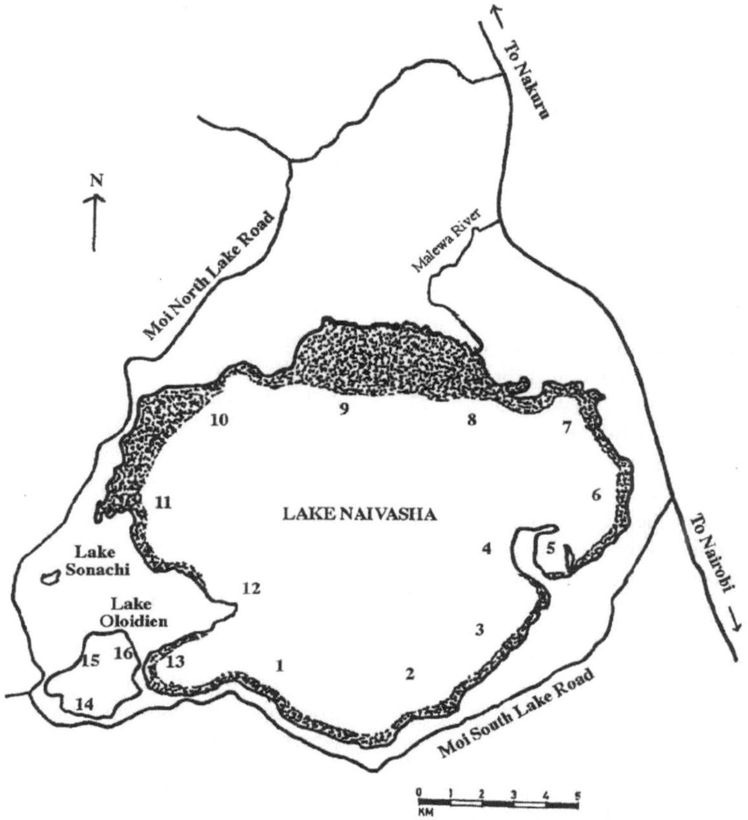

Figure 1. Lake Naivasha population survey map; sources: Hartley, 1984 & Ornithology Dept., National Museums of Kenya; numbers indicate shoreline survey sections.

primarily of two introduced tilapine species (*Oreochromis leucostictus* and *Tilapia zillii*) and the introduced American Black Bass (*Micropterus salmoides*).

These data are believed to represent a random sample of the fish the nestlings were fed during the period when they were between 5 and 8 weeks old. However, because Great Cormorants appear to be opportunistic feeders taking whatever is available, and because 8-week-old nestlings are as large as adults, these data were also used to indicate the diet of adult cormorants during the breeding season.

Results

Cormorant population trends

Between January 1993 and January 1995, the number of Great Cormorants using Lake Naivasha for foraging and resting increased by 56% from 555 to 868, while the number of Long-tailed Cormorants decreased 64% from 1035 to 374 (Table 1).

The population shifts observed in 1993–95 continued through 1996. From a low of 555 in January 1993, the number of Great Cormorants counted using the Lake Naivasha ecosystem for foraging and resting increased by 110% to 1167 individuals in January 1997, after peaking at 2403 individuals in January

Table 1. Counts of cormorants using Lake Naivasha ecosystem* each January, 1993–95

	1993	1994	1995
Great Cormorant	555	861	868
Long-tailed Cormorant	1035	823	374
Both species	1590	1684	1242

*Includes small adjacent lakes Oloidien and Sonachi.

1996. During the same time period, the number of Long-tailed Cormorants using the lakes declined by 77% from the peak of 1035 individuals counted in January 1993 (Fig. 2).

Resource-use comparisons

Prey species and size

Approximately 90% of the Great Cormorant diet during the breeding period consisted of fish weighing ≤150 g. Body shape appears to have had a significant influence on the weight of individual fish eaten and regurgitated (Fig. 3). Mean weights for the more cylindrical-shaped bass (131.6 ± 63.4 g S.D., $n = 73$) were twice as great as for the broader, deeper-shaped *Tilapia* (65.2 ± 39.2 g S.D., $n = 117$; $t = 8.05$, $P < 0.001$; separate variance t-test).

We did not document the Long-tailed Cormorant diet. Only 83 individuals were seen foraging during the 12 censuses and they were rarely observed catching anything. Analysis of stomach contents was not attempted, and examination of pellets cast at their day roosts was unsuccessful.

Foraging distance from shore

The Great Cormorant was found foraging most often (62.8%) in open water >100 m from the edge of the lake, while the Long-tailed Cormorant was found foraging most often (88.0%) <100 m of the lake edge (Fig. 4). Long-tailed Cormorants often foraged very close inshore: 44.6% of foraging individuals were recorded <5 m of the lake edge, compared to just 9.8% for the Great Cormorant. The difference between the two species in foraging distance from shore is highly significant ($\chi^2_1 = 87.39$, $P < 0.01$; χ^2 test).

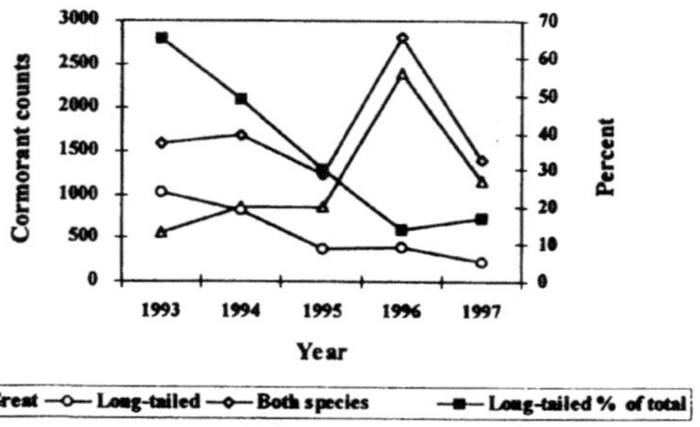

Figure 2. Counts of cormorants using Lake Naivasha ecosystem each January, 1993–97, and the Long-tailed species percentage of the total each year; source: Bennun, 1993; Nasirwa & Bennun, 1994, 1995; Nasirwa personal communication. Includes small adjacent lakes Oloidien & Sonachi.

Foraging methods

The Long-tailed Cormorant was only observed foraging solitarily, while the Great Cormorant usually foraged in flocks. Of 3023 Great Cormorants recorded foraging, only 15% were solitary, the rest being in flocks ranging in size from nine to 407 birds ($n = 28$, mean = 91.7 ± 96.6 S.D.).

Effect of water transparency on Great Cormorant foraging method

When water transparency was <40 cm, Great Cormorant foraging was almost exclusively communal (95.8%). However, when water transparency was ≥40 cm, the relative frequency of solitary foraging increased to 37.4%. There was a highly significant association between communal foraging and water transparency <40 cm, and between solitary foraging and water transparency ≥40 cm ($\chi^2_1 = 797.7$, $P < 0.01$, χ^2 test). Long-tailed Cormorants were never observed foraging in groups, so water transparency appeared to have no effect on this aspect of their foraging behaviour.

Effect of water transparency on foraging activity

Long-tailed Cormorants foraged as readily in areas where water transparency was relatively low (<40 cm Secchi disk depth), as in those where it was relatively high (≥40 cm Secchi disk depth) (Fig. 5). However, a disproportionate number of Great Cormorants were recorded foraging in waters where transparency was relatively low (<40 cm). There was a highly significant association between Great Cormorant foraging and water transparency <40 cm ($\chi^2_1 = 662.2$, $P < 0.01$, χ^2 test).

Day and night roost locations and habitats

From their night roosts, both species flew to Lake Naivasha to feed just after dawn. Before beginning to forage, however, the Long-tailed Cormorants first flew singly or in small groups of twos and threes to day roosts, large clumps of papyrus or floating water hyacinth located near their foraging areas at various points around the lake. These roosts, which usually accommodated between 10 and 50 birds each, acted as bases for the day's foraging, with birds flying out to forage and then returning to rest throughout the day.

At the end of the day, the Long-tailed Cormorants again gathered at their day roosts before flying off in small groups to their night roost just before sunset. (We were unable to locate their night roost.) The locations of the day roosts varied throughout the

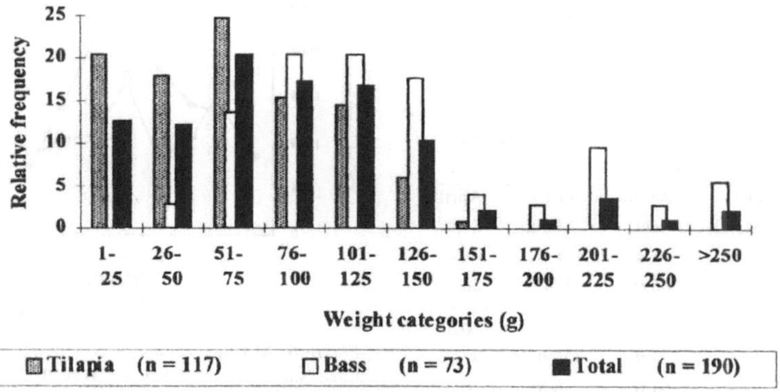

Figure 3. Proportion of Great Cormorant colony diet represented by fish weight category: tilapia (combined species) and American large-mouthed bass, 1995–96.

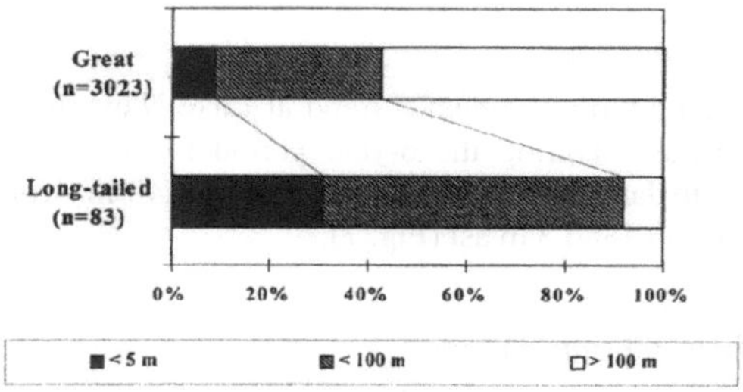

Figure 4. Relative frequency of foraging distance from the lake edge by species.

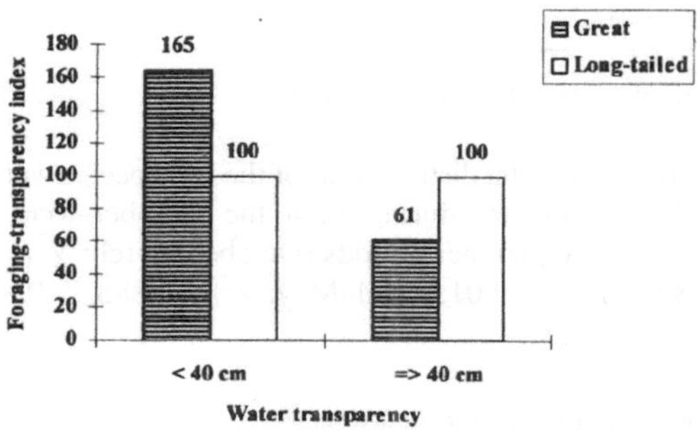

Figure 5. Foraging index (the proportion of individuals recorded foraging at each transparency level divided by the proportion with which that level occurred in the survey). Sample sizes according to water transparency (Secchi disk) depth. Great Cormorant: <40 cm 1611; ≥40 cm 984; total 2595; Long-tailed Cormorant: <40 cm 27; ≥40 cm 45; total 72; survey transparency levels: <40 cm 47; ≥40 cm 78; total 125.

year depending on the foraging locations currently in use. For 6 months of the year (October–March), extensive beds of floating Water Hyacinth *Eichhornia crassipes* became anchored on a large area of submerged macrophytes located north of Crescent Island. During this part of the year, the majority (61.8%) of resting Long-tailed Cormorants were found on these centrally located beds (Table 2). During the remaining 6 months, when these beds were much smaller, the majority (61.0%) rested on sedges around the lake edge (Table 2). The October–March pattern of resting habitat occupation was significantly different from the April–September pattern ($\chi^2_6 = 652.0$, $P < 0.01$).

The Great Cormorants roosted at night in their breeding colony trees at Lake Oloidien. At dawn, they would leave this roost in groups of 15–30 birds, usually flying to Lake Naivasha. Here, they would fly around the edge of the lake. When a group located a shoal of fish, it landed and began foraging. It would soon be joined by other groups of Great Cormorants, and the resulting flock could quickly build to 300–400 birds. As with the Long-tailed Cormorant, the period of most intense foraging occurred between daybreak and about 09:00 h. After completing foraging, the Great Cormorants also began gathering in ones and twos at day roosts. Like the Long-tailed Cormorants, the Great Cormorants rested in large numbers during the day on beds of floating water hyacinth, either the large central beds north of Crescent Island when these were available (October–March), or the beds around the edge of the lake (April–September). At all times of the year, even larger numbers rested during the day on muddy or sandy shoreline (>40%, Table 2).

The differences between the October–March period and the April–September period in the frequency with which different resting habitats were used (Table 2) were statistically significant, even after combining the two water hyacinth categories ($\chi^2_5 = 46.6$, $P < 0.01$, χ^2 test). Likewise, the differences between

Table 2. Seasonal relative frequency of resting habitat use by species

Habitat	Percentage of Long-tailed Cormorants resting		Percentage of Great Cormorants resting*	
	Oct.–March (n = 1930)	April–Sept. (n = 669)	Oct.–March (n = 3743)	April–Sept. (n = 2636)
Central hyacinth beds	61.8	8.2	11.0	0.5
Littoral hyacinth beds	10.5	23.9	22.0	35.8
Sedges	9.9	61.0	0.8	0.2
Mud/sand shore	0.4	1.2	41.5	42.9
Rocks	3.6	4.3	14.1	14.1
Stumps	3.3	1.4	6.9	3.9
Man-made structures	0.5	0.0	3.7	2.6

*Excludes birds in breeding-colony trees.

the resting habitat distributions of the two species were highly significant during both the October–March and April-September periods (October-March: χ^2_6 = 2886.5, P =< 0.01; April–May: χ^2_6 = 2886.5, P < 0.01; χ^2 tests).

Breeding timing and locations

The Great Cormorants nested in a single colony from March to August/September, peaking in April/May. The nest sites of the Long-tailed Cormorants were not found, but this species appeared to breed later in the year, peaking between June and August. This was inferred from the relatively low numbers of adults counted on the lake at that time, presumably being away sitting on nests, and the increase in young birds in immature plumage counted during October–December.

This breeding timing would result in peak prey demand for one species occurring during a period of relatively low prey demand for the other. During 1995–96 the Great Cormorant population in the Lake Naivasha ecosystem was at its lowest between November 1995 and April 1996, the period between the dispersal of the 1995 fledglings and the start of the 1996 breeding effort. This was the same period during which the Long-tailed Cormorant was most numerous (Fig. 6). Similarly, in 1996 the Great Cormorant population peaked with the fledgling of that year's young between May and July, at a time when the Long-tailed Cormorant numbers on the lake were at their lowest level of the year.

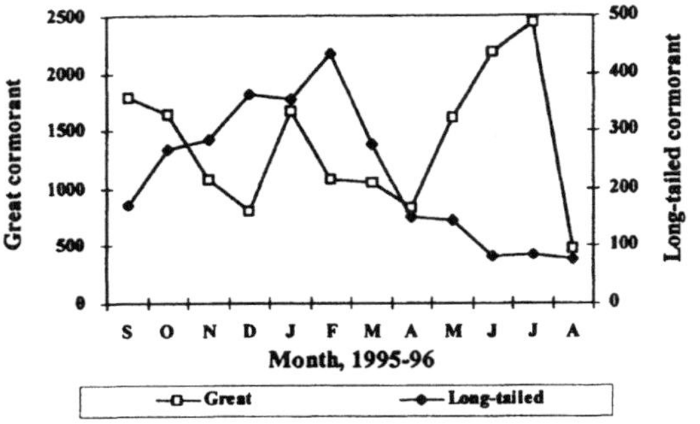

Figure 6. Monthly counts of cormorants (both species) using the Lake Naivasha ecosystem, 1995–96. Includes small adjacent lakes Oloidien and Sonachi.

Environmental changes

Lake water level

The long-term water level trend at Lake Naivasha is downward. During the 6-year period from January 1991 to January 1997, the level fell by 1.48 m, from 1887.9 to 1886.3 m asl (Fig. 7).

Lake water transparency

Between 1990 and 1996, water transparency in Lake Naivasha appears to have declined significantly. We replicated Secchi disk measurements made in 1990 near Hippo Point, the deepest area in the main lake (excluding the small Crescent Island lagoon). The mean decline in transparency at this location between 1990 and 1996 was 44% (Table 3; t-test for differences in means, t = 5.63, P < 0.01, paired t-test). Transparency was consistently lowest in the shallowest areas of the lake along the northern and eastern shores. In 1995–96, mean transparency in this part of the lake (32.5 cm ± 11.5 cm S.D.) was significantly lower than that along the southern, western, and central shores (54.4 cm ± 18.4 cm S.D., t = −8.78, df = 141, P < 0.001; unequal variance t-test).

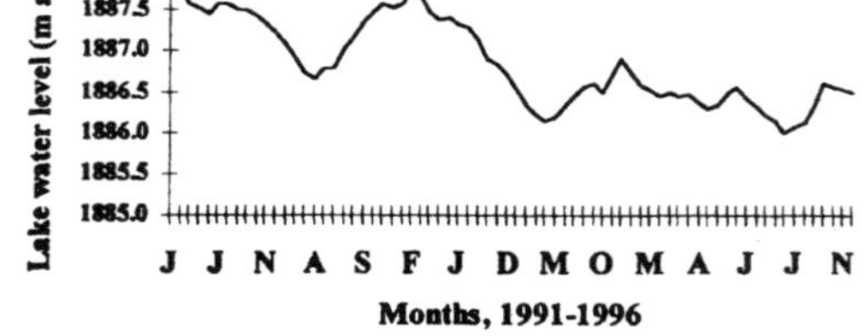

Figure 7. Lake water levels and trend, October 1991–1997; source: Sulmac Company Limited.

Table 3. Comparison of water transparency in centimetres near Hippo Point, Lake Naivasha, in 1996 and 1990. Each monthly value is based on six readings.

Month	Transparency (cm)		
	1990	1996	% Chg.
Feb.	70	55	−21.4%
March	70	44	−37.1
May	80	40	−50.0
June	85	48	−43.5
July	105	53	−59.4
Aug.	115	54	−53.0
Mean	87.5	49.0	−44.0

Effect of lower water level on Long-tailed Cormorant distribution

In 1993, before the substantial reduction in water level and transparency, the shallower northern and eastern shores held over half of the resting and foraging Long-tailed Cormorants (Table 4). By 1997, the population had shifted significantly toward the deeper southern and western sections ($\chi^2{}_2 = 113.3$, $P < 0.01$, χ^2-test). Since the overall Long-tailed population also declined greatly during this period, the southern and western sections in effect lost proportionately fewer birds than the other sections (Table 4).

Discussion

Due to the limited number of suitable prey species in Lake Naivasha, there appeared to be extensive overlap in the diets of the two cormorant species for fish weighing $\leq$75 g. Approximately 45% of the Great Cormorant's diet during the breeding period appeared to consist of fish of this size, and although we were unable to document the Long-tailed Cormorant's diet in this study, two earlier studies using stomach contents analysis (Bowmaker (1963) and Birkhead (1978)), indicated that prey size ranged from 2.20 cm in length (mean =7 cm) and from under 19 g to 75 g in weight.

This overlap might have been expected to result in substantial competition. However, no evidence was found that such competition occurs. The two species tended to forage in different areas of the lake, using different methods, and were rarely seen foraging close together. Furthermore, their breeding timing appeared to be sufficiently different to reduce competition for prey during the periods of maximum demand (immediately after fledging). Thus, despite probably having substantial dietary overlap, the two species seemed to be well separated on other niche dimensions. Previous studies have shown similar ecological differentiation throughout sub-Saharan Africa where the species are sympatric, such as on lakes Victoria and Malawi in East Africa and Lake Kariba in southern Africa (Benson et al., 1971; Linn & Campbell, 1992; Wanink, 1996). Where only one species is present, such as the Great Cormorant on Lake Abiata in Ethiopia (Urban, 1992) and the Long-tailed Cormorant on Lake Bangweulu in Zimbabwe (Bowmaker, 1963), each retains its restricted habitat utilisation.

Between 1990 and 1996, there were substantial declines in lake water level and transparency at Lake Naivasha. The Long-tailed Cormorant's foraging behaviour did not appear to be affected by lower water transparency, but as the water level declined and large areas of mud flats or very shallow turbid water developed in its favourite feeding areas close to shore along the shallower northern and eastern sections of the lake, this species virtually disappeared from these areas. It also largely disappeared from the Crescent Island lagoon, the deepest and least turbid section of the lake, although the reason for the decline in this section of the lake is less clear.

One reason might be increased disturbance due to fishing activity. During the study period, the numbers of Long-tailed Cormorants declined substantially in all sections of the lake. In 1995 and 1996, virtually the entire littoral zone of the lake was occupied on each day by fishermen. Where the water was deep enough, fishermen set small-mesh gill nets close to and parallel to the floating vegetation fringing the lake edge. This fringing vegetation is the Long-tailed Cormorant's preferred foraging habitat, presumably because the fish use it as cover. The fishermen's nets would block

Table 4. Distribution of Long-tailed Cormorant by lake section, 1997 vs. 1993; data from the Dept. of Ornithology, National Museums of Kenya, annual January waterbird censuses

Lake census sections	Frequency		%	Relative frequency	
	Jan 1993	Jan 1997	chg.	Jan 1993	Jan 1997
North and east	571	62	−89.1	53.2	34.7
South and west	164	95	−41.5	15.3	53.6
Central (macrophyte beds and Crescent Island lagoon)	339	21	−93.8	31.5	11.7
Total lake	1074	179	−83.3	100.0	100.0

the movement of fish in and out of the vegetation, and thus the ability of the Long-tailed Cormorants to forage effectively. They would also pose a danger to cormorants of both species, which are easily caught and drowned in the nets (RBC, unpubl. data). In those areas where the water is too shallow for gill nets, the fishermen often fished in pairs with seine nets, moving parallel to the fringing vegetation while making great commotion, both to frighten the fish into the nets and to ward off the many *Hippopotamus amphibius* occupying the lake's edge. Long-tailed Cormorants seemed very sensitive to human disturbance: seldom was it possible to approach within 50 m before a roost would empty or a foraging bird would relocate.

It is unknown whether fishing activity was increasing during the period when the Long-tailed Cormorant population was declining. Many fishermen are unlicensed, and there are no records on their activity. Anecdotal information from local residents suggests the fishing pressure has increased substantially, in parallel with a documented increase in the human population around the lake (Goldson Associates, 1993). If this is so, it is likely that this activity, with its focus in the littoral region of the lake, has affected the Long-tailed Cormorant much more than the Great Cormorant.

Acknowledgements

We thank four colleagues – Matt Prescott, Colin Beale, Steven Windels and Joseph Ojambo – for their invaluable assistance in the field, the Zwager family for allowing access to the study site, and the Department of Ornithology, National Museums of Kenya the use of their survey map and population data. Elsamere Conservation Centre kindly provided boats, outboard motors, office and laboratory facilities, and overnight accommodation for field assistants. The Earthwatch Institute provided support for DH, and some volunteer teams assisted with observations. We thank the Office of the President of the Republic of Kenya for written permission to conduct this study, as part of a larger project understanding the ecology of Lake Naivasha.

References

Bennun, L. A., 1992. Summary results of the 1992 January waterfowl counts. The National Museums of Kenya, Department of Ornithology, Nairobi.

Bennun, L. A. & O. Nasirwa, 2000. Trends in waterbird numbers in the southern Rift Valley of Kenya. Ostrich 71: 220–226.

Benson, C. W., R. K. Brooke, R. J., Dowsett & M. P. S Irwin, 1971. The birds of Zambia. Collins, London.

Birkhead, M. E., 1978. Some aspects of the feeding ecology of the reed cormorant and darter on Lake Kariba, Rhodesia. Ostrich 49: 1–7.

Bowmaker, A. P., 1963. Cormorant predation on two central African lakes. Ostrich March: 2–26.

Brandl, R. & H. Utschick, 1985. Size, ecology and wading birds: a nonparsimonious view. Naturwissenschaften 72: 550–552.

Brown, L. H., E. K. Urban & K. Newman (eds), 1982. The Birds of Africa, Vol. 1. Academic Press, London.

Furness, R. W. & R. T. Barrett, 1985. The food requirements and ecological relationships of a seabird community in north Norway. Ornis Scand. 16: 305–313.

Goldson, J., 1993. A three-phase environmental impact study of recent developments around Lake Naivasha. Lake Naivasha Riparian Owners Association, Naivasha, Kenya.

Harper, D. M., K. M. Mavuti & S. M. Muchiri, 1990. Ecology and management of Lake Naivasha, Kenya, in relation to climate change, alien species' introductions, and agricultural development. Environ. Cons. 17: 328–336.

Hartley, J., 1984. A guide to the Lake Naivasha area. Evans Brothers (Kenya) Ltd., Nairobi.

Johnsgard, P. A., 1993. Cormorants, Darters and Pelicans of the World. Smithsonian Institution Press, Washington, DC.

Lack, D., 1945. The ecology of closely related species with special reference to cormorant (*Phalacrocorax carbo*) and shag (*Phalacrocorax aristotelis*). Ecology 14: 12–16.

Lack, D., 1971. Ecological Isolation in Birds. Blackwell Scientific Publications, Oxford.

Linn, I. J. & K. L. I. Campbell, 1992. Interactions between white-breasted cormorants *Phalacrocorax carbo* (Aves: Phalacrocoracidae) and the fisheries of Lake Malawi. J. appl. Ecol. 29: 619–634.

Nasirwa, O. & L. A. Bennun, 1994. Waterbirds in the southern Kenyan Rift Valley, July 1993 and January 1994. Centre for Biodiversity Research Reports: Ornithology, No. 17: July 1994. Department of Ornithology, National Museums of Kenya, Nairobi.

Nasirwa, O. & L. A. Bennun, 1995. Monitoring of waterbirds in central Kenya, July 1994 and January 1995. Centre for Biodiversity Reports: Ornithology, No. 19: October 1995. Department of Ornithology, National Museums of Kenya, Nairobi.

Schoener, T. W., 1974 Resource partitioning in ecological communities. Science 185: 27–39.

Urban, E. K., 1992. Seasonal and opportunistic nesting of great cormorants *Phalacrocorax carbo* in Ethiopia. pp 475–480. Proceedings of the VIII Pan-African Ornithological Conference.

Wanink, J. H., 1996. Foraging locations of kingfishers and cormorants at Lake Victoria depend on the distribution of harvestable prey. Afr. J. Ecol. 34: 90–93.

Wiens, J. A., 1989. The Ecology of Bird Communities, Vol. 1. Foundations and Patterns. Cambridge University Press, Cambridge.

Zijlstra M. & M. R. Van Eerden, 1995. Pellet production and the use of otoliths in determining the diet of cormorants *Phalacrocorax carbo sinensis*: trials with captive birds. Ardea 83: 123–131.

Hydrobiologia **488:** 171–180, 2002.
D.M. Harper, R. Boar, M. Everard & P. Hickley (eds), Lake Naivasha, Kenya.

Population fluctuations and their causes in the African Fish Eagle, (*Haliaeetus vocifer* (Daudin)) at Lake Naivasha, Kenya

David M. Harper[1,*], Maureen M. Harper[1], Munir A. Virani[1,2], Andrew Smart[1,3], R. Brooks Childress[1,4], Rachel Adatia[1], Ian Henderson[1,5] & Bernard Chege[6]
[1]*Department of Biology, University of Leicester, Leicester LE1 7RH, U.K.*
[2]*Present address: The Peregrine Fund, Africa & Asia Office, Box 45341, Nairobi, Kenya*
[3]*Present address: Duchy College, Rosewarne, Camborne, Cornwall, TR14 0AB, U.K.*
[4]*Also at: The Wildfowl Trust, Slimbridge, Gloucestershire, U.K.*
[5]*Present address: The British Trust for Ornithology, The Nunnery, Thetford, Norfolk, U.K.*
[6]*Department of Ornithology, National Museums of Kenya, Box 40658, Nairobi, Kenya*
**Author for correspondence. E-mail: dmh@le.ac.uk*

Key words: submerged water plants, *Eichhornia*, *Salvinia*, *Procambarus*, *Fulica cristata*

Abstract

The population of (*Haliaeetus vocifer* (Daudin)) at Lake Naivasha, Kenya, was censused over a 13-year period (1987–99). During this time it declined by over half, to a low of 70 birds, by 1996. Breeding effectively ceased, with no courtship or nesting observed for most of the 1990s. The reason seemed to be a food shortage, which provided enough to stay alive but not enough to breed. Declines in the main prey items – fish species and red knobbed coot (*Fulica cristata*), occurred during the early 1990s and feeding conditions were worsened by a combination of increased turbidity in the lake, floating mats of exotic vegetation and the loss of shallow lagoons behind fringing *Cyperus papyrus*. This explanation was confirmed when heavy rains, following the strong ENSO event in 1997–98, caused a rapid lake level rise of 3 m vertically, flooding new lagoons behind formerly stranded *C. papyrus*. Breeding re-commenced, leading to 17–24 fledged juveniles in a population exceeding 100, by 1999. This population of *H. vocifer* is a valuable index of both the health of the lake and of its littoral ecotone.

Introduction

The African fish eagle (*Haliaeetus vocifer* (Daudin)) is a species of aquatic ecotones: its food items are aquatic fish and birds and its preferred perching and nesting habitats are horizontal branches of trees close to or overhanging the water (Brown, 1980). It is a top predator of the aquatic food web, as its fish prey may be herbivorous or carnivorous while its bird prey may be herbivorous or piscivorous (Brown & Hopcraft, 1973). Rarely do individuals feed on terrestrial prey and when they do, it is likely they are immature and scavenging (Krueger, 1996). It occurs at alkaline lakes, such as lake Bogoria, Kenya where it feeds upon lesser flamingos (*Phoeniconaias minor*), but it is much more common in freshwater lakes or rivers (Brown, loc cit.).

The population at Lake Naivasha was reported as the densest population in Africa, based on studies in the late 1960s and early 1970s (Brown, loc cit.). The lake was very full at this time (Becht & Harper, 2002) and the density represented approximately one bird km^{-2} of lake or, more realistically, a maximum density of just under 3.5 birds km^{-1} shoreline (maximum count 224 birds (Brown, loc cit.), shoreline estimated as 65 km from lake level in Becht & Harper (loc cit.)).

Birds of prey have been used as indicators of environmental health for several decades (Watson, 1991; Ormerod & Tyler, 1993). Initially, this was because the amplification of contaminants such as pesticides or heavy metals in food chains had led to heavy mortalities of predators in the temperate zone in the 1960s (Peakall, 1993), which still continues in the tropics under certain pest-control operations (e.g., Bruggers et al., 1989). Heavy metals have caused certain concern; for example lead in the north temperate zone (from lead shot) and mercury in aquatic ecosystems (because of use in industry discharging wastes to water (Peakall, 2002). A study of the contamination of 42 eggs of *H. vocifer*, and tissues from two adults, at lake Kariba in Zimbabwe 1989–90, found that DDT residues affected hatching success but suggested that the population size was more likely limited by nest site availability (Douthwaite, 1992). The two samples from adults also contained high levels of mercury, which were considered to pose a future problem as the risk from pesticides was expected to decrease with cessation of DDT use for tsetse fly control (Douthwaite, loc. cit.).

More recently, birds of prey have been used as 'umbrella species' in the promotion of conservation because of the effects of land use change upon them (Fanshawe & Bennun, 1991; Sorley & Andersen, 1994). The *H. vocifer* population of lake Kariba is believed to be limited more by the availability of trees for nesting than contamination (Douthwaite, loc. cit.) and so indicates aspects of land use changes there.

At Lake Naivasha, the lake and the land-water ecotone have been degraded in the last two decades by human intervention. Arrival of the exotic floating species: *Salvinia molesta* in 1970 and *Eichhornia crassipes* in 1988 have created floating rafts in the shallows (Adams et al., 2002) which have probably impeded *H. vocifer* fishing. Intensive horticultural development has directly accelerated *C. papyrus* and tree (mainly *Acacia xanthophloea*) clearance along the littoral edge (Boar et al., 1999) and application of fertilizers and pesticides at or just behind it (Ntayia & Ngatia, 1995; Donohue & Thenya, 1998), could drain directly into the lake where the buffering effect of *C. papyrus* has been lost by clearance. The lake is now moderately eutrophic as a consequence of the loss of buffering which has resulted in the former 'North swamp' at the entry of the two rivers, being reduced to a riparian fringe (Kitaka et al., 2002). Any one of these changes could impact the population of *H. vocifer* through food, nesting or breeding success.

The population of *H. vocifer* has been studied from 1987 to the present, with the aim of using the species and its health as an indicator of lake health as well as better understanding the population dynamics and ecology of the species. This paper analyses the population changes between 1987–99 and their implications.

Methods

Fish eagles are very territorial birds, holding linear territories along the shoreline: at lake Naivasha they make very few forays on land because of intra-specific hostility largely from augur buzzards (*Buteo augur*) (personal observations). Their brown and white colouring makes them very conspicuous and they spend most of the day perching overlooking water. For these reasons it is feasible for the total population of an aquatic ecosystem to be counted. In this study, counting was carried out by boat and 25 HP outboard motor, with observation using binoculars and location using hand-held GPS (Garmin, USA) and compass (bird location was compass bearing and estimated distance from the GPS location of the boat). The whole lake counts usually commenced at 07:30 h. Usually one-third of the lake was counted per day, in three successive days. In the later years of the study, the census repeated three times within a 2-week period.

Juveniles and immature birds can be distinguished from adults by the pattern of white, and males from females by difference in white plus the tone of their calls (Brown, loc. cit.). In this paper we just separate adults from juveniles plus immatures, and paired from single adults.

The whole population was counted, rather than a smaller sample of the whole and there was minimal human error in counting birds, because they are seen singly rather than in a flock. Any error is caused by the birds being absent from view. Territory-holding birds were not usually absent; when they were, it was either because they were not perched near to the lake (unlikely) or because one individual was soaring out of sight. Non-territory holding immature and adult birds represent a more vagrant component of the population. Time of day was varied on one part of the lake to test whether that affected number of birds counted. On three occasions, repeated counts were made on 16 territories known to be occupied, in order to calculate the frequency of absent birds. In 1994 and twice in 1997, counts were repeated for six successive days. The probability of not making a full census from a single count was thus estimated as the counting error. In 1995, counts were repeated monthly for 6 months, and the corrections from the repeat daily counts used to estimate the contribution of vagrant birds to the population pool at the lake.

The land-water edge habitat of the shoreline in each territory was recorded in 1994, 1996 and 1997 on a five-point scale, which described the physical and vegetation combinations. The extent of shoreline tree cover, tree composition, distance of treeline from water, shoreline papyrus cover and any other relevant features were estimated for the location of each eagle seen. In 1997 a video camera was used to film 100 m either side of each fish eagle seen so a visual record of the habitat vegetation could be used to confirm the accuracy of the estimates made from the boat. The categories, were:

1. Rocky shore with continuous canopy (*A. xanthophloea, Ficus spp*) and *C. papyrus* lake frontage.
2. Flat shore with ± continuous canopy cover close to lake edge of scattered mature *A. xanthophloea* trees with intermittent *C. papyrus* clumps;.
3. Open grassland with scattered trees and intermittent *C. papyrus*;
4. Continuous papyrus frontage with flat reclaimed lake bed behind where receding water levels had exposed large areas of land, (see Hickley et al., 2002 for contour map of the lake), trees at distance, along the edge of the former (1960s–70s) higher lake edge
5. Degraded *C. papyrus* frontage (burnt or cleared), flat reclaimed lake bed behind, trees cleared or continuous only at considerable distance.

In each whole-lake eagle count, details of the perch type were recorded. The perch categories were: *Acacia, Ficus* or *Eucalyptus* mature trees, dead trees, tree stumps, fringing papyrus *Eichhornia* clumps, other floating vegetation, or man-made objects.

The food of the birds was determined by analysis of remains from beneath perching trees (area cleared after each visit). These were dissected and examined under a microscope.

Forty-one birds were trapped, between 1997 and 1999, weighed, measured and ringed (banded). A small blood sample was taken for future use. Trapping was by nooses attached to a floating fish, which attached to the bird's legs when it swooped to capture the fish. 34 birds were trapped between April 1996 and July 1997 and five dead birds measured similarly. Five birds were trapped in 1998–9. Weighing was by spring balances to the nearest 1 g.

Results

No significant difference was found between counting early in the morning, mid-day or early afternoon. The lake was usually calmest in the early morning, and so this was the time of day used for the majority of counts.

The remains examined from visits to perch sites of four pairs in the south-east of the lake ($n = 20$) indicated that *H. vocifer* feeds continuously on birds (100% remains contained feathers black, probably coot, 75% contained bird bones) and regularly on fish (50% of remains contained fish and crayfish).

The repeat counts visited the same 16 territories on 17 different days (occasionally not all 16 were visited; Table 1). Thirteen were the same territories between the 3-year gap. There was a total of 268 territory-counts; on 30 territory-counts only a single bird was seen. On 13 territory-counts, neither bird was seen. This means that, on average, a single bird was absent or missed for every 8.9 pairs counted and on average, a pair was absent or missed for every 20.6 pairs counted. Examining successive counts, it took on average four counts (four successive days) to be 100% certain of seeing both birds in every one of the 16 territories. In a whole lake count, many of which were only carried out once (albeit several times a year in some years), the recorded count is likely to have underestimated the total count by approximately one pair every 10, or 10%. This is an acceptable error for a total population census when a single count is undertaken and can be borne in mind when considering the data below. When successive counts were undertaken shortly after each other, the mean and confidence limits represent the lower error of a total population count. These confidence limits, in the years where successive monthly counts were taken (see below), were close to 10%. Single counts are likely to underestimate the total population by 10%, and so in successive monthly counts where a deviation by more than 10% occurs, it is likely to be a result of vagrant birds. The longest period of consecutive monthly counts was 23 months in 1995 and 1996, made by RBC and MZA whilst doing PhD studies on other bird species. The average, 97.4 ± 4, reflects a consistent count for most of the period (Fig. 1) although a slight decline from beginning to end could be seen. The implication is that a population of close to 110 fish eagles existed at lake Naivasha in this period. There are only two months when there was evidence of vagrants: in February 1995, when although the population count was not outside limits, 11 birds were counted in Oloidien lake when its normal complement was four (two pairs). The second was April 1995, when 117 and 120 birds were counted (at beginning and end of the month) compared with 83 before. Counts for the next 2 months were over 100,

Table 1. Raw data from 17 days successive counts in three periods 1994 and 1997. Dash indicates not visited

Territory number	Territory name	Count 1 3/23/94	Count 2 3/26/94	Count 3 3/27/94	Count 4 3/29/94	Count 5 3/30/94	Count 1 12/4/97	Count 2 13/4/97	Count 3 18/4/97	Count 4 19/4/97	Count 5 20/4/97	Count 6 24/4/97	Count 1 19/7/97	Count 2 20/7/97	Count 3 21/7/97	Count 4 23/7/97	Count 5 24/7/97	Count 6 25/7/07
1	Hippo point	2	2	2	2	2	2	2	1	2	1	2	2	2	2	2	2	2
2	Round bay	2	2	1	2	2	2	1	1	2	2	0	2	2	2	1	1	1
3	*Ficus bay*	2	2	2	2	2	2	2	2	2	2	1	0	1	2	1	1	0
4	Bay Point	1	2	2	2	2	2	0	2	2	2	2	2	2	1	1	2	2
5	Barton's	2	2	2	2	2	2	2	1	1	0	2	2	1	1	1	1	2
6	Cunningham	2	1	2	2	2	1	2	0	1	2	1	1	0	2	0	2	0
7	Os Jetty	2	2	1	2	1	2	2	2	2	2	0	1	2	2	2	2	2
8	Djinn Pal	2	2	1	1	1	2	2	1	0	2	1	2	2	1	2	1	2
9	Elsa	2	2	2	2	2	2	1	1	2	2	2	2	2	2	2	2	2
10	P&L	1	2	2	2	2	2	2	2	2	2	2	2	1	1	2	1	2
11	Wallington's	2	2	2	1	2	2	2	1	2	2	2	0	1	2	2	1	1
12	Fish Eag Inn	2	2	2	2	2	2	2	2	1	2	2	2	1	0	0	1	2
13	Fish Camp	2	2	2	2	2	2	2	2	2	2	2	2	2	2	2	2	2
14	Sher	2	2	2	2	2	–	–	2	2	1	2	–	–	2	2	1	2
15	Bunny's	2	1	1	2	1	–	–	2	2	2	2	–	–	1	1	2	2
16	Root's	2	2	2	1	2	–	–	2	1	2	2	–	–	2	1	2	2
17	Safariland	–	–	–	–	–	–	–	2	2	2	2	–	–	2	2	0	2
Total birds		30	30	28	29	29	25	22	24	26	28	25	20	19	25	22	24	26
Percentage of total counted		94	94	88	91	91	78	69	75	81	88	78	63	59	78	69	75	81

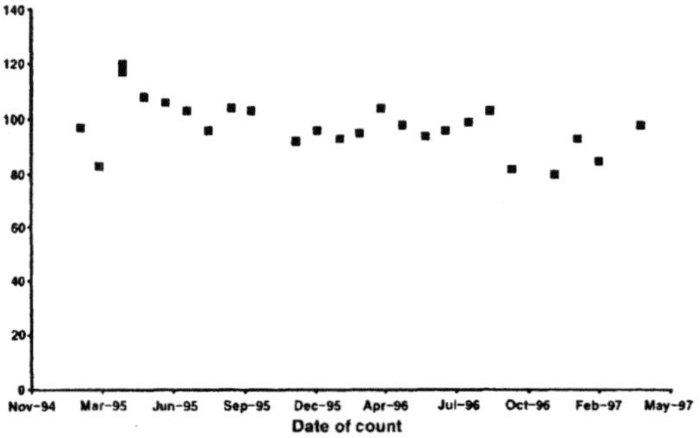

Figure 1. The whole-lake counts recorded at monthly intervals between January 1995 and November 1996.

Table 2. The annual mean percentage of non-adult birds recorded in the whole-lake counts, 1987–1997

Year	Percentage non-adult
1987	6.73
1991	0
1992	0
1993	0
1994	0
1995	1.9
1996	0.9
1997	1.9

before falling back to average. In both examples, a maximum of 10% of the population, 10 individuals, was probably made from vagrant birds.

Forty-nine whole lake counts were made between 1987 and the end of 1999. For 12 years from the commencement of the study, there was a consistent decline in numbers, the population halving from its late-1980s level of around 165 birds (150 counted + 10%). The lowest population was reached in April 1998, when 62 birds were counted, indicating a population total of about 70 (Fig. 2). During this time the percentage of juvenile and immature birds in the population fell to zero, remaining there in the early 1990s, rising slightly in the mid 1990s (Table 2).

A marked change in the fortunes of the population came about in 1998, through to 1999. Total counts increased from the April 1998 low to about 100 birds throughout 1999 (Fig. 3). Moreover the population recommenced breeding in early 1998, with an average percentage of fledged juveniles in the 1999 counts of 20.7 ± 9.8% ($n = 5$). This was the highest recorded in any year since the study began. Throughout the 1990s, only isolated individual juveniles had been seen, until an upsurge of courtship and nest building behaviour was observed in late 1997, which led to

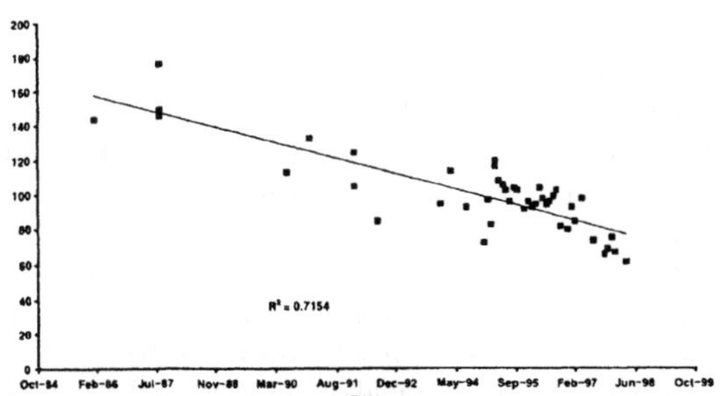

Figure 2. The whole-lake counts of *H. vocifer* between 1987 and 1996 (one count from 1986, Johnson et al., 1986) is included).

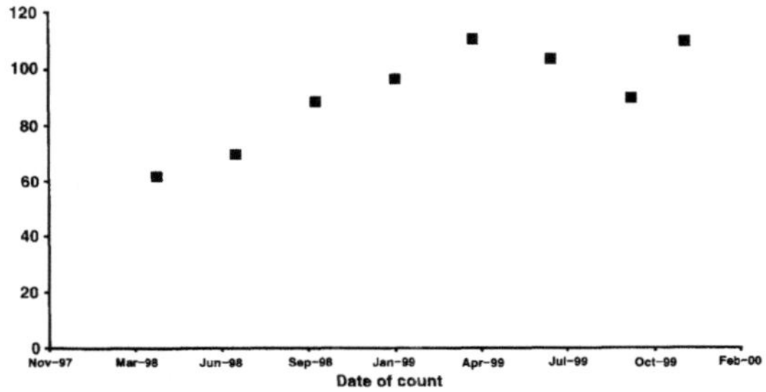

Figure 3. The whole-lake counts of *H. vocifer* between 1997 and 1999.

Figure 4. The percentage distribution of habitats at which *H. vocifer* individuals were seen, 1995 and 1996.

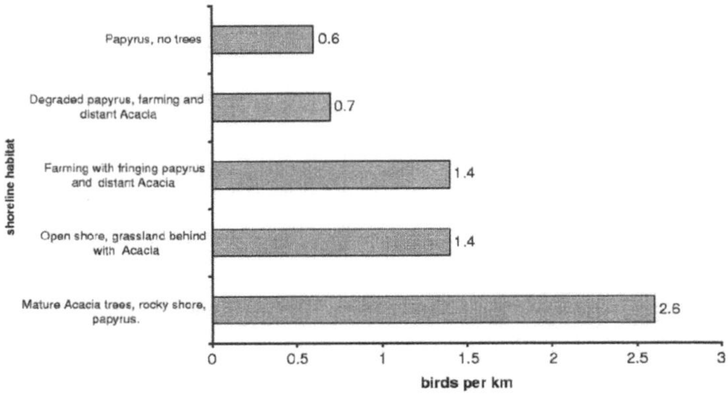

Figure 5. The mean density of birds recorded in five different riparian habitat type, 1995 and 1996.

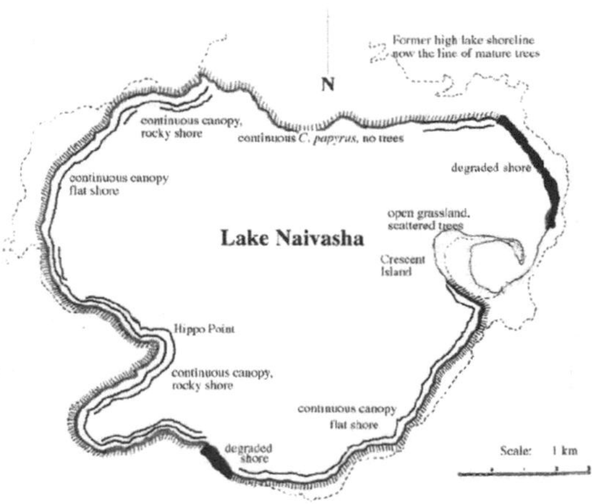

Figure 6. The distribution of the five habitat types along the lakeshore.

Continuous woodland fringed the lake in 1996–7 for 27 km of shoreline, when the whole lake population counted was close to 100 birds. This shoreline accounted for the highest density of birds, rocky shores 2.6 km^{-1} and flat shores 1.4 km^{-1} (Fig. 5), approximately 57 birds. A further 20 km of shoreline, with intact *C. papyrus* edge and trees behind, held 1.4 km^{-1} and accounted for a further 28 birds. Six km of shoreline had trees absent or the lake shore degraded by agricultural clearance, holding 0.6 km^{-1} and accounting for four birds. The difference was significantly higher in 'natural' habitats (1996 χ^2=19.5, 1997 χ^2= 21.2, df=4, p<0.001). The distribution around the lake was thus uneven, with most birds around the western and southern shores and fewest in the north or south-east (Fig. 6).

an estimated half the population (15 pairs) engaged in breeding activity in early 1998, then 38 new nests built by early 1999 (not all active), and finally one-fifth of the population juvenile (17–24 birds) in 1999.

The weight of adult birds captured in 1997 was 3.01 ± 0.3 kg (females, n = 19) and 2.35 ± 0.02 (males, n = 16). In 1999, out of the five birds caught, one male and one female were each heavier than any of the same sex measured in 1997, although too few were caught to ascribe any statistical meaning to this.

Birds were seen predominantly perched on tree branches (64%) above the lake but also on lower objects overlooking the lake (12%) and floating vegetation in the lake (21%). Only for 3% of the time were they seen flying (Fig. 4).

Discussion

When it became apparent that the population was declining in the early 1990s, three reasons were sought, grouped as habitat loss, food loss and pesticide (or other poisoning).

Suitable riparian habitat could be lost in two ways. Naturally, it can be lost by lake level decline, which on the one hand shortens the total length of shoreline and on the other hand results in papyrus clumps becoming stranded on dry land as the shore progressively moves further away from fringing mature *A. xanthlophloea* trees. The lake level during the time the *H. vocifer* population was first studied by Brown, 1968–71, was increasing after a very rapid rise in 1966; in the 5 years to 1971 it rose 5 vertical m. The population counted by Brown increased by half in this time (Fig. 7). The ri-

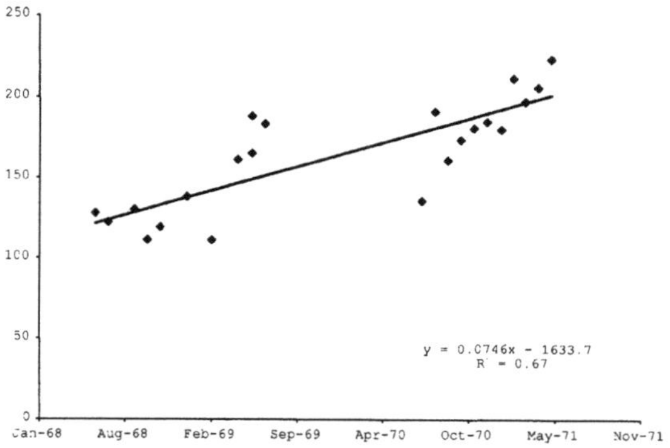

Figure 7. The whole lake population census of *H. vocifer* in 1968–1971 by Leslie Brown and colleagues (Brown, 1980).

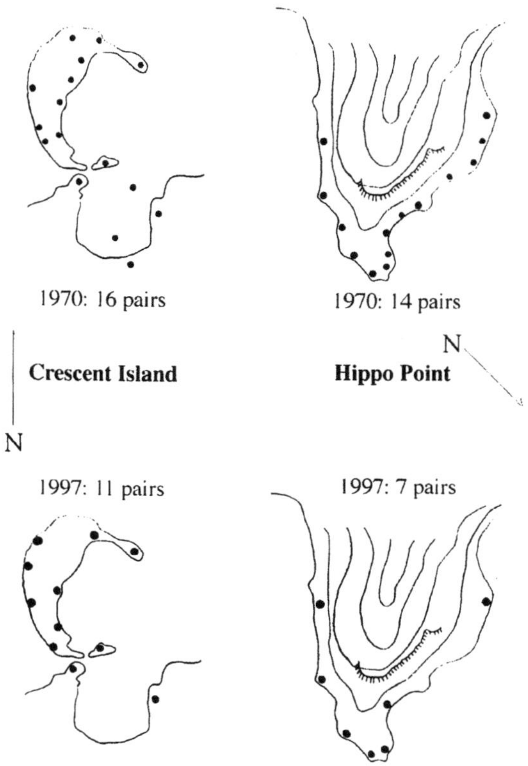

Figure 8. The territory location of *H. vocifer* pairs on two locations of 'prime habitat' in 1968 (Brown 1980) and 1997. See Figure 6 for the location on the main lake.

parian land use at this time was low-intensive ranching and small-scale vegetable growing, so it is likely that the population responded to an increase of lake area making new shoreline territories available.

The lake level decreased about 2 m from June 1990 to 1996 (Becht & Harper, 2002) with some small short-term rises of 50–80 cm in 1992 and 1994. The lake had declined 4 m vertical between its maximum level in the late-1960s and its level in 1996, losing 20% of shoreline length in the flattest topography. This loss could only explain a loss of 15 birds from the maximum recorded at high lake levels.

The lake in the late 1970s was probably surrounded almost totally by *A. xanthophloea/Ficus* spp. woodland. At 1996 densities for riparian woodland, the loss in habitat accounts for another 15 birds. A combined loss of 30 birds as a result of anthropogenic and natural habitat change does not fully explain the population fall.

Even in two areas of 'prime' habitat studied by Brown (1980) – rocky wooded shore at Crescent Island and Hippo Point – there have been absolute declines, from 16 to 14 and 11 to 7 pairs respectively between 1970 and 1996/7 (Fig. 8). This implies that the additional causes of population decline lie away from the riparian habitat. There may have been at least three reasons for this:

(a) contamination of lake by pesticides leading to breeding failure;
(b) decline in absolute availability of prey – coot or fish;
(c) loss of feeding opportunity – as the lake has become infested with floating alien weeds (*Salvinia* and *Eichhornia*) and there is a decrease in transparency of lake water due to eutrophication.

The first reason is unlikely, as Gitahi et al. (2002) found no evidence for pesticide amplification in lake food chains. Contamination by organophosphate pesticides, not detected as tissue residues, or other poisons, would logically be expected to lead to more mortalities, which has not been detected by resident naturalists (J. Root, pers comm.). Lincer et al. (1981) earlier measured DDT in a number of organisms from Naivasha and the other Rift Valley lakes Baringo, and Nakuru. They too found that there was limited contamination in Naivasha. In Zimbabwe, several workers culminating in Douthwaite (1992) had found levels of DDT (and mercury from unknown sources) in birds around Lake Kariba Zimbabwe sufficently high to pose a risk to the population, although censuses in 1987 and 1990 showed that hatching success was greater than 72%. The area investigated had been widely treated for tsetse fly control using organochlorine pesticides, the use of which was declining, and so it is unlikely that in Naivasha, where there is no known

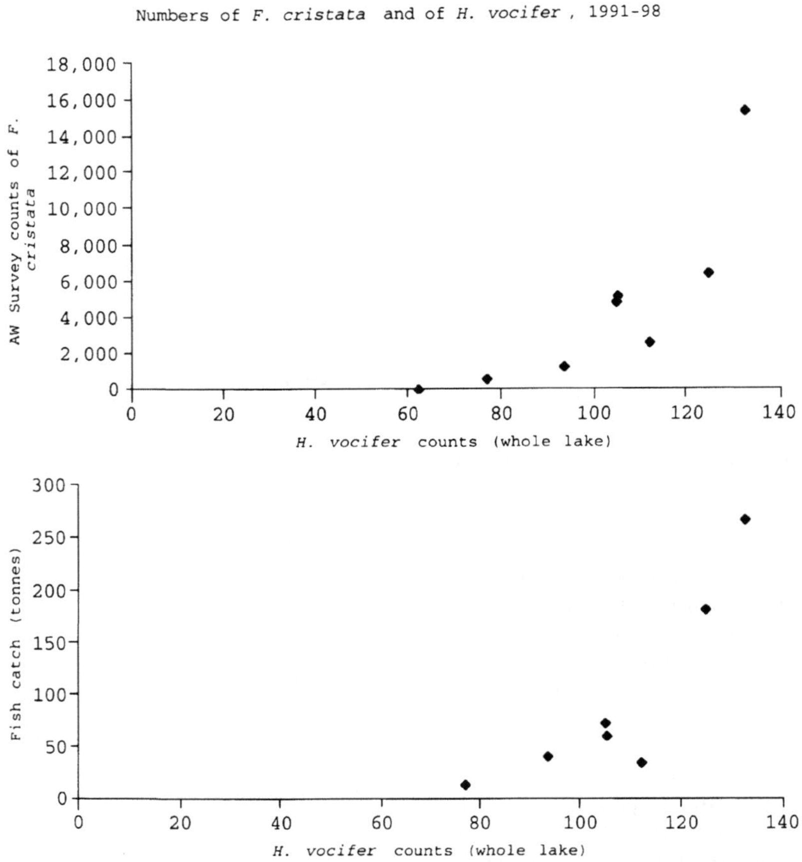

Figure 9. The correlation of numbers of coot (*F. cristata*) and commercial fish catches at lake Naivasha with the numbers of *H. vocifer* counted. (*F. cristata* numbers from the African Waterfowl Census data, every January).

extensive use of organochlorine pesticides on the scale of tsetse fly control, there has been contamination sufficient to cause breeding failure.

The second reason is a distinct possibility. Fish catches have declined (if measures of commercial catches is an accurate indication of populations and availability to *H. vocifer*) as a consequence of over-fishing (Hickley et al. 2002). At the same time, coot have declined as a consequence of the loss of their main food, submerged plants, which have themselves disappeared as a result of consumption by the exotic crayfish *Procambarus clarkii* (Harper et al., 2002). The declines of both fish and coot show good correlations with the decline in *H. vocifer* numbers through the 1990s (Fig. 9).

The feeding habitat of *H. vocifer* has changed markedly. The lake shallows are now dominated by floating exotic vegetation which curtail the area of open water and the absence of a submerged plant littoral brings the turbidity of the open, eutrophic, lake, to the shore (Harper, 1992). Leslie Brown described the best feeding habitat at the end of the 1960s as the wide lagoons with water lilies, clear water full of fish, and other water birds. Food was optimal. The fish eagles fed on fish, young water birds and pirated prey from pelicans and herons. This sheltered habitat no longer exists. Today the lake level is low, there are no water lilies, and the water is turbid. The eagles are forced to fish in open water, a habitat that Leslie Brown described as poor (even in 1968). Now, fish eagles with no lakeside trees utilise the *Eichhornia crassipes* for perching. This must be a poor substitute for a 25-m high perch in an *Acacia* or *Ficus* tree with branches overhanging the water where an eagle can

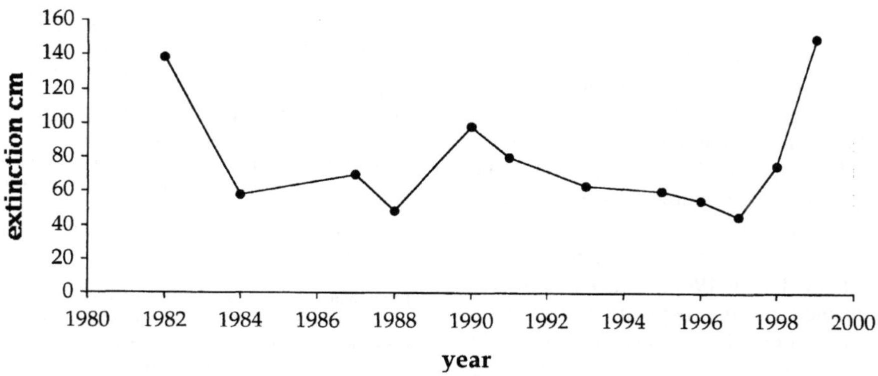

Figure 10. Transparency in the water of Lake Naivasha, 1982–1999.

swoop with great speed and dexterity to snatch a fish by surprise. Fish eagles in the past have also taken coot and flamingo, but it would be difficult to achieve surprise from a low perch, and an eagle catching a bird from this position was never observed.

The birds trapped and weighed were lower in weight than the range given by Brown (1980), the female average was at the lower end of the range (3–3.63 kg) and the male average closer to the middle (1.99–2.5 kg). This lends support to the hypothesis that food limitation prevented breeding, since the female needs to put more energy into egg production. Final support is provided by the continuous breeding throughout the 1990s of one pair of *H. vocifer*, fed daily on a turkey head per individual, by the owner of the land on which they held their territory (S. Higgins, pers. comm.).

Studies elsewhere lend support to the conclusion that the population was limited by food, both absolute quantity and availability. A study of ospreys in Oregon in 1992 (Henny, pers. comm. to M. Virani) found that when the ospreys arrived 2–3 weeks late at their nesting sites, in poor body condition, most did not lay eggs. Following a treatment of a lake with rotenone to kill the fish, osprey breeding was reduced and prey delivery rates were about half the rate observed at a nearby lake not treated with rotenone. Eltringham (1975) reported a density of 3.5 *H. vocifer* km^{-1} and Sumba & Hebrard (1988) 2.4 km^{-1} along the Kazinga Channel in Uganda. Examination of present-day densities in more detail (Kreuger, 1995) showed similar densities to these earlier studies. Along the Kazinga channel in optimal biotopes he found that density varied from 1.75 to 3.25 individuals km^{-1}. This compares with 1.4–2.6 km^{-1} in the best habitats at Naivasha. In the Kyambura Gorge and the bush grassland, both less suitable biotopes, densities fell to between 0.45 and 0.58 km^{-1}. At Naivasha in degraded vegetation these were similar. The density found at either end of the channel, around Lake Edward was 1.9–2.6 km^{-1} of shore and at Lake George 1.0–1.1 *H. vocifer* km^{-1} of shore. These are similar waters, and Lake George in particular is highly turbid with algal growth and continually wind-mixed (Burgis & Morris, 1988). Brown (1980) found densities at Lake Baringo that were variable but low, with long stretches of the shoreline apparently suitable but not occupied by eagles. Baringo is a highly turbid lake whose catchment has suffered badly from soil erosion since the 1950s. There is thus some evidence from elsewhere that turbid lake waters support lower populations of *H. vocifer*, but where the water is in a physically suitable environment (e.g., Lake George water flowing down the Kasinga channel, which is wide, shallow and with littoral lagoons and *C. papyrus*) combined with suitable tree densities, increases *H. vocifer* density. Eltringham (1975) also found distribution related to availability of trees for perching.

The final 2 years of this study gave unexpected support to these conclusions. The courtship, breeding and increase in numbers followed a sudden rise in lake level (Becht & Harper, 2002). That rise flooded several tens of km^2 land, behind papyrus which had become rooted in the former shallow sediments. In these new flooded lagoons, tilapias bred rapidly, submerged macrophytes germinated and attracted coot, and the lagoons remained clear. They did not connect directly to the lake, being behind *C. papyrus* where that existed, but even where they did, the lake itself had become much clearer by the dilution with river water whose sediment settled (Fig. 10). For a short time, in 1998, littoral conditions returned partly to those described as 'optimum' by Brown (1980) and the population of *H. vocifer* returned towards the level of breeding recorded by him.

In conclusion the study showed the population of *H. vocifer* was limited in the 1990s by lack of available food, which enabled the adult population to maintain territories but not to breed. Probably birds were not replaced on territories that had been vacated because of severe unsuitability for feeding or one bird's death. The species does give a summary of the overall health of the lake and thus provides both a good indicator (a census only takes 3 days) and a 'flagship' for raising environmental awareness.

Acknowledgements

This study formed past of the work of the University of Leicester research project at Lake Naivasha, which since 1987 has been authorised by the Office of the President, Government of Kenya under research permit to Dr D. M. Harper no. OP 13/001/12C 46. The project was funded by the Earthwatch Institute, Boston, USA, and Oxford, England. The data collection would not have been possible without the assistance of numerous Earthwatch volunteers. Our sincere thanks go to the numerous colleagues in Kenya in particular Jill & Angus Simpson and Velia Carn.

References

Adams, C. S., R. R. Boar, D. S. Hubble, M. Gikungu, D. M. Harper, P. Hickley & N. Tarras-Wahlberg, 2002. The dynamics and ecology of exotic tropical species in floating plant mats: Lake Naivasha, Kenya. Hydrobiologia 488 (Dev. Hydrobiol. 168): 115–122.

Becht, R. & D. M. Harper, 2002. Towards an understanding of human impact upon the hydrology of Lake Naivasha. Hydrobiologia 488 (Dev. Hydrobiol. 168): 1–11.

Boa R. R., D. Harper & C. Adams, 1999. Biomass allocation in *Cyperus papyrus* in a tropical wetland, Lake Naivasha, Kenya. Biotropica 31: 411–421.

Brown, L. H., 1980. The African Fish Eagle. Bailey Bros., Folkestone.

Brown, L. H. & J. B. D. Hopcraft 1973. Population structure and dynamics in the African fish eagle Haliaeetus vocifer (Daudin) at Lake Naivasha, Kenya. East African Wildlife Journal, 11: 255–269.

Bruggers, R. L., M. M. Jaeger, J. O. Keith, P. L. Hegdal, J. B. Bourasa, A. A. Laitigo & J. N. Gillis, 1989. Impact of fenthion on non-target birds during Quelea control in Kenya. Wildlife Society Bull. 17: 149–160.

Burgis, M. & P. Morris, 1988. The Natural History of Lakes. Cambridge University Press, Cambridge.

Donohue, C. & T. Thenya, 1998. Development and Implementation of a Water Quality Monitoring Programme for Lake Naivasha. Lake Naivasha Riparian Association: 33 pp.

Douthwaite, R. J. 1992. Effects of DDT on the fish eagle *Haliaeetus vocifer* population of Lake Kariba in Zimbabwe. Ibis 134: 250–58.

Eltringham, S. K., 1975. Territory size and distribution in the African fish eagle. J. zool. Soc. London 175: 1–13.

Fanshawe, J. H. & L. A. Bennun, 1991. Bird conservation in Kenya: creating a national strategy. Bird Conserv. Int. 1: 293–315.

Gitahi, S. M., D. M. Harper, S. M. Muchiri, M. P. Tole & R. N. Ng'ang'a, 2002. Organochlorine and organophosphorus pesticide concentrations in water, sediment, and selected organisms in Lake Naivasha (Kenya). Hydrobiologia 488 (Dev. Hydrobiol. 168): 123–128.

Harper, D. M., 1992. The ecological relationships of aquatic plants at Lake Naivasha, Kenya. Hydrobiologia 232: 65–71.

Hickley P., R. Bailey, D. M. Harper, R. Kundu, M. Muchiri, R. North & A. Taylor, 2002. The status and future of the Lake Naivasha fishery, Kenya. Hydrobiologia 488 (Dev. Hydrobiol. 168): 181–190.

Kitaka, N., D. M. Harper, K. M. Mavuti & N. Pacini, 2002. Chemical characteristics, with particular reference to phosphorus, of the rivers draining into Lake Naivasha, Kenya. Hydrobiologia 488 (Dev. Hydrobiol. 168): 57–71.

Kreuger, O., 1995. Population density and intra- and inter-specific competition of the African fish eagle in Kyambura Game Reserve, southwest Uganda. Ibis 139: 19–24.

Lincer, J. L., D. Zalkind, L. H. Brown & J. B. Hopcraft, 1981. Organochlorine residues in Kenya's rift valley lakes. J. Appl. Ecol., 18: 157–72.

Ntayia, R. O. & J. Ngatia, 1995. Monitoring of pesticide residues in and around Lake Naivasha. Crop Protection Special Report, Kenya Agricultural Research Institute, National Agricultural Laboratory, Nairobi.

Ormerod, S. J. & S. J. Tyler, 1993. Birds as indicators of change in water quality. In Greenwood, J. J. & R. W. Furness (eds), Birds as Monitors of Environmental Change. Chapman & Hall, London: 179–216.

Peakall, D. B., 1993. DDE-induced eggshell thinning: an environmental review. Environ. Rev. 1: 13–20.

Peakall D. B., 2002. Poisoning in wild (free-living) raptors. In Cooper, J. E. (eds), Birds of Prey, Health and Disease. Blackwell, Oxford: 163–170.

Sorley, C. S. & D. E. Andersen, 1994. Raptor abundance in south-central Kenya in relation to land-use patterns. African J. Ecol. 32: 30–38.

Sumba, S. J. A. & J. J. Hebrard, 1988. Influences of habitat factors on the population distribution and structure of the African fish eagle in Queen Elizabeth National Park, Uganda. Proceedings of the VII Pan African Ornithological Congress: 503–511.

Watson, R. T., 1991. Using birds of prey as an environmental conservation tool: The Peregrine Fund's World Programme. Environmental Conservation 18: 269–270.

Hydrobiologia **488:** 181–190, 2002.
D.M. Harper, R. Boar, M. Everard & P. Hickley (eds), Lake Naivasha, Kenya.

The status and future of the Lake Naivasha fishery, Kenya

Phil Hickley[1], Roland Bailey[2], David M. Harper, Rodrick Kundu[3], Mucai Muchiri[4], Rick North[1] & Andy Taylor[1]
University of Leicester, Leicester LE1 7RH, U.K.
[1]*Present address: National Coarse Fisheries Centre, Environment Agency, Arthur Drive, Hoo Farm Industrial Estate, Worcester Road, Kidderminster, DY11 7RA, U.K.*
[2]*Life Sciences Department, King's College London, 150 Stamford Street, London SE1 9NN, U.K.*
[3]*Present address: Fisheries Department, P.O. Box 47066, Nairobi, Kenya*
[4]*Present address: Department of Fisheries, Moi University, P.O. Box 3900, Eldoret, Kenya*

Key words: fishery yield, introduced species, *Oreochromis leucostictus*, *Tilapia zillii*, *Micropterus salmoides*

Abstract

Lake Naivasha is a freshwater lake situated in the eastern rift valley of Kenya. Only five species of fish are present, all of which have been introduced. They are *Oreochromis leucostictus*, *Tilapia zillii*, *Micropterus salmoides* (largemouth bass), *Barbus amphigramma* and *Poecilia reticulata* (guppy). The first three of these form the basis of an important gill net fishery and bass are also taken by rod and line for sport. *Barbus* are occasionally caught by dip net. Actual and potential yields for the Lake Naivasha fishery are discussed and the fishery is shown to be under-performing. The feeding regimes of the commercially exploited fish were examined in the context of available food supply, in particular the benthic fauna. Small bass depend heavily on *Micronecta* and large bass mostly take crayfish. Detritus predominates in the diet of *Oreochromis leucostictus* and *Tilapia zillii* but the former also eats algae and the latter, *Micronecta* and macrophyte. Various food resources, especially the benthos, appear to be under-utilized and so it is possible that further species could be introduced to enhance the commercial fishery.

Introduction

Lake Naivasha is a shallow (3–6 m) freshwater lake, approximately 160 km^2 in area, situated in the eastern rift valley of Kenya about 100 km north of Nairobi. It lies in a closed basin at an altitude of 1890 m above sea level, receives 90% of its water from the perennial River Malewa, and is subject to considerable fluctuations in water level. Dominant vegetation types are belts of papyrus (*Cyperus papyrus* L.) around the margins, stands of submerged macrophytes, of which the principal species is *Najas pectinata* (Parl.), and mats of floating plants comprising *Salvinia molesta* Mitch. and *Eichhornia crassipes* (Mart.). An overview of the lake and its changing ecology can be found in Harper et al. (1990).

Resulting from a probable history of occasional drying out, Lake Naivasha when first studied (c.1900) had only one species of fish present, the endemic *Aplocheilichthys antinorii* (Vinc.) which was last recorded in 1962 (Elder et al., 1971). Since 1925, various fish introductions have been made, some successful and some not (Litterick et al., 1979; Muchiri & Hickley, 1991). Present today are *Oreochromis leucostictus* (Trewewas), *Tilapia zillii* (Gervais), *Micropterus salmoides* Lacépède (largemouth bass), *Barbus amphigramma* Blgr. and *Poecilia reticulata* Peters (guppy). Since 1959, the two tilapias and the bass have formed an important fishery (Muchiri & Hickley, 1991) with all three species being commercially exploited using gill nets and the bass also being taken by rod and line for sport.

The purpose of this paper is to relate the current status of the commercial fishery to information on fish feeding regimes and potential yields so as to identify whether or not there is scope for better management and utilisation of the resource.

Methods

Commercial catch data were obtained from the Fisheries Department of the Kenya Government. Fishermen use multifilament gill nets with a minimum stretched mesh size of 100 mm and are limited to a maximum of 10 nets of not more than 100 m in length. All catches must be landed at a single station near Naivasha town where fisheries personnel record the weight of the catch from each canoe. Data collection commenced in 1963. In addition, some *B. amphigramma* were taken downstream of a weir on the River Malewa.

Gill net catch statistics were fitted to a version of the Schaefer model (Ricker, 1975; Pauly, 1983): $C = a + bE$ where C = catch per unit effort, E = effort (with effort being the number of canoes licensed to fish in a given year), a and b are constants. Maximum sustainable yield (MSY), optimum effort and equilibrium yields were then calculated: MSY = $a^2/4b$; optimum effort = $a/2b$; equilibrium yield = $aE - bE^2$.

The theoretical annual yield (Y) of fish was estimated using a range of models suitable for use with non-temperate data. These were based on various measures including lake surface area (A), morphoedaphic index (MEI; the ratio of total dissolved salts (TDS) to mean lake depth (Z)), primary productivity ($P = gO_2\ m^{-2}\ d^{-1}$, $X = gC\ m^{-2}\ yr^{-1}$), air temperature (T) and effort (E) as listed below by author:

(a) Crul (1992): $Y = 8.32\ A^{0.92}$ ($Y = t\ yr^{-1}$);
(b) Henderson & Welcomme (1974): $Y = 8.7489\ MEI^{0.3813}$ ($Y = kg\ ha^{-1}$)
(c) Melack (1976): $\log Y = 0.113\ P + 0.91$ ($Y = kg\ ha^{-1}$);
(d) Oglesby (1977): $Y = 0.00038\ X^{2.21}$ (Y = mg dry wt $m^{-2}\ yr^{-1}$);
(e) Schleisinger & Regier (1982):
(i) $\log_{10} MSY = 0.061\ T + 0.043$ ($Y = kg\ ha^{-1}$);
(ii) $\log_{10} MSY = 0.044\ T + 0.482 \log_{10} MEI + 0.021$ ($Y = kg\ ha^{-1}$);
(iii) $\log_{10} Y = 0.051\ T + 0.358\ E_{Low} + 0.161 \log_{10} MEI - 0.383$ ($Y = kg\ ha^{-1}$);
(iv) $\log_{10} Y = 0.050\ T + 0.349\ E_{High} + 0.146 \log_{10} MEI - 0.367$ ($Y = kg\ ha^{-1}$).

In the foregoing equations the following values were used: $A = 163\ km^2$; $Z = 3.35$ m; TDS = 231; MEI = 68.95; $P = 8.98\ gO_2\ m^{-2}\ d^{-1}$; $X = 99.16\ gC\ m^{-2}\ yr^{-1}$; $T = 21$ °C; $E_{Low} = 1$; $E_{High} = 2$. Sources of data for mean values of TDS and primary production were Melack (1979), Harper (1987), Harper (1991) and original (this study) and, for temperature, Litterick et al. (1979) and Muchiri (1990). Area is for 1991, the year in which the average depth was measured (see below).

In order to determine the above figure for average depth the lake was surveyed during August 1991 using a Lowrance X-15 chart recording echo-sounder with a 20° transducer beam. The soundings were carried out from a small boat driven at constant speed between two defined positions and, using known travel times and distances, the chart recordings were transcribed into depth profiles. Åse et al. (1986) had used a similar echo-sounder in 1983 but had concentrated on the north of the lake. After confirmation of the 1983 findings, the 1991 survey concentrated on previously unmeasured areas.

From 1987, annual collections of bass and tilapia specimens for gut analysis were obtained by gillnetting (11–50 mm bar mesh) in the major habitat types of open water, submerged macrophyte beds, rocky shoreline and the marginal vegetation fringe. Stomachs were removed from representative samples across the size range captured, for each habitat type from each sampling year, and analysed as described by Muchiri (1990) and Hickley et al. (1994).

Results

Lake depth

A depth contour map based on both actual readings from the 1991 echo-sounder traces and interpolation from the map published by Åse et al. (1986) is given in Figure 1. Note that the lake level was approximately 1.75 m lower in 1991 than in 1983. Two examples of depth profiles, from west to east and south to north, are given in Figure 2.

Commercial fish catches

The annual catches for the gillnet fishery are shown in Figure 3. Over the years there have been great fluctuations in both the amount of fish landed (21–1150 t yr^{-1}) and the number of fishing canoes (6–104). Catch per canoe ranged from 0.73 to 92.8 t yr^{-1}. The average species composition of the catch (1987–1998) was *O. leucostictus* 67.6%, *T. zillii* 10.2% and *M. salmoides* 22.2%. The mean sizes (fork length) of fish landed were *O. leucostictus*, 192 mm (S.D. = 23 mm), *T. zillii*, 168 mm (S.D. = 17 mm), and bass, 262 mm (S.D. = 18 mm).

Three separate phases in the development of the fishery were identified:

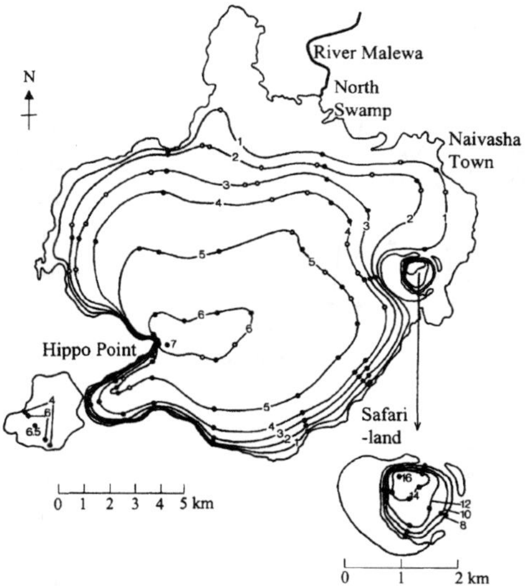

Figure 1. Bathymetric map (m) of Lake Naivasha based on 1991 transects (solid circles) and Åse et al. (1986, open circles).

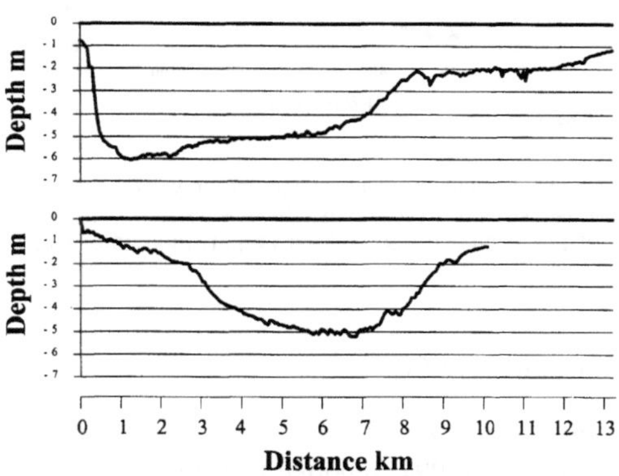

Figure 2. Depth profiles of Lake Naivasha. The upper graph shows approximately West to East, from Hippo Point to Naivasha Town, and the lower graph approximately South to North, from Safariland to North Swamp. Locations are shown in Figure 1.

(1) The period of development of the new fishery from 1959 to the time when the fishery collapsed in the early 1970s. During this 'boom and bust' phase the maximum recorded catch of 1150 t yr^{-1} was attained.

(2) The middle phase from 1974 to 1988 during which the fishery recovered and was then typified by fish catches which fluctuated but were still relatively high.

(3) The recent period of a poorly performing fishery with low reported catches.

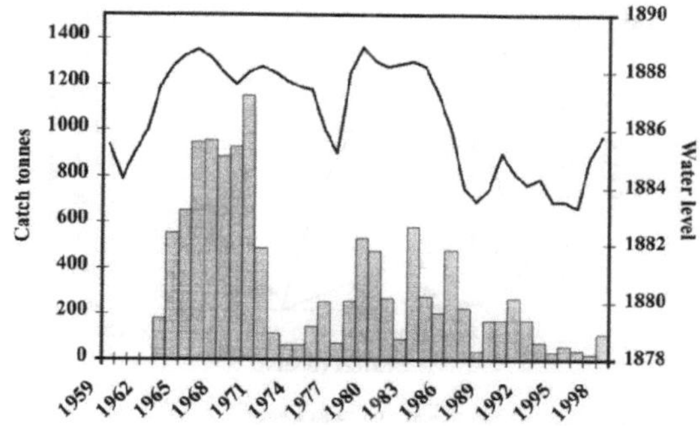

Figure 3. Total catches (bars) from the gill net fishery and water level changes (solid line) for Lake Naivasha during the period 1959–1998 inclusive. Commercial fishing started in 1959 and records kept from 1963.

Muchiri & Hickley (1991) attributed some of the fluctuations observed during the first two phases to fishing pressure, changing lake levels and the loss of submerged macrophytes. Although there were restrictions imposed on the maximum number of gill nets allowed under each fishing licence, the number of licensed canoes was at times very large. There is a considerable demand for fish in the neighbouring urban centres and horticultural estates whose populations recognise the importance of including fish protein in their diet. Fish catches appeared to be related to trends in water level changes with a rise in lake level generally followed by increased catches and a fall in level followed by a corresponding decline in fish catch. Submerged macrophytes almost disappeared during the early 1980s and this appeared to severely affect recruitment of new individuals to sustain the fishery.

Fish yields

Figures 4(a)–(d) show regression plots of computed catch per unit effort on effort, together with the resulting equilibrium yield curves, for the whole period from 1962 to 1998 alongside separate plots for the three identified phases of the fishery.

Estimates of actual fish yields are shown in Table 1(a). For the entire period 1962–1998, the computed maximum sustainable yield (MSY) was 641 t yr^{-1} with an optimum effort of 36 canoes. MSY for the development phase (1962–1973) was 731 t yr^{-1} and optimum effort was 20 canoes whereas in the 15 year long middle phase (1974-88) MSY was much lower at 341 t yr^{-1} and optimum effort higher at 51 canoes. These changes were reflected in a consequential decrease in CPUE from 47 t yr^{-1} to 9 t yr^{-1}. In

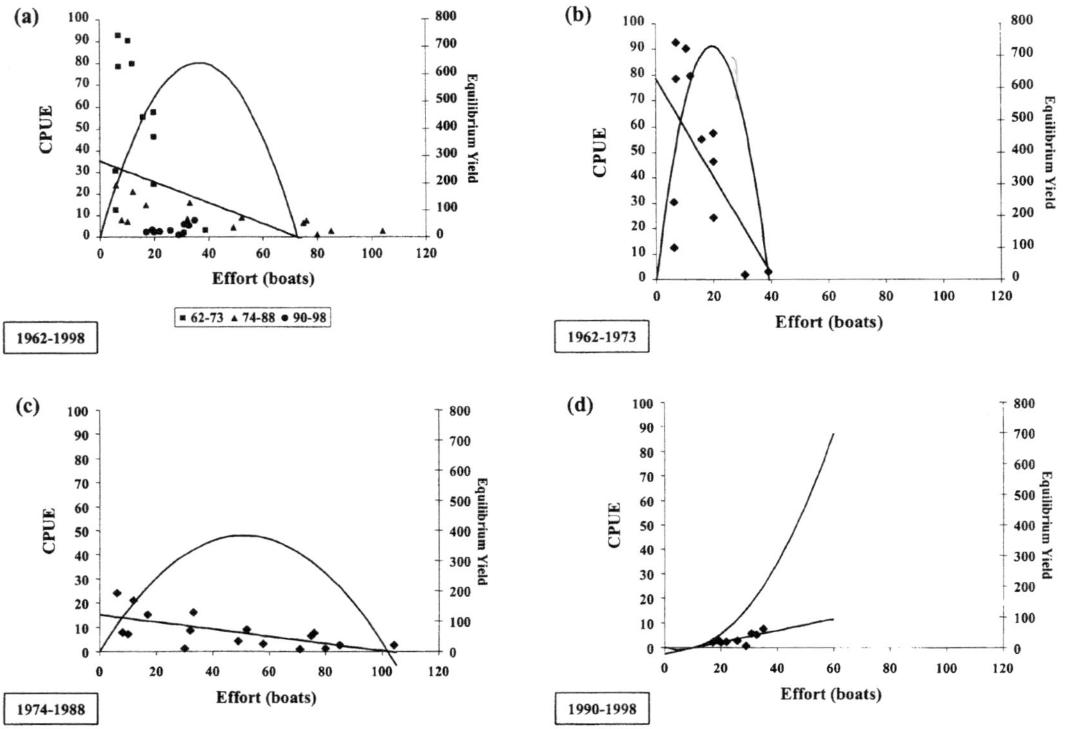

Figure 4. Regression plots of computed catch per unit effort (CPUE) on effort (E), together with the resulting equilibrium yield curves, for the whole study period alongside separate plots for the three development phases of the fishery. (a) 1962–1998; CPUE = 35.29 − 0.485E (r^2 = 0.196) (b) 1962–1973; CPUE = 78.48 − 1.897E (r^2 = 0.358) (c) 1974–1988: CPUE = 15.06 − 0.148E (r^2 = 0.435) (d) 1990–1998; CPUE = 2.59 − 0.237E (r^2 = 0.491).

recent times, the average CPUE (3.5 t yr^{-1}) has been at an all time low and MSY could not be calculated.

Theoretical annual fish yields are presented in Table 1(b). The average estimate shows the potential fishery yield to be 889 t yr^{-1}, more than previously recorded MSY values and considerably higher than the 384 t yr^{-1} for the relatively stable 1974–1988 period. The lowest estimates of theoretical yield equate to the 1974–1988 MSY value whereas estimates based on primary production are twice the value of the MSY for the development phase of the fishery (1962–1973).

The River Malewa fishery

The exploitation of the *B. amphigramma* spawning migration up the River Malewa by fishermen using dip-nets was short-lived. Fishing commenced in 1983, catches peaked at 62.9 t yr^{-1} in 1986, decreased to 26.1 t yr^{-1} in 1987 and thereafter only a small number of fish have been caught.

Feeding habits of fish

Micropterus salmoides

Hickley et al. (1994) reported that, until fish reach a size of 260 mm fork length, bass in Lake Naivasha were almost totally dependent upon free-living invertebrate food organisms. Thereafter, additions to the diet included crayfish, fish and frogs. The most important invertebrate prey species for the smaller bass were the water boatman, *Micronecta scutellaris* (Stal.) and dipteran pupae. For bass in the larger length category, the crayfish, *Procambarus clarkii* (Girard), was the preferred food. *Micropterus salmoides* in Lake Naivasha can, therefore, be classed as a generalized macro-predator, principally feeding on free living animals of a kind most likely to be found in the littoral zones.

Table 1. Estimates of actual and theoretical fish yields calculated for Lake Naivasha: (a) Summary statistics for catch per unit effort (CPUE) on effort (E) and the resultant maximum sustainable yield (MSY) (b) Theoretical yields based on various predictive models. For formulae and data sources see Methods

(a)

Year	E No. canoes (range)	Average CPUE t canoe^{-1}	Optimum E No. canoes	MSY t yr^{-1}
1962–1998	30 (6–104)	20.52	36.34	641
1962–1973	16 (6–39)	46.88	19.63	731
1974–1988	45 (6–104)	8.94	50.96	384
1990–1998	26 (17–35)	3.52	(5.46)	(−7)

(b)

Model	Parameters used	Theoretical annual yield t yr^{-1}
Crul (1992)	Surface area	902
Henderson & Welcomme (1974)	Morpho-edaphic index (MEI)	716
Melack (1976)	Gross primary production (g O_2m^{-2} d^{-1})	1371
Oglesby (1977)	Gross primary production (g Cm^{-2} yr^{-1})	1599
Schlesinger & Regier (1982)	Air temperature	344
Schlesinger & Regier (1982)	Air temperature & MEI (for depth <25 m)	1105
Schlesinger & Regier (1982)	Air temperature, MEI & effort (low)	358
Schlesinger & Regier (1982)	Air temperature, MEI & effort (high)	727
Mean of estimates		889

Oreochromis leucostictus

Detritus was found to be the principal component of *O. leucostictus* diet (Muchiri, 1990; Muchiri et al., 1994). It is the most abundant food material available to fish in Lake Naivasha and its importance has been previously noted by Malvestuto (1974) and Siddiqui (1977). Of the other dietary constituents, various algae, especially planktonic forms, were predominant although some chironomid larvae and oligochaete worms were also taken. In terms of feeding category classification, *O. leucostictus*, whilst in the main a detritus feeder, can also be described as a micro-herbivore.

Tilapia zillii

As for *O. leucostictus*, the diet of *Tilapia zillii* was reported as being principally detritus (Muchiri, 1990; Muchiri et al., 1994) but this species consumes a significant amount of macrophyte (mostly *Najas* and *Potamogeton*), rather than algae, and relatively large quantities of insect material. *T. zillii* is, therefore, a detritus feeder like *O. leucostictus* but with a secondary classification of omnivorous browser.

Barbus amphigramma

Muchiri (1990) indicated that *B. amphigramma* is primarily an insectivore, although zooplankton was an important dietary component.

Poecilia reticulata

As the guppy was introduced in an attempt to control mosquito larvae (Litterick et al., 1979), and has been used elsewhere in the tropics for this purpose (Costa, 1985), it is likely that the fish will take any similarly sized aquatic invertebrates to be found in the littoral zone.

Invertebrate fauna

The benthic fauna of Lake Naivasha comprises principally oligochaetes and chironomid larvae (Clark et al., 1989; Muchiri, 1990; Muchiri et al., 1994) a typical biota of many tropical lakes (e.g. Burgis et al., 1973; Green, 1979; Marshall,1982). The worms most commonly found in samples were the tubificids *Limnodrilus hoffmeisteri* (Clarepede) and *Branchiura sowerbyi* (Beddard). The predominant chironomid was *Chironomus formosipennis* Keiffer. Recent work (Raburu, 2002) estimated biomass and production of these benthic organisms to be 4.07 g dry wt m^{-2} yr^{-1} and 8.08 g dry wt m^{-2} yr^{-1}, respectively.

The littoral macro-invertebrate fauna shows greater species richness with approximately 50 species being present at any one time (Clark et al., 1989; Muchiri, 1990). By far, the most abundant organism recorded in all habitats, ranging from flooded farmland to open water, was the hemipteran *Micronecta scuttellaris* (Stal.) at approximately 400 m^{-3} Clark (1992).

Food web

From the information that they obtained on fish diets, Muchiri et al. (1994) constructed a basic food web. In order to facilitate recognition of any under-utilised resource, they restricted the pathways to only the most important of the feeding relationships. Possible under-utilisation was identified in relation to zooplankton, phytoplankton, fish and, in particular, the benthos.

Discussion

Since commercial fishing started, fish have been taken with gill nets which, in theory, makes regulating the fishery easier than if gears such as seine nets and trawls are used. As gill nets are selective, desired mesh sizes can be imposed and effort controlled by limiting numbers and lengths of nets in addition to regulating the numbers of fishing vessels. Unfortunately, the task of enforcing such regulations is often the difficult part for the fisheries manager. The most serious problems to be addressed are those of ensuring conformity with the gill net specifications and size limits by licensed operators and of curbing illegal fishing by unlicensed fishermen who have no regard for regulations. The latter issue is, however, as much a social concern as a fisheries one and requires a combined effort by the Fisheries Department, riparian interests and other stakeholders in the Naivasha region in order to educate and rehabilitate those people involved. A code of conduct has been drawn up by the Lake Naivasha Fisheries Department (1998) in an attempt to foster responsible fishing practices and to promote compliance with the legislation (Kenya Fisheries Act – Cap 378).

Although fish caught by unlicensed fishermen are unrecorded there was no loss of information on catches brought to the official landing station. Therefore, in spite of difficulties in obtaining accurate data, it could be considered that the reported commercial catches provide a good representation of the pattern of exploitation. Unfortunately, however, in recent years it has been observed that there has been an increasing tendency for more fish to be traded other than at the official location. It is possible, therefore, that the presumed poor performance of the fishery in the 1990–98 period could in part be due to under-reporting rather than reduced catches. Nonetheless, the fact that the lake level and catch show a very recent upturn (Fig. 3), and that in Figure 4(d) CPUE increases with effort, could indicate that the fishery is about to enter a recovery phase.

Calculations of MSY, although based on a simple version of the Shaefer model (Ricker, 1975; Pauly, 1983), are likely to be reliable. Muchiri et al. (1995) also worked on the fish catches for the 15 year period 1974–1988 inclusive. Using sophisticated catch and effort data analysis software (MRAG, 1992) to fit the data to various models of population growth, MSY estimates of 365–446 t yr^{-1} were obtained, as compared with 384 t yr^{-1} given in Table 1.

The theoretical fish yields given in Table 1 were calculated from limnological and primary production data. Hanson & Legget (1982), using a data set of 26 temperate lakes, suggested that total phosphorus and macrobenthos biomass were superior to MEI as predictors of fish yields but reported that for two tropical lakes, George and Naivasha, predictions from total phosphorus fell below observed values. Reasons suggested for this were that for tropical lakes, as against temperate ones, fish production is higher and most commercially important species are herbivores. The same situation of underestimation appears to be true for the use of macrobenthos biomass because when the current estimate of 4.07 g m^{-2} (Raburu, 2002) is used in the various formulae proposed by Hanson & Legget (1982) fish yield estimates of only 70–225 t yr^{-1} are obtained. Recently, Welcomme (1999) considered methods for estimating the potential output from fisheries and concluded that there is no system as yet to predict the symptotic level at which Y_{max} will occur other than the generalised predictors such as MEI. Accordingly, the theoretical fish yields for Lake Naivasha were calculated in such a way.

The potential fish yields, with a mean of 889 t yr^{-1}, suggest that the overall MSY for the Naivasha fishery could be higher. In particular, the estimates based on primary production models (Melack, 1976; Oglesby, 1977) and on the Schlesinger & Regier (1982) model for shallow lakes produced yield estimates (1105–1599 t yr^{-1}) around double the overall MSY of 641 t yr^{-1}. These results suggest that Lake Naivasha may have a potential for a larger fisheries output than is realised at the present time.

Tropical lakes typically support many species of fish, the number being broadly correlated with basin area (Welcomme, 1999). Albeit considerable variance according to altitude and degree of isolation, from figures derived from Vanden Bosch & Bernacsek (1990) for 160 tropical lakes (Number of species = 5.9 lake area $(km^2)^{0.2684}$), a lake such as Naivasha could be expected to contain in the order of 23 species. Such species would be then be likely to undertake a wide range of feeding strategies (Lowe-McConnell, 1987).

By modifying the classification scheme proposed by Matthes (1964), Hickley & Bailey (1987) identified five major feeding categories for common species of fish found in the large, shallow, papyrus-fringed lagoons which form the perennial waters of the Sudd swamps (R.Nile). Whilst not necessarily fully applicable to a lake like Naivasha, such a classification serves to indicate the types of fish that could be expected in a shallow water tropical system. The categories are mud-feeders, microherbivores, macroherbivores, omnivores and carnivores. Carnivores can be further subdivided into zooplanktivores, bottom-feeders, browsers in vegetation, generalised macropredators and piscivores. In Lake Naivasha the tilapias, *Oreochromis leucostictus* and *Tilapia zillii*, with their large uptake of detritus, represent mud-feeders. *O. leucostictus* also partly fulfils the role of microherbivore by taking green algae and diatoms, although this species alone is unlikely to fully exploit phytoplankton production. Of the larger plant material, some is eaten by *T. zillii* but *Procambarus clarkii*, which forms the basis of a commercial crayfish fishery (Lowery & Mendes, 1977), is capable of cropping the macrophytes to a major extent (Harper et al., 1990). Omnivores are represented by *T. zillii*, *B. amphigramma* and the guppy. Although zooplankton featured in the diet of both *B. amphigramma* and juvenile bass, neither species fulfils the role of zooplanktivore to a degree compatible with reports that offshore secondary production is high (Mavuti & Litterick, 1981; Harper, 1984, 1987). The largemouth bass is a generalized macropredator.

It appears, therefore, that in terms of the above classification the Naivasha fish fauna is not sufficiently species rich. Indeed, it would be difficult for any fish population comprising only five species to fully exploit all potential food resources. This view is supported by Moreau (1997) who, taking account of modeling by Mavuti et al. (1996) using ECOPATH (Christensen & Pauly, 1992), concluded that Lake Naivasha is a good example of trophic under-exploitation. As early as the mid 1970s it was suggested (Siddiqui, 1977) that new species should be introduced to exploit the remaining niches. Such a proposal has now been carried forward into the Lake Naivasha Management Plan (LNRA, 1999) with a view to increasing commercial catches and providing local employment opportunities.

Based on prospective feeding guilds and the actual food web, Muchiri et al. (1994) identified four areas in terms of food and space with respect to the potential for stocking additional species of fish. The most convincing case, and that which would best be addressed in the first instance, is the bottom feeder. Given that it is the benthic oligochaetes and chironomid larvae which are under-utilized, one of the species of *Mormyrus* is likely to be a suitable candidate for introduction. With regard to introducing fish to consume more phytoplankton and zooplankton than is currently taken by the existing species, there is scope but the proposal is less robust. Species such as *Limnothrissa* or *Alestes* could be stocked to occupy the open water and take zooplankton but, as these species would not be susceptible to the existing gill net fishery, new fishing methods would have to be authorised. *Heterotis niloticus* (Cuvier) could be used to target the phytoplankton. *H. niloticus* has been introduced to a number of waterbodies and has established self-reproducing populations without seriously threatening the pre-existing fish fauna (Moreau, 1997) although, in Naivasha, the long term decline in *C. papyrus* (Boar et al., 1999) could impact upon the availablity of nest sites. With regard to the introduction of a dedicated piscivore, it might be best to leave all the cropping of fish to be carried out not by a top predator but by the activities of the commercial fishermen.

Clearly, the necessary feasibility appraisals of each potential species must address all aspects of suitability or otherwise. The EIFAC code of practice for species introductions (Turner, 1988) recommends taking into account feeding habits, growth rates, reproductive strategy, environmental conditions required, competition with resident species and the potential implications of the fishery for the new species. Nonetheless, when assessing the balance of risks associated with further introductions into Lake Naivasha, the overall species composition and present ecological nature of the lake is such that any unforeseen detrimental effects of stocking could be considered to be less serious in the long term than if they occurred elsewhere.

Due to the relatively unstable environmental conditions of Lake Naivasha (Harper et al., 1990), there is need for continuous appraisal of the fishery. Catch records need to be of high quality and so adequate enforcement of the fisheries regulations is essential (Lake Naivasha Fisheries Department, 1998). In particular, the key enforcement areas to be addressed are the non-compliance with the minimum (100 mm) mesh size, the trading of undersized fish, the setting of nets closer than 100 m from the shoreline, and the landing of fish other than at the official landing station. The establishment of more landing and recording stations (LNRA, 1999) would comprehensively improve the robustness of the catch statistics. Also, the number and length of gill nets fished would be a better measure of effort than the number of licensed canoes.

Fishery recommendations

It is concluded that, at the present time, the fish population of Lake Naivasha does not fully exploit the available food resources and the fishery is not as productive as it could be. It is recommended that there be acceptance of the proposal to enhance fish production by the introduction of additional species (LNRA, 1999). Potential species should be selected and full feasibility analysis as to their suitability be carried out as soon as is practicable. In addition, consideration should be given to the possibility of re-establishing the *B. amphigramma* of the 1980s. Building fish passes to facilitate spawning migrations might enable the population to recover sufficiently to be exploited once again.

It could be argued that there should be no introductions to a conserved wetland. Even if it were to be considered desirable, it is totally impractical, however, to attempt to reverse the inheritance of past stocking history. The lake cannot be returned to its former status of having only its single endemic fish species. Therefore, given that under certain circumstances social and economic benefits can be considered to be as important as conservation, the concept of further introductions becomes acceptable. Effort should be made, however, to draw from an African fish list when considering potential candidate species.

Any introductions must be part of an overall management package which should also include:

- Conservation measures based on sound ecology such as refuge areas, fish passes on the river weirs and close times;
- Appropriate legislation and the enforcement thereof, such as minimum mesh sizes and a ban on trading in undersized fish;
- Education to improve awareness of both user groups and those in authority;
- Addressing of associated social issues such as redeployment of the poachers.

If introductions are to be made, then this should be done sooner rather than later, with a phased approach to the associated measures such that any new restrictions are accompanied by increasing availability of fish.

Acknowledgements

This study is part of a wider research project at Naivasha undertaken by the Universities of Leicester, U.K., and of Nairobi, Kenya which since 1987 has been authorised by the Office of the President, Government of Kenya under research permit to Dr D. M. Harper no. OP 13/001/12C 46. The project was funded by the Earthwatch Institute, Boston USA and Oxford, England. The data collection would not have been possible without the assistance of numerous Earthwatch volunteers. We are indebted to the Fisheries Department in Naivasha for access to catch data. Our sincere thanks go to the numerous colleagues in Kenya in particular the staff at the erstwhile Elsamere Conservation Centre, Jill & Angus Simpson. The views expressed are those of the authors and not necessarily those of their parent organisations.

References

Åse, L. E., K. Sernbo & P. Syren, 1986. Studies of Lake Naivasha, Kenya, and its drainage area. Naturgeografiska Institutionen Stockholms Universitet, Forsningsrapport 63: 1–75.

Boar, R. R., D. M. Harper & C. S. Adams, 1999. Biomass Allocation in *Cyperus papyrus* in a Tropical Wetland, Lake Naivasha, Kenya. Biotropica 31: 411–421.

Burgis, M. J., J. E. P. C. Darlington, I. G. Dunn, G. G. Ganf, J. J. Gwahaba & L. M. McGowan, 1973. The biomass and distribution of organisms in Lake George, Uganda. Proc. R. Soc. Lond. B, 184: 271–298.

Christensen, V. & D. Pauly, 1992. A Guide to the ECOPATH II Software System. version 2. 1. ICLARM Software, 6.

Clark, F. , 1992. A study of a population of *Micronecta scutellaris*. Hemiptera: Corixidae. in Lake Naivasha, Kenya. Hydrobiologia 248: 115–124.

Clark, F. A. Beeby & P. Kirby, 1989. A study of the macro-invertebrates of Lakes Naivasha, Oloidien and Sonachi, Kenya. Rev. Hydrobiol. Trop. 22. 10: 21–23.

Costa, H. H., 1985. Control of *Culex quinquefasciatus* larvae by the larvivorous fish *Poecilia reticulata* in drains and ditches of Colombo. J. Natl. Aquat. Resour. Agency Sri Lanka. 32: 24–33.

Crul, R. C. M., 1992. Models for Estimating Potential Fish Yields of African Inland Waters. FAO Committee for the Inland Fisheries of Africa. CIFA. Occasional Paper, 16: 22 pp.

Elder, H. Y., D. T. Garrod & P. J. P. Whitehead, 1971. Natural hybrids of the African cichlid fishes *Tilapia spirulus nigra* and *T. leucosticta*, a case of hybrid introgression. Biol. J. linn. Soc. 3: 103–146.

Green, J., 1979. The fauna of Lake Sonfon, Sierra Leone. J. Zool. Lond. 187: 113–133.

Hanson, J. M. & W. C. Legget, 1982. Empirical prediction of fish biomass and yield. Can. J. Fish. aquat. Sci. 39: 257–263.

Harper, D. M., 1984. Recent changes in the ecology of Lake Naivasha, Kenya. Verh. int. Ver. fur Theor. Ang. Limnol. 22: 1192–1197.

Harper, D. M., 1987. The ecology and distribution of the zooplankton in Lakes Naivasha and Oloidien. In Harper, D. M. (ed.), University of Leicester Studies on the Lake Naivasha ecosystem, 1982–1984; Final Report to the Kenya Government: 101–121.

Harper, D. M., 1991. Primary production in Lake Naivasha, Kenya. Verh. int. Ver. Limnol. 24: 1112–1116.

Harper, D. M., K. M. Mavuti & S. M. Muchiri, 1990. Ecology and management of Lake Naivasha, Kenya, in relation to climatic change, alien species introductions and agricultural development. Env. Conserv. 17: 328–336.

Henderson, H. F. & R. L. Welcomme, 1974. The Relationship of Yield to Morpho-edaphic Index and Number of Fishermen in African Inland Fisheries. UN/FAO CIFA Occ. Pap. No. 1: 19 pp.

Hickley, P. & R. G. Bailey, 1987. Food and feeding relationships of fish in the Sudd swamps. River Nile, southern Sudan. J. Fish Biol. 30: 147–159.

Hickley, P., E. North, S. M. Muchiri & D. M. Harper, 1994. The diet of largemouth bass, *Micropterus salmoides*, in Lake Naivasha, Kenya. J. Fish Biol. 44: 607–619.

Lake Naivasha Fisheries Department, 1998. Code of Conduct for the Fisheries Sector of Lake Naivasha. In: LNRA 1999, Lake Naivasha Management Plan: 74–77.

LNRA, 1999. Lake Naivasha Management Plan. Lake Naivasha Riparian Association, P. O. Box 1011 Naivasha, Kenya: 77 pp.

Litterick, M. R., J. J. Gaudet, J. Kalff & J. M. Melack, 1979. The limnology of an African lake, Lake Naivasha, Kenya. Manuscript prepared for the SIL-UNEP Workshop on Tropical Limnology, Nairobi: 73 pp. Mimeo.

Lowe-McConnell, R. H., 1987. Ecological Studies in Tropical Fish Communities. Cambridge University Press, Cambridge.

Lowery, R. S. & A. J. Mendes, 1977. *Procambarus clarkii* in Lake Naivasha, Kenya, and its effects on established and potential fisheries. Aquaculture 11: 111–121.

Malvestuto, S., 1974. The Fishes of Lake Naivasha: Their Biology, Exploitation and Management. Mimeographed report to the Director of Fisheries, Government of Kenya, Nairobi, Kenya: 37 pp, 5 figs.

Marshall, B. E., 1982. The benthic fauna of Lake McIlwain. In Thornton, J. A. (ed.), Lake McIlwain. Monographiae Biologica 49: 144–145.

Matthes, H., 1964. Les poissons du Lac Tumba et de la région d'Ikela. Etude systématique et écologique. Ann. Mus. R. Afr. Centr. Ser. 8vo. Ser. Zool. 126: 1–204.

Mavuti, K. & M. R. Litterick, 1981. Species composition and distribution of zooplankton in a tropical lake, Lake Naivasha, Kenya. Arch. Hydobiol. 93. 1: 52–58.

Mavuti, K., J. Moreau, J. Munyandorero & P. D. Plisnier, 1996. Analysis of trophic relationships in two shallow equatorial lakes, Lake Naivasha, Kenya and Lake Ihema, Rwanda. Hydrobiologia 321: 89–100.

Melack, J. M., 1976. Primary production and fish yields in tropical lakes. Trans. am. Fish. Soc. 105: 575–580.

Melack, J. M., 1979. Photosynthetic rates in four tropical African fresh-waters. Freshwat. Biol. 9: 444–571.

Moreau, J., 1997. Impact of Fish Introductions in African Ecosystems: an Evaluation by Using ECOPATH II. In Remane, K. (ed.), African Inland Fisheries, Aquaculture and the Environment. Oxford: Fishing News Books, Blackwell Science Ltd: 326–350.

MRAG, 1992. Catch-Effort Data Analysis. Unpublished software manual, MRAG Ltd, London and ODA, London.

Muchiri, S. M., 1990. The Feeding Ecology of Tilapias and the Fishery of Lake Naivasha, Kenya. PhD Thesis unpublished, University of Leicester, U.K.

Muchiri, S. M., J. B. Hart & D. M. Harper, 1995. The persistence of two introduced species in Lake Naivasha, Kenya, in the face of environmental variability and fishing pressure. In Pitcher, T. J. & J. B. Hart (eds), Impact of Species Change in African Lakes. London: Chapman & Hall: 299–319.

Muchiri, S. M. & P. Hickley, 1991. The fishery of Lake Naivasha, Kenya. In Cowx, I. G. (ed.), Catch Effort Sampling Strategies: Their Application in Freshwater Fisheries Management. Oxford: Fishing News Books, Blackwell Scientific Publications: 382–392.

Muchiri, S. M., P. Hickley, D. M. Harper & E. North, 1994. The potential for enhancing the fishery of Lake Naivasha, Kenya. In Cowx, I. G. (ed.), Rehabilitation of Freshwater Fisheries. Oxford: Fishing News Books, Blackwell Scientific Publications: 348–357.

Oglesby, R. T., 1977. Fish yield as a monitoring parameter and its prediction for lakes. In Alabaster, J. S. (ed.), Biological Monitoring of Inland Fisheries. Barking: Applied Science Publishers.

Pauly, D., 1983. Some Simple Methods of Assessing Fish Stocks. F. A. O. Rome. Fish. Tech. Pap. No. 234: 52 pp.

Raburu, P., K. M. Mavuti, D. M. Harper & F. L. Clark, 2002. Population structure and secondary productivity of *Limnodrilus hoffmeisteri* (Claparede) and *Branchiura sowerbyi* Beddard in the profundal zone of Lake Naivasha, Kenya. Hydrobiologia 488 (Dev. Hydrobiol. 168): 153–161.

Ricker, W. E., 1975. Computation and interpretation of biological statistics of fish populations. Bull. Fish. Res. Bd Can. 191: 382 pp.

Schlesinger, D. A. & H. A. Regier, 1982. Climatic and morphoedaphic indices of fish yields in natural lakes. Trans. am. Fish. Soc. 111: 141–150.

Siddiqui, A. Q., 1977. Lake Naivasha. Kenya, East Africa. Fishery and its management together with a note on the food habits of fish. Biol. Conserv. 12: 217–218.

Turner, G. E., 1988. Code of Practice and Manual of Procedures for Consideration of Introductions and Transfers of Marine and Freshwater Organisms. FAO European Inland Fisheries Advisory Commission. EIFAC. Occasional Paper, 23: 44 pp.

Vanden Bosch, J. P. & G. M. Bernacsek, 1990. Source Book for the Inland Fishery Resources of Africa. FAO Committee for the Inland Fisheries of Africa. CIFA. Technical Paper, 18, Vol. 1: 411 pp., Vol. 2: 240 pp., Vol. 3: 219 pp.

Welcomme, R. L., 1999. A review of a model for quantitative evaluation of exploitation levels in multi-species fisheries. Fish. Manage. Ecol. 6: 1–19.

Hydrobiologia **488:** 191–203, 2002.
D.M. Harper, R. Boar, M. Everard & P. Hickley (eds), Lake Naivasha, Kenya.

Concluding remarks

Towards the sustainability of the Lake Naivasha Ramsar site and its catchment

Mark Everard[1,*] & David M. Harper[2]
[1]*The Natural Step UK, 9 Imperial Square, Cheltenham, Gloucestershire GL50 1QB, U.K.*
[2]*Department of Biology, University of Leicester, Leicester LE1 7RH, U.K.*
**Present address: Environment Agency, King's Meadow House, King's Meadow Road, Reading, Berkshire RG1 8DQ, U.K. E-mail: mark.everard@environment-agency.gov.uk*

Abstract

Sustainable development remains a pressing priority for the Lake Naivasha Ramsar site, as a matter both of long-term protection of the lake ecosystem and of the economic and social progress that depends upon it. This must be addressed at catchment-scale to protect the 'ecosystem services' that constitute the region's primary resources. Sustainability principles highlight energy, fertilisers, pesticides and other aspects of chemical use, plastics, erosion and sedimentation, wildlife, water and socio-economic issues as central to sustainable development strategy, influenced by both intensive and subsistence farms across the catchment. Market forces and long-term self-interest provide compelling arguments for sustainable development, the delivery of which will depend upon education and the spreading of good practice, perhaps backed up by stewardship schemes. Alternatively, the adverse consequences of continued unsustainable practice are highly significant for the region.

Introduction

Fertile volcanic soils, relatively constant climate, high levels of solar radiation, a near-constant day-length and ready access to fresh water are features of the shores of Lake Naivasha, Kenya 0° 45′ S, 36° 26′ E. These features, in combination with an inexpensive labour force, have attracted a large number of intensive horticultural enterprises to the flat lands of the lake shore and on its inflowing river floodplains. Cut flowers and out-of-season vegetables are the predominant outputs, flown from Nairobi on a daily basis to provide luxury products for European markets. This form of development has proven extremely profitable and contributes the bulk of the nation's foreign exchange from horticulture, which was the third most-important after coffee and tourism in 1999 (Grant, 1999). It comprises a major element of the economic development of the Naivasha region, together with geothermal energy production. Tourism, sport and commercial fisheries are of lesser importance (LNRA, 1999). The wider lake catchment, by contrast, comprises land tilled for subsistence and cash crops with rough pasture. There is evidence that land use practices in this wider catchment may compromise the quality of the lake and rivers as well as the long-term viability of this form of small-scale farming (Everard et al., 2002a; Kitaka et al., 2002).

Some problems have developed with this proliferation in the extent and intensity of lake shore and catchment farming, notwithstanding their obvious economic and social benefits in the short term. Collectively, they reveal the challenges to sustainable farming in the Naivasha region. These problems are likely to be common to all such vulnerable tropical wetland areas, which are nevertheless priority sites for development due to their high productivity (Everard et al., 1995). Such wetland systems have in the past suffered from unsympathetic intensive development, which has led to their destruction or degradation with the loss of wetland 'services' and biodiversity (Dugan, 1990; Denny, 1991 & 1994; Everard, 1997). Naivasha was designated a Ramsar site in 1995 in recognition of its global importance to wildlife (LNRA, 1999), heightening the political and ecological imperative of sustainable development. A lake management plan has begun to address lakeside issues through a process of consensus-building (Enniskillen, 2002) but extension of planning to the catchment scale is required to approach sustainability of the lake and catchment systems, and the livelihoods that depend upon them (Everard et al., 2002a). Such information about both the lake itself and its catchment is essential to gain an understanding of the behaviour and the factors influencing the state of the lake (Everard, 1999a). They are also critically important in determining the contributions of lakeside and catchment development, both urban and agricultural, to changes in lake properties and ecology.

Sustainable development

Rising populations and declining environmental 'headroom' mean that the substantial challenge of sustainable development is unavoidable for governments and for business (WCED, 1987; Hawken, 1993). This challenge is heightened in developing nations characterised by fast-growing populations and increasing expectations of material quality of life (WCED, 1987). The challenge is nowhere more intense than in vulnerable yet productive wetland areas, such as Ramsar sites (Denny, 1991; Everard et al., 1995). Part of the conditions of Ramsar Designation is that wetland management plans be devised to ensure Wise Use, defined as "*sustainable utilisation for the benefit of mankind in a way compatible with the maintenance of the natural properties of the ecosystem*" (Ramsar, 1971). The Ramsar Commission consider Wise Use to be synonymous with sustainable development (Ramsar, 1996).

Everard (1998, 1999b) has discussed the widely-misunderstood, but nonetheless fundamental, distinction between true sustainable development and mere compliance with obligations, or with peripheral 'greening' activities. Sustainable development is the *process* of achieving the *state* of sustainability; the capacity for indefinite continuance through innovative development patterns in balance with the earth's supportive capacities. Critically, it includes the business sector – both agricultural and industrial – which provides wealth-creation and other societal services through exploitation, sustainably or destructively, of natural and human capital (Everard, 1998, 2000). Indeed, the capacity of big business to influence the sustainable development agenda, either positively or negatively, now exceeds that of national governments (Hawken, 1993). A commitment to sustainable development by a business or a government has necessarily to imply far more than mere altruism or 'public relations', enabling it strategically to place itself 'ahead of the game' in a fast-changing world of rising population and diminishing resources. For the small-scale enterprise, such as the subsistence or cash crop farmer, the principle of protecting the natural and human capital upon which future livelihood and profitability depends is more immediate and relevant, although the communication of this concept must necessarily differ. Sustainable development is therefore a mechanism by which both tomorrow's market and tomorrow's 'shocks' – food shortages, other resource scarcities, rising costs, more stringent regulation, soil loss, public perception, etc. – can be pre-empted by strategic action today (Everard, 1998).

Since publication of the 'Brundtland Report' (WCED, 1987) and the signing by 150 nations of the *Declaration of Principles for Sustainable Development* at the 'Earth Summit' in Rio de Janiero in 1992, the evidence for significant progress with sustainable development remains elusive (Everard, 1999b). Indeed, a third of the natural world was lost between 1970 and 1995, a sixth of the world's land area was degraded by poor agricultural practices, and 25% of global fish stocks are currently depleted with a further 44% are being fished at their biological limit (UNDP, 1998; WWF, 1998). The process is continuing at an increasing rate. Wetlands in developing countries are under particular threat (Dugan, 1990; Denny, 1991, 1994; Everard et al., 1995, Thenya, 2001).

Part of the problem has been in deducing what sustainable development actually means in robustly scientific terms that are also generically applicable to day-to-day decisions in business and government (Hawken, 1993; Holmberg et al., 1996). For the purpose of addressing priorities in the Naivasha region, definitions must also be workable from the national to the farm scale. The Natural Step (TNS) provides a systems model for articulating what sustainability means, and how practical progress with sustainable development may be achieved, fitting these needs (Holmberg et al., 1996). TNS principles are being used increasingly world-wide for various sustainable development initiatives (Everard, 1999a). At the core of TNS is a model of the sustainable cycling of matter and energy in the biosphere of this planet, and an emphasis on transition from unsustainable linear resource flows to cyclic processes that prevent the systematic accumulation of pollutants and the degradation of natural and social capital. The breaches of the four first-order 'system conditions' deriving from this systems model enable identification of unsustainable practice and, in conjunction with other TNS tools, help formulation of plans leading enterprises – both large and small – incrementally towards sustainability.

An important aspect of this TNS approach to sustainability is that it supports decisions made on the basis of self-interest (Holmberg et al., 1996), at all scales from the local to the global. This aspect of the TNS approach means that solutions are more likely to be developed and accepted at national and regional government levels, and by businesses, and thereby to become integrated into core decisions rather than merely peripheral 'greening' or public relations exercises that tend not to endure (Everard, 1998, 1999b). Altruism alone is a luxury often beyond the means of those closest to basic subsistence. It also assumes relatively sophisticated understandings of cause and effect, and is not the most durable quality amongst even the most affluent. Everard (1998, 1999b) has discussed the relevance of the TNS approach to the water cycle, and TNS principles consequently provide the background to thinking about sustainable agriculture in this paper. Further elaboration of these high-level principles is provided by Holmberg et al. (1996).

Sustainable development challenges

It is possible to make preliminary assessments based upon the state of the Lake Naivasha ecosystem revealed in this volume. The first TNS system condition relates to the fact that, in a natural system, elements including nutrients are cycled with no net accumulation of waste and that, over geological time scales, any such substances (e.g. fossil carbon, nutrients, heavy metals) have become 'locked away' from the biosphere within the Earth's crust. Many human activities today however break this natural cycle by importing materials from the Earth's crust back into the biosphere where they then accumulate, resulting in problems such as pollution or eutrophication. Part of the challenge of sustainable development is to reduce the linear movement of these, originally mined, materials to levels that natural systems can reintegrate without accumulation (Table 1).

The second TNS system condition relates to human-manufactured substances not formerly found in nature. In a sustainable natural system, substances do not accumulate as they are broken down and reintegrated into natural solar-powered cycles. Many substances produced by society, including pesticides, CFCs, elemental chlorine and plastics, are not readily broken down and tend to accumulate, with unforeseen future potential impacts on human health (Table 2).

The third TNS system condition addresses physical degradation of life support systems. Natural resources such as water, forests and other ecosystems, and soil structure, as well as their supportive processes which provide life-support services that also contribute to economic activities and 'quality of life'. Procurement, usage and disposal/reuse of all such resources should be reviewed by all larger businesses within the context of a wider sustainable development strategy. For smaller subsistence farming, dependence upon the goods and services provided by healthy ecosystems is often critical to self-sufficiency (Table 3).

The fourth TNS system condition relates to socio-economic factors essential to long-term sustainability. Sustainable livelihoods depend upon ecosystem services within catchments, equitable share of their benefits, and collective stewardship. Without an equal share of potential economic and 'quality of life' benefits, there is likely to be no shared concern for, nor strategic approach to, problems relating to the shared ecosystem services. Sustainable development of the Naivasha region will therefore need to involve all stakeholders (Table 4).

Issues relevant to the achievement of sustainable agriculture at Naivasha

Recognising that we are not starting from the perfect position of undamaged habitat nor equitable wealth distribution, there is clearly a pressing need to address sustainable development if the unsustainable trends observed at Lake Naivasha are not to be perpetuated. Indeed, given the fast-rising population – the population of Kenya stood at 35 million in 1997 and is rising by an estimated million people per year – and increasing environmental and political pressures on the Lake Naivasha region which currently forms one of most important economic resources of the country, sustainable development is the over-riding priority.

The assessment using the TNS model outlined above provides a helpful and holistic framework of interconnected issues – including physical resources, ecosystems and ethical elements – relevant to the achievement of sustainable agriculture at Lake Naivasha.

Table 1. TNS System condition 1 in relation to Lake Naivasha

System condition 1: *In the sustainable society, nature is not subject to systematically increasing concentrations of substances extracted from the Earth's crust.* This relates to release of materials immobilized in the lithosphere (e.g., heavy metals, nutrients, radionuclides, fossil carbon) which an unsustainable society reintroduces into the biosphere at rates substantially exceeding nature's reintegration or redeposit rates (Holmberg et al., 1996).

The primary risk from the intensive agriculture at the lake edge are:

- Energy intensity in agricultural practice, particularly from fossil fuel sources
- Haulage of produce to Nairobi
- Reliance on European markets, hence primary dependence on aviation fuel
- High inputs of fertilizers which tend to leach into adjacent water bodies and groundwater

The primary risk from extensive catchment agriculture are:

- Loss of nutrients to waterways through soil loss
- Threats from possible increases of agricultural and energy inputs

Table 2. TNS System condition 2 in relation to Lake Naivasha

System condition 2: *In the sustainable society, nature is not subject to systematically increasing concentrations of substances produced by society.* This relates to synthetic substances alien to nature, and other substances produced by society at high rates, which therefore cannot rapidly be broken down and reintegrated in the biosphere (Holmberg et al., 1996).

The primary risk from intensive agriculture around the lake are:

- Intensive use of pesticides, some of which have been banned in developing countries
- Incautious reuse or disposal of plastics in growing tunnels, packaging and other processes
- Disposal of other chemicals used in laundry, processing, factory systems.

The primary risks from extensive catchment agriculture are:

- Use of pesticides where crop pest predators have been lost through habitat destruction
- Threats from possible increases of agricultural and energy inputs

Energy

The energy-intensive nature of the larger intensive agricultural activities around the Lake Naivasha shore are offset at least in part by the need for only brief periods of artificial lighting to extend day-length and the avoidance of heating, as compared to the requirements of agriculture in the European countries where the major markets currently exist. However, since the Kyoto Protocol in December 1997, and the subsequent re-negotiation of national greenhouse gas targets and trading agreements in Buenos Aires in the middle of 1998, momentum is growing at the international scale for controls on all sources of greenhouse gases. There are a number of energy impacts that agricultural concerns around Naivasha would be advised to review. The potential for a full transfer to photovoltaic, passive solar, geothermal and other technologies is significant in Kenya, and mature techniques such as natural ventilation and 'intelligent building' controls can further contribute to energy efficiency on site.

The energy entailed in transportation of produce from large commercial growers, including both road and aviation, is significant. Vehicle emission standards in Kenya are less stringent than those in Europe. However, the potential for costs savings through increasing fuel efficiency are significant given the distances (goods are hauled approximately 100 km from the lake to Nairobi's Jomo Kenyatta International Airport and along steep-gradient roads). Furthermore, as European retailers take more account of the ethical and environmental impacts along their entire supply chains in response to consumer demand and public perception, the more pressing will come the issue of tackling resource wastage whether compelled by domestic legislation or not.

The dependence of the major parts of the large agricultural businesses in the Lake Naivasha catch-

Table 3. TNS System condition 3 in relation to Lake Naivasha

System condition 3: *In the sustainable society, nature is not subject to systematically increasing degradation by physical means.* This relates to the 'engines' of processing in the biosphere; the quantity of productive surfaces, and the adaptable and efficient pathways provided by diverse ecosystems (Holmberg et al., 1996).

The primary risks from intensive agriculture around the lake are:

- Direct conversion of functionally important and rare habitat into intensive agricultural land
- Fencing that disrupts migratory behavior in animals (daily and seasonal)
- Water use, over-harvesting from groundwater and surface water sources
- Use of geothermal energy, the impacts of which on groundwater are not known
- Loss of native crop species and varieties, and essentially sustainable agriculture practices
- Degradation of soil structure and loss of topsoil through over-grazing or poor tillage practice
- Displacement of native crops and agricultural practices by intensive practices
- Increasing invasion by non-native weed species

The primary risks from extensive catchment agriculture are:

- Increasing habitat loss through more intensive land use due to population pressures
- Soil loss through erosion from over-grazing or unwise tillage practices
- Loss of ecosystem services, particularly wetland services through riparian development
- The spread of non-native weed species
- Threats from possible increases of agricultural and energy inputs, including tillage

Table 4. TNS System condition 4 in relation to Lake Naivasha

System condition 4: *In the sustainable society, human needs are met worldwide.* This relates to social considerations permitting compliance with sustainable resource use, including equity and resource efficiency (Holmberg et al., 1996).

The primary risks from intensive agriculture around the lake are:

- Disenfranchisement of local people through land ownership patterns
- Centralisation of economic benefits by large businesses and state companies
- Displacement of local nomadic people using lake shore habitat
- Disturbance of traditional tribal practices
- Loss of retreating and other itinerant forms of subsistence agriculture and cash cropping
- Inequitable wages
- Inadequate protection from the risks of agriculture (i.e. pesticide usage)
- Marginalisation of the poorest people to the poorest land
- Provision of luxury goods for developed world markets in place of subsistence for local people

The primary risks from extensive catchment agriculture include:

- Loss of farmland and productivity for future use by poor current land use practices
- Loss of 'quality of life' by extensive habitat and wildlife loss
- Threats from possible increases of agricultural and energy inputs
- Cash-cropping at scales that compromise the need for local food production

ment on aviation to supply European markets should be an issue of great concern in the light of trends in international policy and consumer pressure regarding taxation and controls on greenhouse gas emissions. Aviation fuel is currently not taxed, and is exempt from controls on greenhouse gas emissions under the Kyoto Protocol, rendering aviation artificially cheap. However, there is growing international pressure for taxation of aviation fuels. Indeed, since the Intergovernmental Panel on Climate Change estimated that air traffic accounts for nearly three times its previously-estimated contribution to global warming, there is widespread and growing pressure from NGOs (non-governmental organisations) worldwide for appropriate taxation. Considerable research is already under way in the aviation industry about how to tackle this issue, which is widely regarded as inevitable in the longer term. Such measures are likely to affect the profitability of this form of farming in the medium to long term.

Furthermore, public pressure in Europe is squeezing big retailers to reduce the 'food miles' of produce on their shelves, with a move to local production for staples. All of the above factors have the potential to impact on the 'bottom line' economic performance of farms in the Naivasha region. They serve as classic examples of the ways in which unsustainable environmental and ethical practice will lead to decreased economic sustainability of businesses in a changing world, making more urgent a review of sustainable development policies across the farming sector.

There is a burgeoning population of small-scale subsistence and cash crop farmers throughout the catchment of Lake Naivasha, right up to the high gradient regions in the Nyandarua mountains, where energy use is a far less important issue at present. This scale of agriculture, including both tillage of land and grazing by stock (mainly goats and cattle), is performed on a largely sustainable basis at least with respect of energy inputs, which is substantially reliant upon manual labour. Where produce is transported for trade, this is largely done by bicycle, donkey cart, or people. However, increasing energy intensity is a trend that must be observed carefully, be that increasing usage of farm machinery (for example the use of tractors in the flat uplands of the Kinankop) for more intensive wheat and maize production, or the increasing use of mechanised transport. The low energy intensity of small-scale production in the catchment is arguably an obstacle to freeing people in the region from more productive uses of their time, and must be balanced carefully with other social aspects of sustainability. Local hydro power, photovoltaic and wind generation schemes are perhaps important areas for further consideration, as may be sustainable plantation of fuel-wood.

Fertilisers

Phosphorus and nitrogen are the main contributors to eutrophication in water bodies. Lake Naivasha, naturally nutrient-poor owing to the thin volcanic soils in its catchment, is particularly vulnerable (Harper et al., 1993; Kitaka et al., 2002). Agriculture worldwide is one of the key activities implicated in the transfer of nutrients into water bodies (Environment Agency, 1998).

Examples of best practice in some intensive farms, and included in advice to members of the Lake Naivasha Growers Group (LNGG), focus on a range of control measures. These include education of key staff as a means for preventing accidental losses of fertilisers, secure bunded storage of fertiliser stocks, metering of targeted application through drip feed, and delivery only of sufficient nutrient as determined by soil condition as a strategy both for reducing chemical costs and preventing leakage to the wider environment. One farm has also implemented treatment wetlands to reduce nutrient release from laundry facilities and surface run-off. These examples of best practice provide one element of an integrated and holistic sustainable development strategy. Adoption of European 'Best Practice' will be a defensive measure against future customer pressure, and as a wise proactive investment in resource stewardship including cost controls.

Management of intensive lake shore farms progressively to reduce chemical inputs and outputs is essential to sustainable development of the region. However, in addressing the sustainability of the lake, one has to ensure that its interactions with the whole catchment are adequately taken into account. (Kitaka et al., 2002) have found a close association between suspended particulates and phosphorus in the Naivasha catchment, and conclude that inputs from erosion in the catchment are highly significant compared to inputs from the generally tightly-managed lake shore farms. Everard et al. (2002a) present evidence of extensive poaching and erosion right up to the headwaters of most of these river systems. Though large intensive farms are absent from the less fertile and generally highly erosive soils away from the lake shore, Boar & Harper (2002) suggest that the input to the lake of sediments and their associated pollutants from the lake's riparian zone is a consequence of the loss of the buffering capacity of the *Cyperus papyrus* 'ecotone' which has been progressive in the last two decades (Fig. 1).

Pesticides

The accumulation of pesticides in Lake Naivasha is not yet at a serious level (Gitahi et al., 2002) but it potentially has three primary unsustainable impacts:

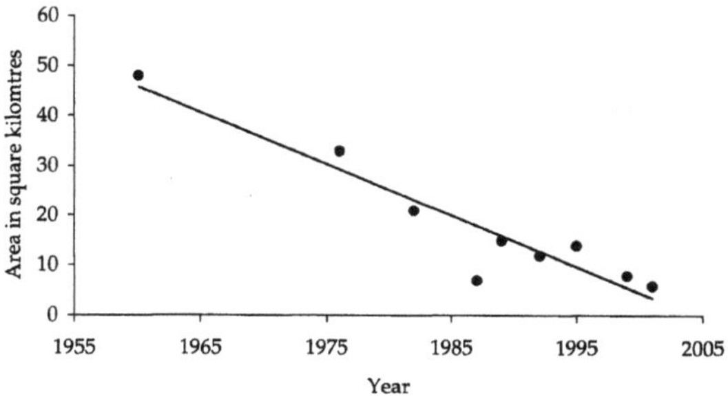

Figure 1. The decreasing area of *Cyperus papyrus* at Lake Naivasha. Data from Boar et al. (1999), Boar & Harper (2002) and Figure 1 of Enniskillen (2002).

- **Ethical** where effective training and protection of staff is not ensured, or where substances banned in developed countries due to known harmful effects continue to be used. There are also ethical implications entailed in the promotion of pesticide sales, and the fostering of reliance on chemically intensive agriculture, by small-scale farmers who can least afford it and who are likely to benefit least (if at all);
- **Ecological** where the pesticide substances are not fully contained on the larger intensive farms, and particularly where they upset the balance of nature by accumulating up food chains, impair critical links in the food chain, or damage vulnerable species or ecosystems. For the smaller-scale farmers, soil loss is a particular problem near watercourses and, where they are used incautiously, pesticides are likely to enter the river systems adsorbed to sediment particles; and
- **Market** stemming from obvious breaches of the above two, or where procedures or residue levels exceed those required by legislation in major markets (for example by European Union Directives). Equally, where used by small-scale farmers acting as 'outgrowers' (supply chain for the bigger growers), pesticide residues are likely to present problems for the final customers in Europe. Inputs deriving from any one of the larger intensive farms at the lakeside or in the catchment, or from a number of small-scale farms in the catchment, are likely to prejudice the reputation of all farms in the region in the eyes of their major markets.

One of the keys to understanding how pesticide usage may become sustainable lies in looking to how they influence, or might potentially influence, nature. This focus is important since it is the natural systems that are the ultimate source of all life-support and economic 'services' such as water and air purification, soil formation, mineral cycling. In the Lake Naivasha context, this includes factors such as clean fresh water, soil fertility and structure, biogeochemical cycling, and natural predators of crop pests. The ultimate aim is therefore not necessarily to avoid the use of pesticides; this may not be technically or economically feasible in the short term, and benign alternatives such biological control measures may not always be effective. It is instead to ensure that neither the pesticides nor their residues accumulate in natural systems, through incautious release at rates that nature can not break down and reintegrate.

Examples exist of best practice in some of the larger intensive lakeshore farms. These are focused on education of key staff in pesticide handling, secure and bunded stores, precision application, treatment of wash water, and interception of run-off. Some of this best practice is incorporated in the LNGG *Codes of Good Practice* which is commendable as part of a wider holistic sustainable development strategy. At least one farm is investing heavily in biological control agents and practices to bypass reliance on pesticides and to reduce the accumulation of pesticide residues in produce at levels unacceptable to European customers. Again, compliance with European 'Best Practice' is both a defensive and proactive measure.

Plastics

Plastics are used widely in intensive agriculture on the lake shore, both as growing tunnels and in packaging. Measures should be in place to ensure not only their collection and safe disposal, but also their efficient recycling. In a rural/urban region of hundreds of thousands of people, the quantity of plastic in litter is substantial. Present issues are:

- Accumulation of litter around the lake may attract condemnation from tourists, and adversely affect customer perception of supply chain stewardship by European retailers;
- Reuse of potentially contaminated waste chemicals drums, etc., that might cause health problems;
- Breakdown products from plastics, such as potentially hazardous plasticisers and stabilisers, might accumulate in local ecosystems with harmful results such as endocrine disruption, stimulation of cancers, or ecological imbalances. A full phase-out from persistent and bioaccumulative additives

is recommended, as implied by TNS system condition 2.
- Potential reuse of drums and other plastic packaging for domestic purposes such as carrying water, which risks contamination of users by substances such as pesticides.

A full re-evaluation of dependence on plastics, and their sustainable use, reuse and disposal, should feature as one component of an integrated sustainable development strategy. The LNGG *Code of Good Practice* guidance already addresses aspects of the issues, and small local enterprises are developing which recycle plastics either into moulded fence posts or woven domestic matting.

Erosion and sedimentation

Erosion is a significant issue in the catchment of Lake Naivasha, contributing to increasing loads of sediments and attached substances to the lake. The adoption of the principles, in the Lake Naivasha Riparian Association (LNRA) guidelines, of lakeshore buffer zones of 100 m from the shoreline, or 50 m behind a papyrus fringe, is a welcome measure to prevent sediment entry from lakeside farms. Boar & Harper (2002) provide data to support the fact that the papyrus fringe around the lake is of significance in preventing the entry of suspended sediments by diffuse run-off into the lake. Their recommendation for its conservation, first made a decade ago (Harper et al., 1990), has been taken up in the management plan (LNRA, 1999: Enniskillen, 2002).

Erosion and sediment input is a significant issue throughout the catchment, in addition to having a serious impact on the lake because of the loss of the buffering papyrus. In colonial times, there was a prohibition upon cultivation within 15 ft (approximately 5 m) of the river, and conversations by the first author with older catchment residents suggest that this contributed to richer riparian vegetation, greater habitat structure, and water and sediment retention throughout the river system. Re-instatement of these important riparian wetland features, and of their ecosystem services, is therefore important, whether achieved by voluntary or compulsory means. The former is much more likely to succeed, and a strategy for tackling this through community involvement is essential.

Wildlife

The physical 'footprint' of farms, as indeed of all forms of human development, inevitably displaces wildlife. Furthermore, many examples of unacceptable degradation of critical riparian habitat are readily observed around the lake (Enniskillen, 2002, Fig. 1). This habitat loss is a consequence of agricultural development, even if the land is subsequently managed sympathetically. Consequently, it is necessary to mitigate in some way this habitat loss in ways that genuinely address the needs of critical species that may have been displaced or disturbed, and of ecosystem 'services' lost.

Examples of good practice occur in the form of 'buffer zones' between the lake and river systems and some riparian farms. This is helpful not merely for good ecological reasons, but also as a matter of good sense in the light of the significant fluctuations of lake level such as that following the 1997/8 'El Nino' rains.

In the upper catchment, agreements on critical areas for wildlife and the protection of wetland services are urgently required (Bennun & Njoroge, 1999) – including for example 'buffer zones' along watercourses – as an element in the sustainable development of the Naivasha catchment. However, it is also important to recall that sustainable development for purely altruistic reasons is unlikely to be persuasive, or indeed a morally defensible approach, to present to those closest to subsistence. The 'enlightened self-interest' aspect of sustainable development is of far more direct relevance to small-scale farmers, and is one that is more likely to be adopted and maintained. Benefits derived from wildlife and ecosystem services, such as revenues from tourism, soil conservation, refuges for predators of crop pests, and flood protection should therefore form the basis for their protection.

Water

Like all life-supporting cycles, the water cycle *per se* is truly sustainable. However, current human exploitation of it is not. Furthermore, water is a vital resource, and will not only be limiting to development globally in the next millennium – it already is so in many countries, including Kenya and parts of developed nations such as the UK, Australia and the USA – but also touches all aspects of human life and is therefore affected by all sectors of society (Everard, 1999b). Any serious sustainable development policy, whether at farm/enterprise, regional, national or supra-national level, must therefore address water, since it underpins and is impacted by all facets of human life.

Efficient water usage and re-usage in the arid Kenyan environment is essential to avoid damaging sensitive wetland habitats and denying others access to a fair share of water (Orie, 1996). For the intensive farms around the lake, it is important for their commercial success to prevent the perception by European customers of being irresponsible with the use of water and other resources. The designation of Lake Naivasha as a Ramsar Site accentuates the high global profile of the lake in the eyes of the public and the media. The consequences of irresponsible water usage – indeed anything less than an agreed 'best practice' guide – is therefore likely to attract bad press and therefore to damage market share.

The requirement to meter water abstraction in riparian farms (LNRA, 1999) highlights the need to review the efficient and equitable usage of water as a key component of sustainable development. In Naivasha, such eco-technologies include:

- Rainfall harvesting to reduce demand on abstraction from sensitive habitats. A further benefit of this approach is the desynchronisation of floodwater arising from hard surfaces, abating erosion, and also the loss of suspended and dissolved substances to the lake and rivers during heavy rainfall;
- 'Greywater' reuse and other methods of water reuse;
- Intermediate water treatment technologies, such as the use of treatment wetlands, and management and construction of these systems in line with best practice guidance such as that provided by Nuttall et al. (1997); and
- Further research and policy change is required with respect to the regional water abstraction licensing system which, in the absence of any robust quantified hydrological model for the lake, is unlikely to bear any relationship with sustainable yield from Lake Naivasha or its rivers or groundwater systems.

Damming is common in the catchment for small-scale water supply, and indeed is advocated as best practice by some of the larger growers with farms remote from the lake. Dams bring with them a range of possible adverse consequence, including disruption of flows of water and sediment, creation of centres of waterborne disease and exotic species, disadvantaging those downstream and possibly resulting in inequitable distribution of resources in the region.

Aside from local dams, small-scale farmers remote from watercourses in the Naivasha catchment have little direct influence on water bodies. However, there is a real risk of cumulative impacts deriving from farming close to rivers via sedimentation, inputs of contaminants, and from the loss of habitat and ecosystem services provided by riparian habitat, impacting upon river systems and the lake.

Any exceedence of the 'carrying capacity' of the catchment will inevitably lead to adverse ecological effects and decreasing longer-term sustainability. Water allocation, both on the lakeshore and throughout the catchment, therefore requires holistic review with respect to ecosystem needs and equitable distribution to the human population.

Socio-economic issues

Social considerations are integral to the achievement of sustainability. The reason for ensuring the inclusion of all sectors of society in matters relating to the sustainable development of the Lake Naivasha catchment goes way beyond a vague notion of 'equity', howsoever construed. It relates to the issue of collective stewardship. If sectors of society are not party to understandings of their interdependence with the catchment, or feel that they have no stake in its future, or if the benefits others derive from catchment services marginalise the majority, then there is no obvious reason for changing unsustainable behaviour. Without an equal share of potential economic and 'quality of life' benefits, there can be no shared concern and strategic approach to the problems that compromise them. Consequently, the sense of self-interest will, for entirely understandable reasons, lack the 'enlightenment' that provides for consideration of the impacts of decisions and practices upon others, on natural resources, or for one's own longer-term interests.

The continual exclusion of the majority from economic success, social development and decision-making can only perpetuate their unwitting contribution to continual ecological decline and lack of social cohesion. There is no political incentive to stop the practice of 'land grab' (enforced government repossession and reallocation of land ownership common throughout Kenya) and to reduce cultural instabilities, and so the way forward is through private and communal initiatives.

Reputation

Reputation, even in a distant part of the world relative to the European markets, is of direct relevance to the financial 'bottom line' of the intensive lake-shore growers. Examples abound of multinational enterprises, and of national retail chains and manufacturers, that source from developing nations and have suffered adverse reputation and loss of market share as a consequence of one form or another of bad practice down their supply chains. Reputation is therefore a facet of sustainability that the proactive company will manage, not merely for altruistic reasons but as a matter of 'enlightened self-interest' (Everard, 1999b, 2000). The importance of a commitment to sustainable development based on one's own best long-term interests is an issue of great importance in focusing thinking about sustainable development, as it provides a basis for policies and actions that are durable and which contribute to the best outcome for the region as a whole.

The case for collective action

From the perspective of the outside world, including critically those that comprise the final retail markets for farmers around Lake Naivasha, all economic activities in the lake catchment will be blamed for degradation of the health of the lake. Market forces, or 'common sense', will not prevent damage to the lake at an unacceptable scale. The tendency is exemplified in the *Tragedy of the Commons*, wherein all individually seek competitively to maximise their take of what is generally assumed to be a limitless shared resource. A good example is provided by major sea fisheries in international waters, a significant proportion of which have now collapsed due to individuals competing to maximise individual yields which collectively go well beyond the sustainable limits of the fishery. The case for concerted and proactive collective action by the whole community of Lake Naivasha is persuasive. There are a number of such groups representing common interests around the lake, including the Lake Naivasha Riparian Association (LNRA), the Lake Naivasha Growers Group (LNGG) and the Kenyan Flower Council (KFC).

There is clearly advantage to be sought in linking the work of these groups to the direction and output of scientific research programmes on the lake. These groups therefore need to expand to include all stakeholder groups across the catchment. Better understanding of eco-hydrological links between catchment land use and the lake is also necessary, together with a common approach to stewardship that reflects the central importance of ecosystem integrity to sustainable social and economic wellbeing.

Education and influence within the catchment

The quality and character of water bodies are inseparable from processes within their wider catchments, including the activities of the human population within living landscapes (Everard, 1999a). Appropriate adjustment of social/political and economic activities is therefore critical to overall sustainable development of catchments, upon which long-term protection and/or restoration of ecosystems depends.

Everard et al. (2002a,b) highlight the habitat and biological diversity of river corridors in the catchment of Lake Naivasha, noting their vulnerability to excessive or unwise land use. Everard et al. (2002a) and Kitaka et al. (2002) also note that poaching and manuring by stock and erosion through over-grazing and tillage were common phenomena throughout the catchment. Discussions by the authors with teachers and subsistence and cash-crop farmers in the upper catchment revealed that many were unaware where the river near which they farmed drained, or that their activities could contribute to problems downstream and in the lake itself. Education and other forms of influence on land use practices throughout the catchment therefore form an important component of any holistic lake management plan aimed as achieving Wise Use of the Lake Naivasha Ramsar site (i.e. sustainability of the catchment serving the lake).

Such education and awareness programmes need not be expensive. Indeed, viewed at the regional scale, they may be highly cost-effective methods for delivering a range of social, public health and environmental messages, and can attract sponsorship from national funds, international aid sources, charitable trusts and businesses. In Kenya, for example, the National Environmental Action Programme (NEAP) proritises issues requiring awareness-raising, and particularly via schools and public education which may provide a framework within which to prioritise, seek further funding and implement the proposed land/water catchment stewardship strategy. On a bigger scale, the UNDP/UNEP Global Environmental Facility funding process adds money to national support in ways that protect biodiversity within a broader participatory initiative that also enhances social and economic development, including the facility to apply for seed Project Development Funding.

The key to successful education programmes would appear to be effective local management from stakeholders, and the tailoring of delivery to the needs of local people. Strategies have a better chance of success if local communities, with their wealth of traditional management techniques, are involved from the first stages of development, and with their own needs clearly prioritised, rather than being patronised with 'developed world' solutions (IIED, 1994). Nirarita (1999) describes a successful schools programme in Indonesia focusing on Wise Use of wetlands, and integrating issues of demography and environment, family planning, pollution and conservation, with an emphasis on 'local content and local context'.

The message that sustainable development is a matter of 'enlightened self-interest', implicit in The Natural Step approach, is helpful in focusing on the needs of communities and developing awareness. It can help address not merely conceptual issues, such as the wider geography of the catchment and the potential for harm downstream, but the long- to medium-term benefits accruing to farmers and their families at the local scale in terms of sustaining their basic needs. Simple examples of self-interest include:

- Prevention of soil erosion as a means to protect fertile soils and rich grazing over time. This can be effected by simple buffer zoning techniques, contour tillage and drainage, and appropriate stock density rules. Moisture content and soil structure are also protected where erosion is avoided through rainwater infiltration. Advise must be tailored to local culture and need.
- Protection of natural ecosystems ensures the continued availability of natural resources such as fresh water, timber, bush meat and lianas through low-frequency cropping, etc.
- Natural ecosystems also provide a refuge for the predators of common crop pests, averting the need for future reliance on expensive and potentially dangerous agrochemicals.
- Trees, if cropped at a sustainable rate, help maintain soil temperatures, promoting optimal grass growth rates, and also help prevent erosion. Regulatory controls are required to address unsustainable clear-felling and other extensive destruction of ecologically important forest areas.
- Targeted and appropriate use of agrochemicals (particularly fertilisers and pesticides) can save substantial costs.
- Prevention of pollution, both directly and through diffuse route, will ensure the continual supply of water that is both clean and free from disease.

Local champions using best practice on their own shambas (small subsistence farms) in the catchment could potentially serve as demonstration sites. These have advantages over the use of remote educators, as they are 'hands on', local to those one seeks to influence, connected directly to their needs and daily realities, and proving the principle of 'enlightened self-interest' in practice.

The potential benefits of a stewardship scheme

The development of a regionally specific stewardship scheme is a logical extension of the ideas in this paper, contributing to a business-led protocol for sustainable development that will yield a range of direct benefits.

Industries dependent directly upon ecosystem health have been amongst the first to recognise that the sustainable exploitation of primary natural resources is fundamental to their economic sustainability (Everard, 2000). Examples of this recognition include the enormous strides taken by the paper industry over recent years towards the goal of sustainable profit through sustainable resource management and manufacturing (Porritt, 1998). Forest industries and those using forest products have joined together in Europe to establish sustainable stewardship schemes for both wood production and forest growth, including the Forest Stewardship Council (FSC), the UK Woodland Assurance Scheme (UKWAS) and the nascent Pan-European Forest Certification scheme. For marine fisheries, the parallel Marine Stewardship Council (MSC) scheme has been led by WWF and Unilever, whilst there are plans for the establishment of a related Land Stewardship Council (LSC) for sympathetic land usage in agriculture in the UK (Tickell, 1999).

Furthermore, the stewardship idea is one with which developed world corporations are comfortable, and particularly the larger European manufacturers and retailers (Unilever, 1997). Lake Naivasha would provide an ideal case study for the development of such a scheme in Kenya. Whether this takes the form of an internal certification marque within a company, or one for which external auditing was required for any grower from the Naivasha region, across Kenya, or across tropical Africa, the benefits for stimulation of trade through contribution to the wider sustainability of the region are clear.

The *Kenya Flower Council* (KFC) is already a significant step down the road to the type of independently verified stewardship scheme suggested here. For good business and competitive reasons, further progress is strongly recommended. The new business direction for many multinationals is sustainable development, although many openly acknowledge the struggle entailed in getting to grips with the meaning and measurement of the concept in practice (for example, Unilever, 1997; Wessex Water, 1998). However, sustainable development is for them a pressing priority and, in the light of diminishing environmental capacity and increasing population across the world, is one that is not likely to disappear.

Conclusions

Global society is a long way from sustainability (or, in other words, the capacity for indefinite continuance). The pressure to tackle sustainability will intensify over the coming years, and the challenge will be no more intense than in developing nations where the pressures of growing population, increasing expectations of material 'quality of life', and decreasing environmental capacity will be felt most directly. Tropical wetlands offer fertile land for development, both agricultural and industrial, that may act as a source of basic commodities for society as well as a source of foreign revenue. The challenge of achieving their sustainable development is therefore pressing, ensuring that the ecological productivity and diversity of natural systems – soil formation, biogeochemical processing, wildlife, etc. – continue to be sustained and to underpin the economy in the longer term. Sustainable farming on sensitive and important wetlands such as the shores of Lake Naivasha, and its increasingly populated catchment, is at the forefront of this challenge, ultimately delivering Wise Use as required under the Ramsar Convention.

Tackling sustainable development in a concerted and serious manner is a mechanism by which the social development and economic benefits of the Naivasha regions can be secured on a sustainable basis. The achievement of business with 'zero-impact' on the systems from which life-support, economic and 'quality of life' benefits derive is not likely to be quick or easy. However, the potential for loss of unique ecosystems, global market share, and social wellbeing and stability is so great that the challenge of sustainable development must be grasped as a matter of urgency.

Sustainable development of Lake Naivasha can only seriously be conceived at the whole-catchment scale, and the challenges, whilst differing between the lake shore and in the catchment, must be addressed in harmony within a unified regional plan. Aid may well be required, and indeed is available from a number of national and international sources, to realise this goal. Nevertheless, it remains a pressing priority to protect the long-term ecological, social and economic well-being of this vulnerable and productive region of Kenya.

Naivasha is a resource of global significance, and very much in the public eye. The message that sustainable development is a matter of 'enlightened self-interest' will therefore resonate with residents and local businesses reliant on foreign markets attuned to the adverse press reaction, and loss of market share, generated by apparent ecological or social exploitation along supply chains. The consequences of continued non-sustainability are as significant as are the gains to be achieved from seriously addressing sustainable development as a major and strategic issue for local businesses and regional development.

Acknowledgements

The authors would like to thank the Earthwatch Institute for funding the fieldwork at Naivasha over 15 years, from which these observations were derived. Also, ME is grateful to the Freshwater Biological Association for assistance with the fees for the conference *Shallow Tropical Waters and Humans* at the Kenya Wildlife Services Training Institute, Naivasha, 1999 at which the ideas were broadened and further developed. Gary Tomlins of Homegrown Ltd gave helpful guidance on technical aspects of horticulture at the lake shore.

References

Bennun, L. & P. Njoroge, 1999. Important Bird Areas in Kenya. Nature Kenya, Nairobi: 318 pp.

Boar, R. R., D. M. Harper & C. Adams, 1999. Biomass and its allocation in *Cyperus papyrus* in a tropical wetland, Lake Naivasha, Kenya. Biotropica 31: 411–421.

Boar, R. R. & D. M. Harper, 2002. Magnetic susceptibilities of lake sediment and soils on the shoreline of Lake Naivasha, Kenya. Hydrobiologia 488 (Dev. Hydrobiol. 168): 81–88.

Denny, P., 1991. African wetlands. In Finlayson, M. & M. Moser (eds), Wetlands. International Wetlands and Wildfowl Research Bureau, Slimbridge: 115–148.

Denny, P., 1994. Biodiversity and wetlands. Wetlands Ecol. Manage. 3: 55–61.

Dugan, P. J., 1990. Wetland Conservation: A Review of Current Issues and Required Action. IUCN, Gland, Switzerland.

Enniskillen, 2002. Introduction: The Lake Naivasha Management Plan – consensus-building to conserve an international gem. Hydrobiologia 488 (Dev. Hydrobiol. 168): ix–xii.

Environment Agency, 1998. Aquatic Eutrophication in England and Wales: A Proposed Management Strategy. Environmental Issues Series, Environment Agency, Bristol.

Everard, M., 1997. Development of a British wetland strategy. Aquatic Conservation 7: 223–238.

Everard, M., 1998. Application of The Natural Step to water management. J. Inst. environ. Sci. 76: 10–13.

Everard, M., 1999a. Towards sustainable development of still water resources. Hydrobiologia 395/396: 29–38.

Everard, M., 1999b. The Natural Step: a framework for sustainable development. FBA News 7, Spring 1999: 16–18.

Everard, M., 2000. The place of aquatic ecology in the sustainability agenda. Freshwat. Forum 13: 31–46.

Everard, M., P. Denny & C. Croucher, 1995. SWAMP: A knowledge-based system for the dissemination of sustainable development expertise to the developing world. Aquat. Conservation 54: 261–275.

Everard, M., J. A. Vale, D. M. Harper & H. Tarras-Wahlberg, 2002a. The physical attributes of the Lake Naivasha catchment rivers. Hydrobiologia 488 (Dev. Hydrobiol. 168): 13–25.

Everard, M., A. Kuria, M. Macharia, J. A. Vale & D. M. Harper, 2002b. Aspects of the biodiversity of the rivers in the Lake Naivasha catchment. Hydrobiologia 488 (Dev. Hydrobiol. 168): 43–55.

Gitahi, S. M., D. M. Harper, S. M. Muchiri, M. P. Tole & R. N. Ng'ang'a, 2002. Organochlorine and organophosphorus pesticide concentrations in water, sediment, and selected organisms in Lake Naivasha (Kenya). Hydrobiologia 488 (Dev. Hydrobiol. 168): 123–128.

Harper, D. M., K. M. Mavuti. & M. Muchiri, 1990. Ecology and management of Lake Naivasha, Kenya in relation to climate change, alien species introductions and agricultural development. Environ. Conservation 17: 328–336.

Harper, D. M., G. Phillips, A. Chilvers, N. Kitaka & K. Mavuti, 1993. Eutrophication prognosis for Lake Naivasha, Kenya. Verh. int. Ver. Limnol. 25: 861–865.

Hawken, P., 1993. The Ecology of Commerce: A Declaration of Sustainability. Harper Collins, New York.

Holmberg, J., K.-H. Robért & K.-E. Eriksson, 1996. Socio-ecological principles for sustainability. In Costanza, R., S. Olman &. J. Martinez-Alier (eds), Getting Down to Earth – Practical Applications of Ecological Economics. International Society of Ecological Economics, Island Press, Washington.

IIED, 1994. Whose Eden? Empowering Local Communities to Manage their Wildlife Resources. International Institute for Environment and Development Perspectives 13: 3–5.

Kitaka, N., D. M. Harper, K. M. Mavuti & N. Pacini, 2002. Chemical characteristics, with particular reference to phosphorus, of the rivers draining into Lake Naivasha, Kenya. Hydrobiologia 488 (Dev. Hydrobiol. 168): 57–71

LNRA, 1999. The Lake Naivasha Riparian Association LNRA, Kenya. Lake Naivasha Riparian Association, Naivasha.

Nirarita, E., 1999. Community-based environmental education in Danau Sentarum. Wetlands 6: 10–11.

Nuttall, P. M., A. G. Boon & M. R. Rowell, 1997. Review of the Design and Management of Constructed Wetlands. CIRIA Report 180, CIRIA, London.

Orie, K. K., 1996. Proposed amendment to Kenya's water legislation: some missing sustainable water resources management principles. Water Resour. Dev. 12: 79–88.

Porritt, J., 1998. Foreword. Paper Focus Enviro 98, Paper Publications, Hertfordshire.

Ramsar, 1971. Convention on Wetlands of International Importance Especially as Waterfowl Habitat. 2/2/71, Ramsar, Iran.

Ramsar, 1996. The Ramsar 25th Anniversary Statement. Resolution VI.14, 6th Meeting of the Conference of the Contracting Parties, Brisbane, March 1996.

Tickell, O., 1999. First the forests, now the fish. Green Futures, May/June 1999: 38–42.

Thenya, T., 2001. Challenges of conservation of dryland shallow waters, Ewaso Narok swamp, Laikipia District, Kenya. Hydrobiologia 458: 107–119

UNDP, 1998. Human Development Report 1998. Oxford University Press, Oxford. Unilever, 1997. Sustainability: Unilever's Approach. Unilever, Port Sunlight.

WCED, 1987. Our Common Future. Oxford University Press, Oxford, England.

Wessex Water, 1998. Striking the Balance 1998. Wessex Water plc, Bristol.

WWF, 1998. Living Planet Index. Worldwide Fund for Nature, Godalming.

MIX
Papier aus verantwortungsvollen Quellen
Paper from responsible sources
FSC® C105338

If you have any concerns about our products,
you can contact us on
ProductSafety@springernature.com

In case Publisher is established outside the EU,
the EU authorized representative is:
Springer Nature Customer Service Center GmbH
Europaplatz 3, 69115 Heidelberg, Germany

Printed by Libri Plureos GmbH
in Hamburg, Germany